Lecture Notes in Mechanical Engineering

Lecture Notes in Mechanical Engineering (LNME) publishes the latest developments in Mechanical Engineering—quickly, informally and with high quality. Original research reported in proceedings and post-proceedings represents the core of LNME. Also considered for publication are monographs, contributed volumes and lecture notes of exceptionally high quality and interest. Volumes published in LNME embrace all aspects, subfields and new challenges of mechanical engineering. Topics in the series include:

- Engineering Design
- Machinery and Machine Elements
- Mechanical Structures and Stress Analysis
- Automotive Engineering
- Engine Technology
- Aerospace Technology and Astronautics
- Nanotechnology and Microengineering
- Control, Robotics, Mechatronics
- MEMS
- Theoretical and Applied Mechanics
- Dynamical Systems, Control
- Fluid Mechanics
- Engineering Thermodynamics, Heat and Mass Transfer
- Manufacturing
- Precision Engineering, Instrumentation, Measurement
- Materials Engineering
- Tribology and Surface Technology

More information about this series at http://www.springer.com/series/11236

Sümer Şahin
Editor

8th International Conference on Engineering, Project, and Product Management (EPPM 2017)

Proceedings

Springer

Editor
Sümer Şahin
Faculty of Engineering and Natural Sciences
Bahçeşehir University
Beşiktaş, Istanbul
Turkey

ISSN 2195-4356 ISSN 2195-4364 (electronic)
Lecture Notes in Mechanical Engineering
ISBN 978-3-030-08920-7 ISBN 978-3-319-74123-9 (eBook)
https://doi.org/10.1007/978-3-319-74123-9

Printed on acid-free paper

This Springer imprint is published by the registered company Springer International Publishing AG part of Springer Nature
The registered company address is: Gewerbestrasse 11, 6330 Cham, Switzerland

Preface

International Conference on Engineering, Project, and Production Management (EPPM) conference series has a well-established scientific tradition. EPPM has started in 2010 as an autonomous, self-organized event by a group of independent scientists. The sprite is to unify engineering management, project management, and production management on academic and commercial grounds. It was evident that the field of "strong interactions" still has a broad way for further developments to create a synergy for more effective use of academic and industrial assets. EPPM offers a unique platform to inspire new thinking by merging and/or combining different approaches from diverse fields. The engineering, project, and production fields are plenty of opportunities, and it is the spirit of human vocation to explore those domains where interactions can improve some essential aspects of our world and the quality of life.

EPPM Conference has been established in 2010 by David Chua at National University of Singapore and Thanwadee Chinda at Thammasat University and continued annually in different parts of the Globe, see Fig. 1. Previous EPPM Conferences have been held in Pingtung, Taiwan, 2010; Singapore 2011; Brighton, United Kingdom, 2012; Bangkok, Thailand, 2013; Port Elizabeth, South Africa, 2014; Gold Coast, Australia, 2015 and Bialystok, Poland, 2016.

EPPM2017 was organized by Al Zaytoonah University of Jordan in Amman, Jordan, under the Patronage of HRH Princess Sumaya Bint El Hassan. The conference chair was Prof. Dr. Wejdan Abu Elhaija. The conference has received 95 papers, and 45 papers have been accepted for presentation. Table 1 shows the number of accepted papers by topics. In parallel, the special postgraduate poster session was conducted, where 30 student projects were presented.

Professor Wejdan Abu Elhaija, chair of EPPM2017, and Prof. Turki Obaidat, president of Al Zaytoonah University of Jordan, have delivered welcome and opening speeches. The High Guest was HRH Princess Sumaya Bint El Hassan, the patron of the event, and has delivered the "Keynote Speech" on the close relation of science and engineering and their importance for the benefit and convenience of man.

Fig. 1 Geographical view of EPPM conference venues

Table 1 EPPM2017 accepted papers per theme

EPPM2017 themes	Number of accepted papers per theme
Risk management	7
Project and process management	12
Engineering management	7
Operational management and decision support systems	2
Transportation systems	1
Cost analysis and financial management	1
Environmental science and management	1
Quality and reliability management	2
Construction management	10
Maintenance management systems	2

EPPM2017 has been attended by 44 participants out of 24 countries, depicted in Fig. 2. The number of participants by country is shown in Fig. 3 with major participations from Jordan being the host country, followed by South Africa, Poland, Australia, and UK.

Fig. 2 EPPM2017 participant countries

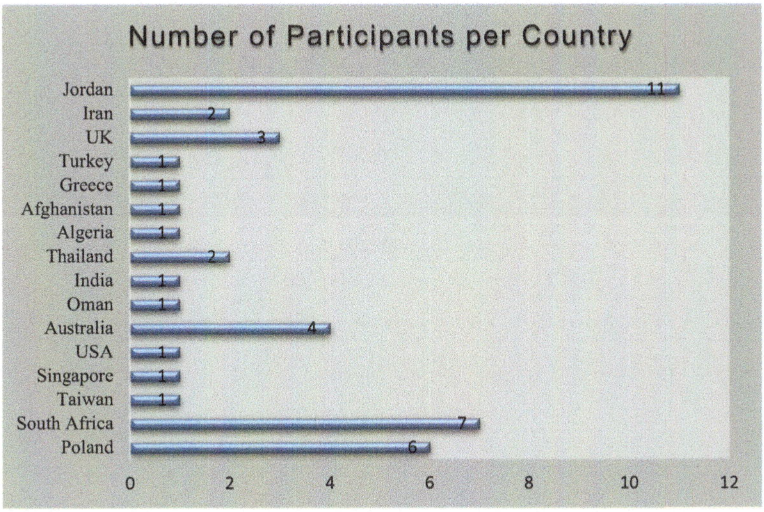

Fig. 3 EPPM2017 participants per country

Finally, 35 papers were accepted to be included in the Proceedings of "The 8th International Conference on Engineering, Project, and Production Management (EPPM2017)," Amman, Jordan.

Istanbul, Turkey
October 2017

Sümer Şahin

Organizing Committee

Conference Chair

Prof. Wejdan Abu-Elhaija

Organization

Scientific Committee

Chair
Prof. Subhi Bazlamit, Al-Zaytoonah University of Jordan, Jordan
Co-Chair
Prof. Erwin Pesch, Professor, Management Information Systems, University of Siegen, Germany
Co-Chair
Dr. Zoubir Hamici, Al-Zaytoonah University of Jordan, Jordan
Co-Chair
Dr. Saeb Al-Ganideh, Al-Zaytoonah University of Jordan, Jordan
Dr. Ernest Abbott, National University of Singapore, Singapore
Dr. Joanna Ejdys, Bialystok University of Technology, Poland
Dr. Katarzyna Halicka, Bialystok University of Technology, Poland
Dr. Alicja Gudanowska, Bialystok University of Technology, Poland
Dr. Julia Siderska, Bialystok University of Technology, Poland
Dr. Arkadiusz Jurczuk, Bialystok University of Technology, Poland
Dr. Nicholas Chileshe, University of South Australia, Australia
Dr. David Chua, National University of Singapore, Singapore
Dr. Agata Czarnigowska, Lublin University of Technology, Poland
Dr. John Smallwood, Nelson Mandela Metropolitan University, South Africa
Dr. Jozef Gawlik, Kracow University of Technology, Poland
Dr. Hsiang-Hsi Huang, National Pingtung University of Science and Technology, Taiwan
Dr. Sittimont Kanjanabootra, University of Newcastle, Australia

Dr. Oleg Kaplinski, Poznan Technical University, Poland
Dr. Kassim Gidado, University of Brighton, UK
Dr. Sherif Mohamed, Griffith University, Australia
Dr. Lukasz Nazarko, Bialystok University of Technology, Poland
Dr. Yiannis Nikolaidis, University of Macedonia, Greece
Dr. Poorang Piroozfar, University of Brighton, UK
Dr. Miroslaw Skibniewski, University of Maryland, College Park, USA
Dr. Jonas Saparauskas, Vilnius Gediminas Technical University, Lithuania
Dr. Jolanta Tamosaitiene, Vilnius Gediminas Technical University, Lithuania
Dr. Hsi-Hsien Wei, Hong Kong Polytechnic University, Hong Kong
Dr. Ker-Wei Yeoh, National University of Singapore, Singapore
Dr. Chris Impey, College of Science University of Arizona, USA
Dr. Jirapon Sunkpho, Thammasat University, Thailand
Dr. Mousa Bani-Baker, Al-Zaytoonah University of Jordan, Jordan
Dr. Ismael Jannoud, Al-Zaytoonah University of Jordan, Jordan
Dr. Mohammad Masoud, Al-Zaytoonah University of Jordan, Jordan
Dr. Yousef Jaradat, Al-Zaytoonah University of Jordan, Jordan
Dr. Moshrik Hamdi, Al-Zaytoonah University of Jordan, Jordan
Dr. Rana Alhorani, Al-Zaytoonah University of Jordan, Jordan
Dr. Fawzi Gharagheer, Al-Zaytoonah University of Jordan, Jordan
Dr. Nabeel Abu Shaban, Al-Zaytoonah University of Jordan, Jordan
Dr. Najm A. Najm, Al-Zaytoonah University of Jordan, Jordan
Dr. Abdel GhafourSaidi, Princess Sumaya University for Technology, Jordan

Organizing Committee

Chair
Dr. Hesham Ahmad, Al-Zaytoonah University of Jordan, Jordan
Co-Chair
Dr. Chien-Ho Ko, National Pingtung University of Science and Technology, Taiwan
Eng. Amal Qassed, Al-Zaytoonah University of Jordan, Jordan
Mr. Ahmad Al-Abbadi, Al-Zaytoonah University of Jordan, Jordan
Dr. Najm A. Najm, Al-Zaytoonah University of Jordan, Jordan
Dr. Loai Dabbour, Al-Zaytoonah University of Jordan, Jordan
Mr. Hani A. Abu Elrub, Al-Zaytoonah University of Jordan, Jordan
Mr. Malik Sarayreh, Al-Zaytoonah University of Jordan, Jordan
Dr. Mousa Bani-Baker, Al-Zaytoonah University of Jordan, Jordan
Dr. Mohammed M. Yassin, Al-Zaytoonah University of Jordan, Jordan
Dr. Nabeel Abu Shaban, Al-Zaytoonah University of Jordan, Jordan
Dr. Abdul Sattar Yousif, Al-Zaytoonah University of Jordan, Jordan
Dr. Ahmad Alaboushi, Al-Zaytoonah University of Jordan, Jordan
Mr. Ayman Alkhadar, Al-Zaytoonah University of Jordan, Jordan
Arch. Dana Shreim, Al-Zaytoonah University of Jordan, Jordan
Dr. Yousef Jaradat, Al-Zaytoonah University of Jordan, Jordan

Dr. Sami Al-Dalahmah, Al-Zaytoonah University of Jordan, Jordan
Mr. Thaer Abu Sharar, Al-Zaytoonah University of Jordan, Jordan
Mr. Mohammad Al-Fada'an, Al-Zaytoonah University of Jordan, Jordan
Dr. Abdallah Khader Atieh, Al-Zaytoonah University of Jordan, Jordan
Dr. Mohammad Masoud, Al-Zaytoonah University of Jordan, Jordan
Dr. Saleh Al-Jazzar, Al-Zaytoonah University of Jordan, Jordan
Eng. Dema Zaidan, Al-Zaytoonah University of Jordan, Jordan
Eng. Khulood Atalla, Al-Zaytoonah University of Jordan, Jordan
Arch. Dua'a Sandouqa, Al-Zaytoonah University of Jordan, Jordan
Dr. Mohammed Ashour, Al-Zaytoonah University of Jordan, Jordan
Dr. Ismael Jannoud, Al-Zaytoonah University of Jordan, Jordan
Dr. Nafiz Nimer Ali, Al-Zaytoonah University of Jordan, Jordan
Dr. Safwan Al-Qawabah, Al-Zaytoonah University of Jordan, Jordan
Dr. Mahmoud Zaidan, Al-Zaytoonah University of Jordan, Jordan
Dr. Ahmad Manasreh, Al-Zaytoonah University of Jordan, Jordan
Dr. Hana'a Jaradat, Al-Zaytoonah University of Jordan, Jordan
Dr. Qeethara Al-Shayea, Al-Zaytoonah University of Jordan, Jordan
Dr. Maha D. Ayoush, Al-Zaytoonah University of Jordan, Jordan
Dr. Moshrik R. Hamdi, Al-Zaytoonah University of Jordan, Jordan
Dr. Malik Al-Amaireh, Al-Zaytoonah University of Jordan, Jordan
Dr. Sa'ad Shaker, Al-Zaytoonah University of Jordan, Jordan
Eng. Amani Nawaflh, Al-Zaytoonah University of Jordan, Jordan
Arch. Valantina Al attah, Al-Zaytoonah University of Jordan, Jordan
Arch. Luma Daradkeh, Al-Zaytoonah University of Jordan, Jordan
Ms. Ibtisam Al-Qirem, Al-Zaytoonah University of Jordan, Jordan
Eng. Motasem Seder, Al-Zaytoonah University of Jordan, Jordan
Eng. Ayman Nasir, Al-Zaytoonah University of Jordan, Jordan
Eng. Osama Albadawi, Al-Zaytoonah University of Jordan, Jordan
Eng. Wafa'a Obaid, Al-Zaytoonah University of Jordan, Jordan
Eng. Mohammed Al-A'araj, Al-Zaytoonah University of Jordan, Jordan
Dr. Abdel Aziz Nidwi, Al-Zaytoonah University of Jordan, Jordan
Dr. Nawwaf Al Quseen, Al-Zaytoonah University of Jordan, Jordan
Mrs. Ala'a Aldwik, Al-Zaytoonah University of Jordan, Jordan
Dr. Khalid Al-Omoush, Al-Zaytoonah University of Jordan, Jordan
Eng. Nour Atieh, Al-Zaytoonah University of Jordan, Jordan
Eng. Amal Al-Masri, Al-Zaytoonah University of Jordan, Jordan
Eng. Huda Bazzari, Al-Zaytoonah University of Jordan, Jordan
Eng. Sultan Mosab, Al-Zaytoonah University of Jordan, Jordan
Eng. Dua'a Fawzi, Al-Zaytoonah University of Jordan, Jordan
Eng. Farah Al-Husseini, Al-Zaytoonah University of Jordan, Jordan
Arch. Noor Al-Shoubaki, Al-Zaytoonah University of Jordan, Jordan
Arch. Heba Al-Abbadi, Al-Zaytoonah University of Jordan, Jordan
Arch. Nisreen Abu-Ajamiyeh, Al-Zaytoonah University of Jordan, Jordan
Mr. Rabah Oweidah, Al-Zaytoonah University of Jordan, Jordan
Mr. Mohammad Al-Azaydah, Al-Zaytoonah University of Jordan, Jordan

Mr. Mahmoud Al-Amareen, Al-Zaytoonah University of Jordan, Jordan
Mr. Abd AL Azeez Al-Attrash, Al-Zaytoonah University of Jordan, Jordan

Sponsors

Al-Zaytoonah University of Jordan
Association of Engineering, Project, and Production Management (EPPM)
ARAB BANK
Jordan Engineers Association
Jordan Tourism Board
Conseil International des Grands Réseaux Électriques (CIGRE)

Keynote

Importance of Science and Engineering for Human Society

HRH Princess Sumaya Bint El Hassan, President of the Royal Scientific Society, Amman, Jordan

It is indeed an honor for me to be a part of the 8th International Conference on Engineering, Production, and Project Management 2017, here at Al Zaytoonah University of Jordan, among so many illustrious men and women of science and engineering from all over the world. The theme of the conference, when I first heard of it, brought to mind in sharp focus the new reality that faces our modern world—a reality in which various branches of science, engineering, and humanities are not separate little islands any more, but are rather coalescing ever faster into vast new areas of interdisciplinary studies, especially tailored to deal with the formidable multi-dimensional challenges of our times.

The speed with which science and engineering have transformed our lives in just a few short decades has thrust us all, willingly or unwillingly, into a superfast digital age where every facet of life demands an easy, practical, and ideally instantaneous solution. It is here that engineering and its twin supplemental branches of production and projects come into play. In fact, their footprint is ubiquitous and may be found in every field including power generation, infrastructure, industry, transportation, telecommunications, education, health care, agriculture, and more. Naturally, engineering projects and production being at the very heart of modern life, new and imaginative specializations, vocations, trades, businesses, and professions have mushroomed around them. And, it is agreed that none of them could function efficiently without proper management.

Jordanian universities have recognized the importance and potential impact of science and engineering and now offer an array of specializations in these fields along with several in management and administrative sciences. Traditional fields of engineering no doubt continue to exist as the bases for all engineering programs, but they now have to coexist with the more unconventional branches of study such as mechatronics, biotechnology, biomedical engineering, bioinformatics, networks and information security engineering, enterprise systems engineering, and natural resources engineering, among others. So it can undoubtedly be assumed that the myriad branches of engineering today with their various products and projects are transforming the world in ways we did not envisage perhaps even just a decade ago.

Progress has been extraordinary. Who would have thought that a journey which took months, or even years, could now be made in a few short hours? Who would have imagined that regular surface or sea mail which took days or sometimes months to deliver letters would become practically obsolete, with the ability to chat with our loved ones, friends, and colleagues instantly and even seeing them in real time, while doing so? I doubt whether just a few decades ago medical professionals thought they might be able to conduct consultations, and even complex surgeries, remotely. It is the magic of engineering that has transformed deserts into oases, and the sun, wind, and rain into great sources of power for the benefit of mankind. I could go on…engineering has made possible a world that was never thought or dreamt of, not so long ago. The great American writer Lyon Sprague de Camp rightly observed that "The story of civilization is, in a sense, the story of engineering—that long and arduous struggle to make the forces of nature work for man's good."

This begs the question, where are we heading? If engineering is a means to an end; to what end? Is technology hurtling along toward a goal that we can see, or is it yet unknown? Are we keeping an eye on the possible ramifications? Are engineering breakthroughs meant to simply make life easier, or are they meant to make a real difference to the quality of human life? How affordable and within reach are the solutions that engineering and technology offer? Does engineering offer equality to the masses or ever greater disparity? Are we mindful of the consequences of technological advances on nature and our environment?

To me these questions must indeed be raised because the ultimate goal of any branch of science and technology should, by default, be to improve, develop, and enhance the human experience. While researching the subjects of product and project management, I came across the areas that production and project managers are responsible for. They included strategy, releases, ideation, organizational training, delivery, resources, capacity, problem resolution, deadlines, quality, profit and loss, budget, and more. Nowhere did I find the need to study, for example, the long-term impact of particular engineering projects on resources, the environment (including flora and fauna), and humans. Shouldn't such studies necessarily be a part of engineering production and projects, or even precede them before implementation?

After all, what good is a dam if it has flooded hundreds of villages and displaced thousands of rural families? Of what use is a combined harvester or modern irrigation system, if the food it produces in immense quantities remains too expensive and out of reach for the masses? Wouldn't information technology be more effective and useful if it were available in every remote, underdeveloped corner of the earth rather than just electrified towns and cities? The National Academy of Engineering calls electrification, the greatest engineering achievement of the 20th century," yet today, in the 21st century more than a billion people all over the world still do not have access to electricity. What about medical equipment that saves lives? Does it serve the underprivileged by giving them the same access to health care as everyone else and help save their lives too? Every large city in the world today is dotted with new housing and construction projects, but how many are meant to shelter the homeless? Are new factories and assembly lines providing enough jobs to the unemployed or are they rather creators of unemployment by adopting automation and robotics without first examining the consequences?

I guess my point here is that engineering, production, and project management are definitely tools that are necessary, even imperative, for our development and progression as a human race. However, if these tools are not inclusive but rather serve only a certain privileged group that can afford them, the goal of sharing our technologies and resources together to benefit the entire human family could remain a distant dream.

Also, let not our focus on a profit-making model of doing business, which includes production and project management, blind us to the fact that the massive, steadily growing disparity between the haves and the have-nots today is a testament to our lack of focus on the well-being, prosperity, and success for all humankind. Gordan Stanley Brown, the great professor of Electrical Engineering at MIT, put is rather eloquently when he said, "Engineering is not merely knowing and being knowledgeable, like a walking encyclopedia; engineering is not merely analysis; engineering is practicing the art of the organizing forces of technological change. Engineers operate at the interface between science and society." In other words, engineers are meant to, and must make it their goal to use the tools of science to serve society. What is any progress, if it does not hold the promise of equality, dignity and a decent livelihood for all? I would argue that all progress, including technological progress, cannot be truly successful if it does not recognize that all humans, irrespective of their social status, must have access to its benefits. In fact, I would go so far as to say, engineering, production, and projects, apart from relying on the regular elements of organization, strategies, planning, training, and budget, need to make "empathy" and "social responsibility" a fundamental and essential feature of their studies and proposals.

The 10 top inventions that changed the world as listed by National Geographic are:

1. Printing Press
2. Light Bulb
3. Aeroplane
4. Personal computer
5. Vaccine
6. Automobile
7. Clock
8. Telephone
9. Refrigeration
10. Camera

All of the above are indeed engineering feats that will go down in the annals of history as having completely transformed human existence. Today, I would add to the list the World Wide Web, space technology, quantum computing, artificial intelligence, indeed this amazing list could go on.

But I stop here to laud scientists and engineers who revolutionized the world, and yet others who stood on the shoulders of giants and continued to improve, develop, and transform it through their incredible knowledge, imagination, and innovation.

I reiterate what has been said before that “Engineering is the art of directing the great sources of power in nature for the use and convenience of man.” Indeed, I ask all the illustrious scientists and engineers gathered here today to use this great and noble profession as a tool for bringing about a social renewal that promises equality and dignity to all; to harness its inevitable disruption and make it positively transformative for all people.

Contents

Confirmatory Factor Analysis of Perceived Risk Factors for Crowd Safety in Large Buildings

Mohammed Alkhadim, Kassim Gidado and Noel Painting

1 Introduction

Crowd safety is a major concern in facilities management and to those who attend events in large buildings and at venues such as sport stadiums, concert halls, and religious events (i.e. Hajj). Crowd safety can be achieved when there are no injurious or serious incidents outcomes experienced by any individual in the crowd. In large buildings used by large numbers of people, there are many threats and different levels of risk that require effective management. A flaw or hazard in large buildings or spaces during an event has resulted in many crowd disasters across the world. Two key existing crowd safety models were identified; FIST [5], and six dimensions and loci of crowd disaster [4]. These models include important factors that may cause risk to crowd safety and lead to crowd disaster. The acronym FIST is defined as: Force (F), Information (I), Space (S) and Time (T) while the six dimensions and loci of crowd disaster model involve 6 factors (Stampede, Riot, Structural and Mechanical failure, Terrorist attacks, Explosion (fire, chemical) and Natural disaster). Two more factors have been added from analyses of previous studies and major crowd incidents including user behavior and perceived safety. The research in this paper used the Holy Mosque during Hajj event as a case study in order to test, verify and to measure the reliability of the factors. The Holy Mosque is the largest mosque in the world, at approximately 356,800 square meters

M. Alkhadim (✉) · K. Gidado · N. Painting
School of Environment and Technology, University of Brighton,
Brighton BN2 4GJ, UK
e-mail: M.Al-Khadim@brighton.ac.uk

K. Gidado
e-mail: K.I.Gidado@brighton.ac.uk

N. Painting
e-mail: N.J.Painting@brighton.ac.uk

S. Şahin (ed.), *8th International Conference on Engineering, Project, and Product Management (EPPM 2017)*, Lecture Notes in Mechanical Engineering,
https://doi.org/10.1007/978-3-319-74123-9_1

and has 32 doors. It can accommodate around 1.2 million worshipers at the same time. Hajj is one of the five pillars of Islam which is mainly concentrated in four holy places: The Holy Mosque, the Mina, Muzdalifah and Arafat. It takes place once a year in a period ranging between 4 and 6 days. The Holy Mosque is a large building and has unique characteristics that facilitate an in-depth understanding of risk factors that may affect crowd safety.

2 Perceived Risk Factors to Crowd Safety

2.1 Perceived Risk of Stampede (PST)

Human stampede is a phenomenon that has occurred many times around the world. It has been associated with grave consequences such as loss of life, serious injury, property damage, psychological trauma and distress. Stampede can occur in many types of large gatherings including political rallies, social events, sporting events or religious events (pilgrimages, etc.). Several studies have reported that religious events have seen the worst incidents of human stampede with most incidents occurring in developing countries. Sociological theorists have stressed that individuals lose their sense of responsibility during a stampede situation. Studies on crowd disaster have shown that when the crowds need to turn in order to change the direction (e.g. in corners and stairwells), there is a risk of trampling and/or stampede to occur. They further state that when such restricted passage has sudden changes in the escape direction, it could also trigger trampling and stampede as people rush to flee.

2.2 Perceived Risk of Riot (PR)

Riot is a risky phenomenon, with many possible causes of incidents. The National Disaster Management Authority of India, defined riot as “a form of civil disorder characterized often by what is thought of as disorganized groups lashing out in a sudden and intense rash of violence against authority, property or people”. It has frequently occurred in some part of the world. One example of a riot occurred in 1992 in Los Angeles in which resulted in 52 people dead and 2500 injured as well as at least $446 million in property damage. Riots are often aggressive and violent, they usually start peacefully and then transform into a violent mob. Once they start, it is likely impossible to control them.

2.3 Perceived Risk of Structural Failure (PSF)

The failure of any temporary or permanent structure in a crowded venue can have an overwhelming effect. It has been indicated that structural failure is not uncommon. Insufficient design, poor construction, inadequate codes of practice and overloading have all caused significant failures. According to the National Disaster Management Authority of India (NDMA), structural failures have also been cited as reason for crowd disasters on numerous occasions.

2.4 Perceived Risk of Terrorist Attack (PTA)

In recent decades, terrorism has been increasing worldwide. Most studies on terrorism have lacked theoretical and empirical analysis. Furthermore, accepted definitions of terrorism are unclear but several elements are shared in common. These common elements refer to the violence or threat of actions that result in fatalities and serious injuries. Although many terrorist events seem irrational, these events must have been planned. According to the current definition of the U.S. Department of Defense (US DoD 2015): terrorism is the unlawful use of violence or threat of violence, often motivated by religious, political, or other ideological beliefs, to instill fear and coerce governments or societies in pursuit of goals that are usually political. The Oxford English Dictionary defines terror as 'the state of being terrified or greatly frightened; intense fear, fright or dread'. Nowadays, terrorism has become one of the main risk dimensions which requires safety planning. It is a veritable threat which targets public venues particularly, crowded places including sports and, religious events.

2.5 Perceived Risk of Explosion (Fire/Chemical) (PE)

Fire and explosion are major accidents which are classified as technological disaster. The International Labour Office (ILO), defined major accident as "an occurrence such as a major emission, fire or explosion resulting from uncontrolled developments in the course of an industrial activity, leading to a serious danger to man, immediate or delayed, inside or outside the establishment, and to the environment, and involving one or more dangerous substances". Many technological disasters have occurred around the world, such as: the fire that swept through a tent in Mina, Makkah during Hajj in 1997; the Gothenburg, Sweden, Disco in 1998; the Rhode Island Rock concert in 2003. A number of disaster cases were reviewed by the researcher with respect to the fire and behavior of the people within the fire situation. These disasters included the Beverly Hills Supper Club in 1977; Summerland Woolworth's in 1937; Bradford King's Cross in 1985. In these cases, the fire made people to panic in response to save their lives; and it is the panic that resulted in fatalities. For example, due to behavior of the people who panicked at the Beverly Hills Supper Club event in Kentucky, USA in 1977, 300 people were stampeded.

2.6 *Perceived (Risk of) Natural Disaster (PND)*

Natural disasters are catastrophic events which occur due to natural forces and are not controllable by mankind. Examples include flood, climate change (heat waves or cold waves), strong wind, volcanoes, earthquakes, etc. Asia and the Pacific are the regions most exposed to natural disasters. Most natural disasters result from heavy rains. Another natural disaster threat comes from climate change. Several studies have shown the significant association between climate change (e.g. high temperature) and mortality. Based on the Intergovernmental Panel on Climate Change 7 report in 2014, the increase of heat and decrease of cold due to climate change will result in increase of mortality in some parts of the world. Numerous studies have discovered that exposure to heat waves may cause cramps, fluid loss, fainting, heat exhaustion, dehydration, heat stress, heat stroke and ultimately mortality. Within a short time of exposure to high temperatures, people affected by heat may suffer fatalities. The elderly (aged 60 years or older), particularly women, and those with chronic lung diseases are more affected. High temperatures are likely to affect people physically and psychologically. It can increase aggressive behavior by directly increasing feelings of hostility and indirectly increase aggressive thoughts. This could be worse within large gathering events including sport, religious and political events.

2.7 *Perceived Risk of Force (PF)*

Perceived force refers to the feeling of the individual while within a crowd that may be produced by either hearing, seeing or sensing the force. The force may reach such a high level that it cannot be controlled or resisted because of crowd pressure. It has been emphasized that crowd compression, compressive asphyxia and a subsequent loss of footing or inability to move are the main reasons of deaths during an event (not by trampling). Berlonghi [3], claims that serious injuries and fatalities may occur from suffocation when people in a crowd are being swept along with movement and compressed. Generally, the forces that can be created when density exceeds a certain level may lead to a serious incident.

2.8 Perceived Risk of Poor Information (PPI)

In large buildings, it is crucial to obtain real time information about the crowd condition including crowd action, reactions whether real or perceived. Information communicated to—or withheld from—the crowd can influence their perceived safety. The communicating with the crowd is essential in maintaining order and managing behavior. It was found that poor information prior to or during an event has led to many crowd incidents. Information comprises all means of communication such as signs and announcements. Fruin [5], suggests that actions and training of personnel, sights and sounds all affect group perceptions. Setting up a communication center and a centralized crowd management system is good practice. Experts have highlighted that real-time information and communication are significant factors in minimizing risk of crowd disasters.

2.9 Perceived Risk of Insufficient Space (PIS)

Fruin [5], claimed that architects and engineers typically pay minimal attention to planning people's movement and perceptions but greater emphasis to meeting the local building codes regarding space in large buildings involving physical facilities, seating areas, corridors, stairs, escalators, and lifts. It has been shown that human psychology usually undergoes a change when the capacity becomes high and the venue does not have enough space to accommodate the crowd. Generally, when the individuals within a crowd perceive risk or a possible disaster, they panic and move to an exit ignoring alternative exits made available. Fruin [5], stated that within a high- density crowd it is difficult to describe the psychological and physiological pressure, and individuals may lose their control. Several studies have emphasized that crowd density has an effect on perceived safety and on people's behavior [1]. It has been argued that insufficient or poor use of space is considered a key risk factor to crowd safety.

2.10 Perceived Risk of Poor Real Time Management (PPRTM)

Poor real time information and interventions are key risk factors to crowd safety [5]. Failure to detect the behavior of the crowd at the right time can lead to serious incidents. Time plays an important role, for example, the inflow of the pedestrian compared to the rapid egress is much less while the pedestrian is leaving an event. It has been emphasized that the flow of the pedestrians must not exceed the capacity of the spaces available. It was indicated that lack of consideration is sometimes given to how crowd flow and density can be successfully managed by controlling timings.

2.11 Perceived Safety (PS)

Feeling unsafe during an event can drive people to panic from real or perceived risk through acting unusually by pushing and shoving. Studies in urban design have identified several factors that may have an influence on perceived safety including characteristics of the environment, the physical condition, and the configuration of spaces. The perception could differ from one person to another, for instance women and older people have a more diverse sense of safety compared with others. Crowd studies have defined the perceived crowding as "the psychological counterpart to population density" is closely tied to perceived safety. There is a negative correlation between perceived safety and perceptions of crowding; people's sense of safety declines as perceptions of crowding increase.

2.12 Crowd (Users) Behavior (UB)

Crowd behavior refers to the way in which persons act or behave towards others. Berlonghi [3], suggested important factors that can influence crowd behaviors and play an important role in designing, management and crowd control at events. These factors include location and time of the event, size of the crowd, crowd mobility, demographics of the crowd, schedule of event activities, crowd movement models, weather conditions and density of crowd in different areas. At some events, the crowd can turn into a mob and become aggressive. The aggressive behaviour may be in response to such strictures or emotional triggers such as elation, fear, or anger, and may be exaggerated by impairments such as drug or alcohol intoxication and lack of accurate information. Aggressive behavior may also result from physical discomfort due to environmental conditions such as heat, cold, noise, etc., and may become more likely if others are displaying aggressive behavior and are either rewarded or go unpunished. Psychological experiments, they have shown that when people get a high level of arousal responsibility is diffused, people may act irrationally and not be able to control their own behaviour [3]. Those people may start throwing objects, screaming and pushing people while some may turn into mobs carrying out theft, vandalism, rioting, group violence leading to a potential crowd disaster.

3 Method

Confirmatory Factor Analysis (CFA) is used as statistical technique for this study. This statistical technique does not specify variables to factors instead the factors are determined by the researcher based on the theory being tested prior to any results being obtained. CFA is applied to test the theoretical pattern of the variables loading on specific constructs and to show how well the theoretical specification of the factors can match the reality (the actual data). CFA enables the research to accept or reject the theory that has been studied [8]. 1940 pilgrims (both local and foreign) were surveyed within the zone of Makkah during the Hajj of the year 2016 (1437 Arabic Calendar). The questionnaire covered thirteen sections: section one is background information and sections two to thirteen are designed to test the perceived risk factors to crowd safety. The items included in the questionnaire were adapted from [1, 4, 5]. All the items were measured using a 5-point Likert scale (1 = strongly disagree to 5 = strongly agree; or 1 = never occur to 5 = almost always occurs). Several items were modified to attain the aim of the research.

4 Results and Discussion

4.1 Confirmatory Factor Analysis (CFA)

When undertaking a CFA, it is necessary to assess the uni-dimensionality, convergent and discriminant validity, as well as reliability [2]. The uni-dimensionality should be made first before assessing the convergent and discriminant validity, and reliability. Uni-dimensionality refers to the measurement items that have acceptable factor loading for the latent construct which is 0.60 and above [2]. Figure 1 presents the structural model; some modifications have been made based on Modification Indices (MI). Several items have been deleted one at a time and others have been covarying the error terms with the purpose of achieving the minimum fitness index.

4.2 Convergent Validity

To establish convergent validity, the model fit must be adequate, and the average variance extracted (AVE) must exceed 0.50 [8]. Table 1 provides the result of the model fit measures. Awang [2] recommend a comparative fit index (CFI) ≥ 0.90, standardized root mean square residual (SRMR) ≤ 0.08, and root mean square error of approximation (RMSEA) ≤ 0.06 for acceptable model fit. The values included in Table 1 indicate that the model is fit and all measures of CFI = 0.940, SRMR = 0.046, and RMSEA = 0.045 have achieved the required level. Also, the results of AVE for all constructs as illustrated in Table 2 have achieved the standard minimum required level of 0.50.

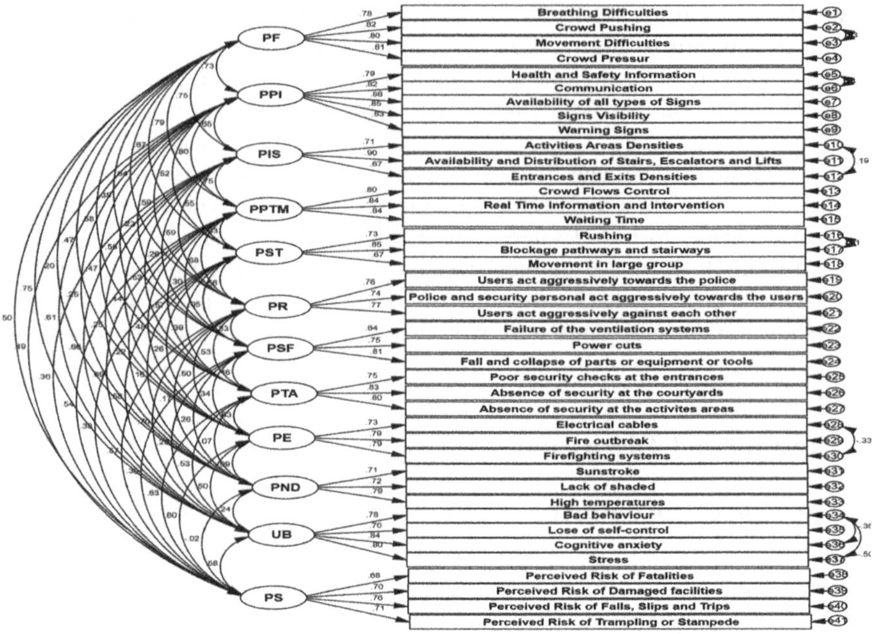

Fig. 1 The path diagram on the confirmatory factor analysis for all variables

Table 1 Fit indices

Measure	Estimate	Threshold	Interpretation
CFI	0.940	>0.95	Good fit
SRMR	0.046	<0.08	Good fit
RMSEA	0.045	<0.06	Good fit

4.3 Discriminant Validity

To establish discriminant validity three criteria must be met [7, 8]. The Fornell-Larcker test needs the square root AVE for each construct to be greater than any inter-construct correlations [6]. All constructs for this study have met this criterion. The square root of the AVE of the construct is greater than its estimates of correlation as presented in Fig. 2.

The other two criteria for discriminant validity that must also be met are the Maximum Shared Squared Variance (MSV) and Average Shared Squared Variance (ASV). Hair et al. [8], recommend that MSV and ASV must be less than the results of AVE (MSV < AVE, ASV < AVE). The results of ASV and MSV as detailed in Table 2 indicate that our measurement model is valid.

Table 2 Reliability and construct validity

	CR (above 0.60)	AVE (above 0.50)	Cronbach (above 0.7)	MSV	ASV	Convergent validity CR > AVE AVE > 0.50	Discriminant validity MSV < AVE ASV < AVE
PF	0.878	0.644	0.886	0.622	0.37	Yes	Yes
PPI	0.920	0.697	0.922	0.643	0.32	Yes	Yes
PIS	0.811	0.593	0.824	0.556	0.31	Yes	Yes
PPTM	0.867	0.685	0.866	0.643	0.38	Yes	Yes
PST	0.796	0.568	0.748	0.448	0.22	Yes	Yes
PR	0.803	0.577	0.804	0.484	0.31	Yes	Yes
PSF	0.779	0.543	0.774	0.210	0.09	Yes	Yes
PTA	0.836	0.630	0.833	0.403	0.27	Yes	Yes
PE	0.811	0.590	0.780	0.403	0.21	Yes	Yes
PND	0.785	0.549	0.784	0.068	0.04	Yes	Yes
UB	0.864	0.616	0.834	0.561	0.33	Yes	Yes
PS	0.805	0.508	0.804	0.401	0.23	Yes	Yes

Construct	PF	PPI	PIS	PPTM	PST	PR	PSF	PTA	PE	PND	UB	PS
PF	**0.802**											
PPI	0.735	**0.835**										
PIS	0.746	0.655	**0.770**									
PPTM	0.789	0.802	0.746	**0.828**								
PST	0.669	0.523	0.547	0.627	**0.753**							
PR	0.637	0.591	0.587	0.684	0.557	**0.760**						
PSF	0.354	0.234	0.260	0.301	0.045	0.331	**0.737**					
PTA	0.578	0.551	0.528	0.573	0.388	0.532	0.458	**0.794**				
PE	0.468	0.468	0.438	0.480	0.257	0.504	0.339	0.634	**0.768**			
PND	0.196	0.253	0.253	0.218	0.164	0.175	-0.260	-0.068	-0.091	**0.741**		
UB	0.749	0.608	0.662	0.692	0.584	0.696	0.292	0.532	0.496	0.236	**0.785**	
PS	0.502	0.492	0.360	0.542	0.381	0.569	0.302	0.633	0.599	-0.019	0.581	**0.713**

Fig. 2 Discriminant validity Fornell-Larcker test

4.4 Reliability and Construct Validity

Table 2 presents the results of the reliability and construct validity test. Two reliability tests have been undertaken for this study: composite reliability (CR) and Cronbach's alpha. Both tests were used to guarantee the reliability of the data before conducting further analysis. CR is more accurate than Cronbach's alpha because it does not assume that the loadings or error terms of the items are equal. The CR test has met the standard minimum threshold of 0.60.

The model also confirms that all Cronbach's Alpha values for the construct as given in Table 2 are above the recommended value of 0.70 [7]. This indicates the acceptability of internal consistency and confirms that all the items used in the model are technically free from the errors [8]. Overall, the result of the assessment of the measurement model shows solid evidence of unidimensionality, convergent validity, discriminant validity, and reliability.

5 Conclusion

This research presents the results of confirmatory factor analysis examining twelve important factors that may cause risk to crowd safety. The theoretical patterns of the variables loading on a developed construct were tested confirming the validity and reliability of the model. It clearly shows that the items on each construct of the study are reliable and the model has got enough measurement properties. Forty-one items were identified with an acceptable factor loading of at least 0.60. The results of CFA show acceptable evidence on the FIST, the six dimensions and loci of crowd disaster, and the two additional items of user behavior and perceived safety. The results of good internal consistency and validity of the constructs support the potential use of these 12 factors for large buildings used for hosting large events. For instance, the model can be used as indicators to evaluate large buildings used for large numbers of people whether it involved any of these risks. Many crowd disasters have occurred across the world where hazards were either not recognized or completely ignored. We concluded therefore that these 12 factors and items must be taken into account for managing large buildings and spaces in order to enhance crowd safety and reduce risk.

References

1. Alnabulsi H, Drury J (2014) Social identification moderates the effect of crowd density on safety at the Hajj. Proc Natl Acad Sci USA 111:9091–9096
2. Awang Z (2015) SEM made simple: a gentle approach to learning structural equation modeling. MPWS Rich
3. Berlonghi A (1995) Understanding and planning for different spectator crowds. Saf Sci 18:239–247
4. Chukwuma A, Kingsley C (2014) Disaster risks in crowded situations: contemporary manifestations and implications of human stampede in Nigeria. Int J Liberal Arts Soc Sci 2:87–98
5. Fruin JJ (1993) The causes and prevention of crowd disasters. Engineering for crowd safety. pp 1–10
6. Fornell C, Larcker DF (1981) Structural equation models with unobservable variables and measurement error: algebra and statistics. J Mark Res 18:382–388
7. Gaskin J (2016) Confirmatory factor analysis. Gaskination's StatWiki. http://statwiki.kolobkreations.com/index.php?title=Confirmatory_Factor_Analysis
8. Hair JF, Black WC, Babin J, Anderson RE (2010) Multivariate data analysis, 7th edn. Pearson, UK

A Comparison of Paid Versus Free Weather Services for Site Specific Weather Forecasts for Construction Projects

Evan M. Lauterbach, Salman Azhar and Amna Salman

1 Introduction

Weather is an undeniable factor that impacts almost every construction project around the world. The science of weather prediction is complex and there may probably never be a way to precisely determine the exact weather forecast well in advance for a specific area to plan day to day activities. However, over the past few decades, meteorologists have been able to refine the prediction accuracy to a pretty astounding level. However, there remains room for improvement and still almost every industry in the world is affected by weather in one way or another.

Weather plays a vital part in most aspects of the construction industry. Historical observations of weather conditions in an area play a pivotal role in construction planning long before the ground is broken. Design parameters are often dictated by the anticipated weather conditions the project will be subjected to over the course of its life. However, what construction managers are most concerned with after the design has been completed is the unexpected weather conditions during the construction execution phase [3].

In the United States, federal projects typically include contractual provisions that mandate a contractor to account for a certain amount of adverse weather days in a given month. These "adverse" days should be added into the weather dependent activity durations, so that when adverse weather happens, the overall completion of the project is not impacted by the event unless the provided amount of days for the

E. M. Lauterbach (✉) · S. Azhar · A. Salman
McWhorter School of Building Science, Auburn University,
Auburn, AL, USA
e-mail: eml0022@tigermail.auburn.edu

S. Azhar
e-mail: salman@auburn.edu

A. Salman
e-mail: azs0072@auburn.edu

S. Şahin (ed.), *8th International Conference on Engineering, Project, and Product Management (EPPM 2017)*, Lecture Notes in Mechanical Engineering,
https://doi.org/10.1007/978-3-319-74123-9_2

month exceeded. The intent of the Federal government's provision is to transfer risk to the contractor to account for a reasonable amount of weather events for the field work so that the planned occupancy date will not be missed. Should that be the case, a contractual liquated damages assessment amount occurs for each day beyond the agreed beneficial occupancy. In order to reduce or eliminate unplanned additional costs due to weather, the contractor must be savvy enough to dynamically plan during construction how weather will impact specific construction activities [3].

A typical construction site will most likely undergo a variety of inclement weather events. Benjamin and Greenwald [1] found that almost 50% of a construction project's activities are sensitive to weather. Probably one of the most significant conditions is extreme high and low temperatures. Temperature can have a broad impact on multiple facets of a job site. Worker efficiency due to fatigue and heat exhaustion from high temperatures may have unfavorable impacts on scheduled productivity. Furthermore, high temperatures impact the placement and curing of concrete. Consideration for hot weather placement may include extra equipment, specialized material and additional procedures, all of which must be in-place prior to execution in order to place and cure concrete according to specifications. Conversely, cold temperatures have just as significant impact. In addition to similar concrete issues such as concerns over slow rate of strength gain, the formation of ice on material and equipment takes time to remove and may manifest into serious site safety issues [1].

Often times the most significant weather condition that comes with little and undependable forewarning, is the *wind*. High wind speeds require careful observation and quick decision making to cease sensitive operations [5]. High wind conditions in excess of 20 mph will almost always shut down crane work and other elevated operations, such as man-lifts, equipment booms and other hosting procedures. Cranes are typically one of the most important pieces of equipment on a jobsite and are often times tied directly to critical path activities. Wind poses a serious danger to crane due to unaccounted moment forces on the crane which could cause overturn. The more accurate forecast to plan for high wind speeds, the better planning can be done in order to minimize delays and the risk of jobsite damage and injuries. Other considerations for high wind include limited visibility and flying loose construction debris throughout the site. Often times contractors are left to their own judgement in responding to high wind situations in their projects [5].

Typically, in the planning phase of laying out the schedule, the contractor provides input as to how activities sequencing should occur, along with a big picture view of when major milestones are completed in conjunction with the time of the year. In locations that are subject to wet seasons, contractors always strive to get buildings "dried in" before the start of the wet season. However, precipitation does occur throughout the year, leaving construction managers with weather related disruptions [7]. In cases of heavy civil construction, a complete jobsite shut down may be required to deal with erosion control measures and overly saturated and unworkable soils [2].

Throughout the construction industry, project managers are continually looking for ways to increase productivity and accurately forecast potential schedule impacts by means of all information and services available. One of the services available is site-specific paid weather forecasts. However, there has to be valid justifications in order to employ a pay-for-use service as opposed to what is freely available. Published research studies have not formally analyzed paid services for site-specific weather forecasts versus free and readily available information from the internet specifically for construction projects. The aim of this research is to compare the reliability and accuracy of "paid" site-specific weather services against readily available "free" information, to justify their use as an effective decision-making tool in construction project management. The rest of the paper is organized as follows: next section discusses the critical role weather plays in project scheduling, it is followed by methodology and major results, and at the end conclusions and recommendations are outlined.

2 Role of Weather in Project Scheduling

Knowing the impact adverse weather conditions may potentially have on a multitude of construction activities, a prudent construction manager will continually monitor upcoming weather for potential issues and revise construction activities to account for their prospective impacts. Continual reevaluation of weather sensitive activity durations is an essential part of project cost control. Moselhi et al. [4] developed an automated decision support system called WEATHER for estimating the impact on construction productivity. The system is a stand-alone interface that requires user-input to define parameters such as temperature, humidity and wind to estimate weathers impact on productivity. The software focused on the human productivity rate as a result of weather. This software application however was under development in 1997 and at the time was limited to electrical work, masonry construction, outdoor manual and equipment tasks and general construction. The current availability of this software is unknown.

Similarly to Moselhi et al. [4], Shahin et al. [6] proposed a framework for simulating weather-sensitive activities under extreme weather conditions [6]. In line with Moselhi et al. [4], Shahin et al. [6] work is based upon human productivity rates. While the model is based upon historical weather observations, the model also relies upon generated weather parameters as opposed to future forecast weather data. However, the simulated weather outputs were statistically analyzed for adequacy in comparison to historical observations.

In 2008, Thorpe and Karan acknowledged that weather delays pose a significant risk in project delivery, and construction managers need a simplistic tool to calculate expected revised durations based on weather delays. They developed a prediction model to help evaluate the effect of weather conditions on the of construction project duration. However, what differentiates their effort from previous research is that instead of being based upon labor productivity to simulate durational impact, their methodology takes a specific activity's sensitivity to a particular weather condition into account. The foundation of their method for calculating schedule delay is based on the following equation:

$$SD = \sum_{i=1}^{n} S_i \times P_t \times D_i \quad (1)$$

SD Schedule Delay
n number of activities
S Sensitivity factor based on its category (e.g. snow, rain, cold temperature)
P Probability of occurrence
D Duration of activity

The equation proposed by Thorpe and Karan [8] provides a revised duration for each activity based upon its sensitivity to a weather activity, its probability of occurrence and the original duration of the activity. The parameters of the equation rely on the judgement and expertise of the construction manager. Thorpe and Karan's intention was to develop future forecast probabilities (P_i) based on historical observations of occurrences broken down into months in which either snow, rain or cold weather would occur. Though recommended in all but none of the cited studies evaluated the accuracy of weather forecast data.

3 Research Design and Methodology

All data collected for this research study are quantitative in nature. The amount of potential weather data available is extensive. Due to limited time available to complete this research, only a small amount of weather forecast data and subsequent observations were recorded. Quantitative weather data was compiled on a daily basis from the National Weather Service (Free) and WeatherGuidance.com (Paid) at latitude and longitude of 30.05466 N, −97.82466 W, outside of Buda, Texas, USA. In order to assess the short and long term accuracies of both weather services, all defined conditions data were collected and recorded for next day, 2-day, 3-day, 4-day and 5-day forecasts. Table 1 depicts the conditions in which data was recorded for a period of 90 days, from May 31, 2016 to August 30, 2016 along with the driving action limit for possible schedule analysis.

In order to keep the comparison of data independent, neither the National Weather Service (NWS) nor WeatherGuidance.com (WG) could be relied upon for the collection of observed historical weather information. Using either service for that purpose could give the appearance of the recorded observations being more favorable to the future forecast data. Traditionally, actual weather observations are

Table 1 Weather data collection parameters

Weather event	Comment	Action limit for schedule analysis
High temperature	Daily maximum was recorded, regardless of time of event	Above 95 °F
Low temperature	Daily minimum was recorded, regardless of time of event	Below 32 °F
Wind	Maximum sustained and peak wind gust was recorded	Above 20 mph
Precipitation	Precipitation potential was recorded	Any recordable amount

limited to regional airports and weather balloons as sources. The problem with collection of actual observed data from these sources is that it may show a much greater variance if the site is farther from the weather station. In this study, the closest station was in San Marcos, TX (29.89275 N, −97.863 W), over 12 miles to the south of site. Hence a neutral and unbiased weather station named KTXBUDA5 residing at 30.065 N, −97.774 W, which is less than 3 miles from the site location was used. KTXBUDA5 is managed by Weather Underground (wunderground.com) which is a world-wide network of personal weather stations providing instantaneous observations and historical data in more locations than reliance on local or federal government weather stations.

After data has been collected for a period of 90 calendar days, statistical analysis on the data was conducted in order to show how accurate the forecasts are in comparison to the actual observed data of the corresponding day. Both descriptive and inferential statistical analysis were used The analysis conducted was overall descriptive statistics, which provided the mean, median, mode and standard deviation of both weather service forecasts for all weather conditions for next day, 2-day, 3-day, 4-day, and 5-day weather forecasts. In addition, inferential statistical analysis was conducted to identify relationships between the two sets of forecasts and actuals. Correlation analysis was performed to determine if a relationship between the weather services and the actual observations exists. In addition, T-tests were conducted to perform hypothesis testing. For this research, the null hypothesis, $H_{o,}$ was that there is no significant difference between the two compared weather services. The alternative hypotheses, $H_{1,}$ is that there is a significant difference. This logic was used to compare the two weather services to each other as well as each individual weather service to the actual observed conditions. Lastly, a direct comparison of accuracy of each weather service against the other was conducted to determine which service provides a more accurate forecast. This was done by examining the standard deviation of each forecast as well as looking at other statistical calculations, include Mean Squared Error (MSE) and Mean Average Percent Error (MAPE).

4 Results

Figures 1 and 2 show the results for maximum (highest) and minimum (lowest) daily temperatures based on 5-day forecast which is found to be the most critical one. Results for 1-day, 2-day, 3-day and 4-day advance weather forecasts are not reported here and can be found in Lauterbach [3].

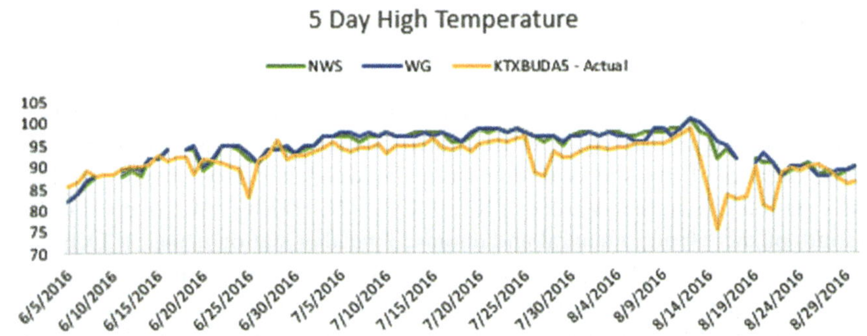

5-Day High Temperature Correlation

5-Day	*NWS*	*WG*	*Actual*
NWS	1.00		
WG	0.98	1.00	
Actual	0.64	0.56	1.00

WG/NWS: Very strong positive correlation
Actual/NWS: Strong positive correlation
Actual/W G: Moderate positive correlation

5-Day High Temperature MSE and MAPEE

5-Day High Temp.	NWS	WG
MSE	18.80466	22.7183
MAPE	3.48%	3.64%

The NWS forecast showed greater accuracy to the actual observed data than the WG forecast did according to the MSE and MAPE.

5-Day High Temperature t-test

4 DAY HIGH TEMP	*NWS*	*WG*	*NWS*	*ACTUAL*	*WG*	*ACTUAL*
Mean	94.17045	94.40909	93.94318	91.55	94.40909	91.60568
Variance	15.91314	15.55486	18.69788	18.92483	15.55486	18.24974
Observations	88	88	88	88	88	88
Hypothesized Mean Difference	0		0		0	
df	174		174		173	
t Stat	-0.39906		3.660092		4.523139	
P(T<=t) one-tail	0.345167		1.67E-04		5.64E-06	
t Critical one-tail	1.653658		1.653658		1.653709	
P(T<=t) two-tail	0.690335		0.000334		1.13E-05	
t Critical two-tail	1.973691		1.973691		1.973771	

NWS v WG: P(T<=t) two-tail is larger than 0.05, fail to reject null hypothesis – no significant difference
Actual v NWS: P(T<=t) two-tail is less than 0.05, reject null hypothesis – significant difference
Actual v WG: P(T<=t) two-tail is less than 0.05, reject null hypothesis – significant difference

Fig. 1 Maximum daily temperatures comparison and statistical analysis

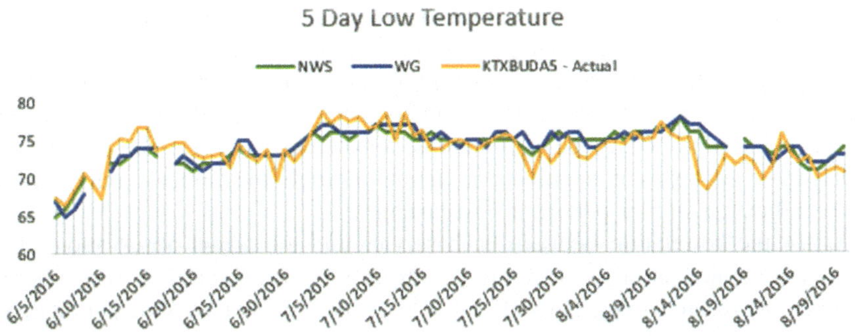

5-Day Low Temperature Correlation

5-Day	*NWS*	*WG*	*Actual*
NWS	1.00		
WG	0.92	1.00	
Actual	0.65	0.63	1.00

WG/NWS: Very strong positive correlation
Actual/NWS: Strong positive correlation
Actual/W G: Strong positive correlation

5-Day Low Temperature MSE and MAPE

5-Day Low Temp.	NWS	WG
MSE	4.075341	4.71625
MAPE	2.17%	2.30%

The NWS forecast showed greater accuracy to the actual observed data than the WG did according to the MSE and MAPE.

5-Day Low Temperature t-test

5-DAY LOW TEMP	*NWS*	*WG*	*NWS*	*ACTUAL*	*WG*	*ACTUAL*
Mean	73.97727	74.14773	73.97727	73.65341	74.14773	73.65341
Variance	4.551202	5.598615	4.551202	6.508724	5.598615	6.508724
Observations	88	88	88	88	88	88
Hypothesized Mean Difference	0		0		0	
df	172		169		173	
t Stat	-0.5019		0.91354		1.332673	
P(T<=t) one-tail	0.308188		0.18113		9.22E-02	
t Critical one-tail	1.653761		1.65392		1.653709	
P(T<=t) two-tail	0.616377		0.36226		1.84E-01	
t Critical two-tail	1.973852		1.9741		1.973771	

Fig. 2 Minimum daily temperatures comparison and statistical analysis

Beginning with the high temperature category, the overall observation was that both weather services continually predicted higher than observed temperatures for all forecasts. The correlation between the National Weather Service (NWS) and WeatherGuidance.com (WG) was a very strong positive correlation, with no less than a correlation coefficient of 0.98 between them. While the correlation between each weather service and the actual observation lessened the farther out the forecast got, with WG showing the weakest correlation compared to the actual over all days. According to the mean squared error, the two services were within a few degrees of accuracy apart. Examining the t-test results is the true indicator of how their performance was in relation to the observed temperature. For all forecast days the null hypothesis was unable to be rejected between NWS and WG, however it was rejected for both services in comparison to the actual observations meaning that the two weather services for high temperature prediction were similar to each other, but neither was similar to the actual temperature. Examining the low temperature category generally showed a much tighter relationship between the forecast data and the observed conditions. The correlations between the two services once again displayed a very strong positive correlation through the forecast range, which would indicate a similar pattern in forecasting.

As for the correlation between each service and the actual observations, both services remained at least a strong correlation. Looking at the mean squared averages and mean average percent error, WG generally showed a greater degree of accuracy for both, however it was within a few degrees of error and percentage points of NWS. Investigating the t-test results showed that in no case was the null hypothesis able to be rejected which means there was no significant difference for either weather service in comparison to each other and versus the actual observed conditions.

The precipitation and wind results for 5-day forecast are shown in Figs. 3 and 4. The precipitation category was a difficult category to analyze due to the fact that only two rain events occurred over the entire period of analysis, with several occasions of both weather services predicating above a 50% chance of precipitation. On those two occasions, both weather service predicted less than 50% chance of precipitation the day before. The correlation analysis indicated that the two weather services maintained a very strong positive correlation with each other throughout the forecast range, while in every case both services had either a very weak, or weak correlation with the actual observations. According to the MSE and MAPE, the WG forecast held a slight edge in accuracy, but not by more than a percentage point. As expected, the t-test results indicated that both weather services were able to reject the null hypothesis, meaning there were significant differences between both NWS and WG and the actual observations. However, in examining the t-test between the two services, the null hypothesis was unable to be rejected, meaning the two services were similar.

By first examining sustained maximum wind, the correlation between NWS and WG remained as a strong positive correlation. According to the MSE and MAPE for sustained winds, the WG forecasts displayed a higher degree of accuracyonly by a very small percentage over the NWS forecasts. Examining the t-tests, the null hypothesis was rejected on all occasions for both weather services in comparison to the actual observations, and failed to be rejected between NWS and WG, with the exception of the next day forecast. This infers that both services were not similar to the actual observations, but with similar to each other with the exception of the next day forecast. As for comparison with the actual observations, the correlations between WG and the actual were a moderate positive throughout and generally very weak or weak positive for NWS. The t-tests results demonstrate that there are

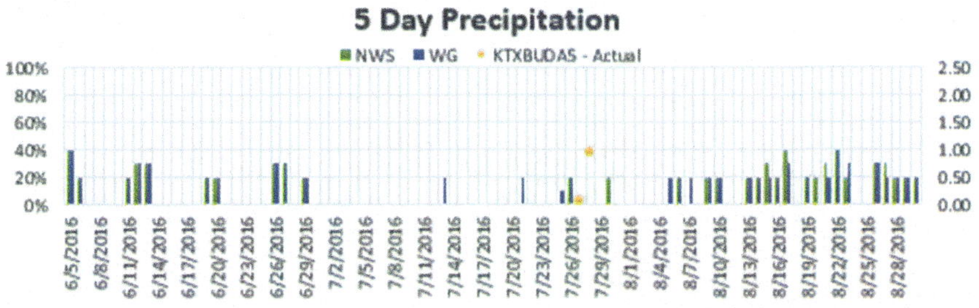

5-Day Precipitation Correlation

5-Day	*NWS*	*WG*	*ACTUAL*
NWS	1.00		
WG	0.87	1.00	
Actual	-0.11	-0.12	1.00

WG/NWS: Very strong positive correlation
Actual/NWS: Very weak negative correlation
Actual/W G: Very weak negative correlation

5-Day Precipitation MSE and MAPE

5-Day Precip	NWS	WG
MSE	0.046818	0.045455
MAPE	5.10%	5.10%

The WG forecast was showed greater accuracy to the actual observed data than the NWS did according to the MSE and equal accuracy according to the MAPE.

5-Day Precipitation t-test

5-DAY PRECIPITATION	*NWS*	*WG*	*NWS*	*ACTUAL*	*WG*	*ACTUAL*
Mean	0.093182	0.093182	0.093182	0.022727	0.093182	0.022727
Variance	0.015585	0.014206	0.015585	0.022466	0.014206	0.022466
Observations	88	88	88	88	88	88
Hypothesized Mean Difference	0		0		0	
df	174		168		166	
t Stat	-7.5E-16		3.388178		3.451308	
P(T<=t) one-tail	0.5		4.38E-04		3.54E-04	
t Critical one-tail	1.653658		1.653974		1.654085	
P(T<=t) two-tail	1		8.77E-04		7.08E-04	
t Critical two-tail	1.973691		1.974185		1.974358	

NWS v WG: P(T<=t) two-tail is larger than 0.05, fail to reject null hypothesis – no significant difference
Actual v NWS: P(T<=t) two-tail is less than 0.05, reject null hypothesis – significant difference
Actual v WG: P(T<=t) two-tail is less than 0.05, reject null hypothesis – significant difference

Fig. 3 Precipitation forecast comparison and statistical analysis

significant differences between the actual observations and both forecast services, however both forecast services were similar to one another.

In a nutshell, both weather services under-predicted sustained wind and maximum wind gusts throughout the forecast range. This is highlighted by universally low sustained wind predictions along with multiple actual observed wind gusts with no forecast from either service.

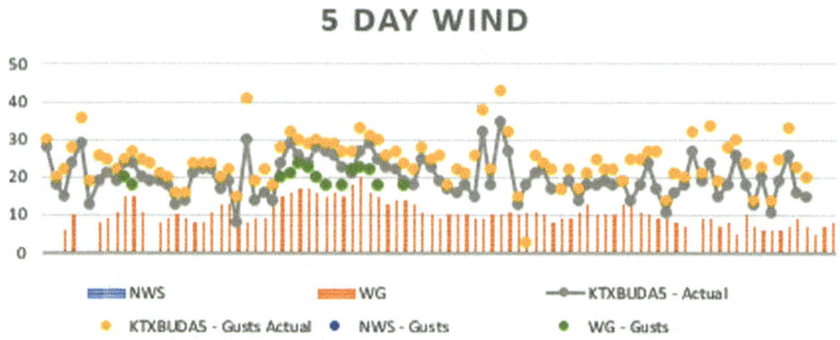

5-Day Wind Correlation – Sustained Wind

5-Day	*NWS*	*WG*	*ACTUAL*
NWS	1		
WG	#DIV/0!	1	
Actual	#DIV/0!	0.3429	1

WG/NWS: No correlation, no forecast
Actual/NWS: No correlation, no forecast
Actual/WG: Weak positive correlation

5-Day Wind Gust Correlation

Gusts

5-Day	*NWS*	*WG*	*Actual*
NWS	1.00		
WG	#DIV/0!	1.00	
Actual	#DIV/0!	0.51	1.00

WG/NWS: No correlation, no forecast
Actual/NWS: No correlation, no forecast
Actual/W G: Moderate positive correlation

5-Day Wind MSE and MAPE

5-Day Sustained Wind	NWS	WG
MSE	425.8864	120.1591
MAPE	100.00%	48.41%

The WG forecast was showed greater accuracy to the actual observed data than the NWS did according to the MSE and MAPE.

5-Day Wind Gust MSE and MAPE

5-Day Sustained Wind	NWS	WG
MSE	614	493.8068
MAPE	100.00%	86.52%

The WG forecast was showed greater accuracy to the actual observed data than the NWS did according to the MSE and MAPE.

5-Day Wind t-test

5-DAY SUSTAINED WIND	*NWS*	*WG*	*NWS*	*ACTUAL*	*WG*	*ACTUAL*
Mean	0	10.55814	0	20	10.55814	20
Variance	0	10.27305	0	26.18391	10.27305	26.18391
Observations	86	86	86	88	86	88
Hypothesized Mean Difference	0		0		0	
df	85		87		147	
t Stat	-30.5483		-36.6652		-14.6214	
P(T<=t) one-tail	6.62E-48		5.46E-55		9.29E-31	
t Critical one-tail	1.662979		1.662557		1.655285	
P(T<=t) two-tail	1.32E-47		1.09E-54		1.86E-30	
t Critical two-tail	1.988268		1.987608		1.976233	

Fig. 4 Sustained wind and gust forecast comparison and statistical analysis

5 Conclusions and Recommendations

Based on the comparison criteria set forth in this study and examining the results for each weather category through statistical analysis, it does not appear that using a "paid" site-specific weather service is more valuable as a project management tool than using "free" weather forecast services. The statistical analysis generally showed that both forecasting services provided very similar results. There were however significant statistical differences between both weather services and the observed actual weather conditions. It implies that the current weather forecasting tools do not provide very accurate information where a construction manager can solely make scheduling decisions based on the provided forecasts. This study recommends not to invest resources in utilizing paid site-specific weather services for use as a decision-support tool. However, it is important to note that the results of any other site-specific weather forecasting services may lead to a different set of conclusions so caution is advised in interpreting these results. This paper presented a brief overview of this research study. Readers interested in more details are recommended to review the full thesis [3].

References

1. Benjamin NB, Greenwald TW (1973) Simulating effects of weather on construction. J Constr Div 99 (Proc Paper)
2. El-Rayes K, Moselhi O (2001) Impact of rainfall on the productivity of highway construction. J Constr Eng Manag 127(2):125–131
3. Lauterbach EM (2016) Viability of site specific weather forecasts for construction. Auburn University, Graduate Capstone
4. Moselhi O, Gong D, El-Rayes K (1997) Estimating weather impact on the duration of construction activities. Can J Civ Eng 24(3):359–366
5. Ratay RT (1989) Wind design problems with building structures during construction. J Aerosp Eng 2(2):102–107
6. Shahin A, AbouRizk SM, Mohamed Y (2010) Modeling weather-sensitive construction activity using simulation. J Constr Eng Manag 137(3):238–246
7. Smith GR, Hancher DE (1989) Estimating precipitation impacts for scheduling. J Constr Eng Manag 115(4):552–566
8. Thorpe D, Karan EP (2008) Method for calculating schedule delay considering weather conditions. In: Proceedings of the 24th annual conference of the Association of Researchers in Construction Management (ARCOM 2008), vol 2, pp 809–818. Association of Researchers in Construction Management (ARCOM)

Use of Ocean Sensors as Wave Power Generators

Rahul Basu

1 Introduction

Solar energy can be harvested from the ocean through ocean currents. Ocean currents being solar and wind generated near the equator are classifiable into marine and tidal currents. Marine currents flow in one direction and are relatively constant. Tidal currents, however, are found near the shoreline due to gravitational forces. Some important currents include the Gulf Stream, Florida Straits Current, and the California Current. The wind and thermal energy get stored in the ocean as currents and surface waves. Wave heights and widths change during the year. Kinetic energy is proportional to the cube of velocity and density. Ocean energy is comparable to wind energy because the two are forms of hydrodynamic and fluid flows. Ocean currents are slower than wind speeds, however, as sea water is 832 times denser than air. Also, ocean currents can be predicted years in advance, since these depend on the sun and the moon. Worldwide, ocean currents of more than five knots or 2.5 m/s (1 knot = 0.50 m/s) exist. Current energy has been estimated greater than 5000 GW, with power densities of up to 15 kW/m^2.

Ocean currents of 4–5 knots occur along the E. African coast and have been studied extensively [11, 16]. The feasibility of using ocean currents off Hawaii was also undertaken [5]. Extractable power from the flowing fluid varies with kinetic energy as

$$P/A = (1/2)\rho V^3 \tag{1}$$

R. Basu (✉)
Mechanical Engineering, Adarsha Institute of Technology,
Kundana, Bangalore 562110, India
e-mail: raulbasu@gmail.com

S. Şahin (ed.), *8th International Conference on Engineering, Project, and Product Management (EPPM 2017)*, Lecture Notes in Mechanical Engineering,
https://doi.org/10.1007/978-3-319-74123-9_3

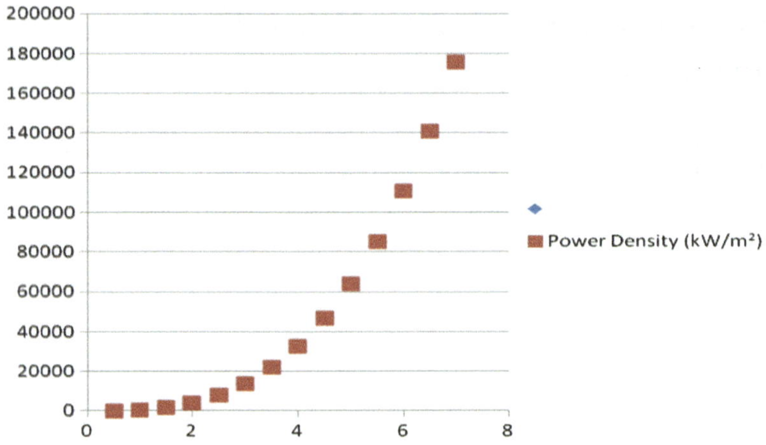

Fig. 1 Power density versus speed of fluid

where A (in square meters), ρ, fluid density (in kilograms per cubic meter; 1000 kg/m^3 for freshwater and 1025 kg/m^3 for seawater), V being flow velocity (meters per second). The velocity is the main factor, also in wind power, since it appears as a cube power. Figure 1 shows power outputs for various speeds. Wave energy also goes as the square of the amplitude and period of the wave. Hence, long periods and large amplitudes are attractive. Various wave energy generators have been used, the Indian experiments have been with oscillating water columns types (150 kWh) at Vizingham fisheries harbor near Trivandrum (Kerala) in 1991 [10]. Successful trials led to improved capacity turbine generators in 1996 of 1.1 MW. Some studies have been done on the spectral components of waves and their effect on the overall energy content [17]. The extractable power available goes as the following formula:

$$P = \left(0.55\,h^2 z\right) \tag{2}$$

where P is in kW per meter length of the wave crest. Where h = average of one-third of the highest waves in meter z = zero crossing periods in seconds. That means a significant wave height of 4 meters, with a zero crossing period of five seconds, will have the wave power of 44 kW per meter length of the wave crest. Along the Indian coast of approximately 7500 km, the average wave energy is around 5–10 kW/m. India has a total potential around 40,000 MW. At 15% utilization this would mean an output of approximately 6000 MW.

1.1 Available Technologies

Since the 1970s, many technologies have been attempted Among these are:

1. Cockerel raft
2. Flexible Bag Energy Converter
3. Submerged circular cylinder converter
4. Clamp wave energy converter
5. Oscillating water column Converter
6. Ocean swell powered renewable energy Converter

The Japan Research and Development Corporation developed wave-powered devices after a study in 1965, resulting in wave powered generators and wave powered light buoys (Wavepowerlab 2016).

1.2 Indian Ocean Networks

In the Indian Ocean, the Indian Meteorological Department (IMD) and other networks provide real-time Seismic data. Indian Ocean Earthquakes larger than Magnitude 6 can be detected within 20 min. The main advance warning method for Tsunami detection is by Deep Ocean Bottom Pressure Recorders (BPRs). In the Bay of Bengal, NIOT (The National Institute of Ocean Technology), has installed 4 BPRs, with 2 BPRs in the Arabian Sea [8]. 30 Tide Gauges to monitor the progress of tsunami waves were placed by the NIOT and Survey of India (SOI). The Integrated Coastal and Marine Area Management (ICMAM), has validated the Tsunami Model for five historical earthquakes and inundations. Another study suggests that high concentrations of mangroves help protect the shoreline from wave action [acm.org]. In Indonesia, the UN Information Management Service (UNIMS) worked with local agencies for planning and recovery activities (Springerlink 2017). Tidal and wave energy development has been attempted in Bay of Bengal and Vizingam. The economics of the project will determine if the investments are worthwhile in the long run. Similar cooperative networks exist in the Pacific region [13].

1.3 Equations

Taking Investment for the WBREDA Sundarbans Project at 40 crores [10], and by using the annualized cost or amortized cost over the life of the project, gives an annual cost A, with P being the present value, "i" the interest rate and n the no. of years as

$$A = P * [i(1+i)^n]/[(1+i)^n - 1] \tag{3}$$

At a market rate of interest i = 10%, for an Output 3.75 MW, at 10 h per day 365 days per year, the annualized output comes to

$$3.75 * 10 * 365 = 13687.5 \text{ MWh.} \tag{4}$$

The INCOIS Centre (Indian National Centre for Ocean Information Services) operates 24 by 7. Satellite data is received from six ocean buoys—equipped with water pressure sensors. Six back-up buoys should also now be ready. Two faults in the Indian Ocean can cause a tsunami and the Bay of Bengal, are potentially hazardous [1].

1.4 The potential for Incorporating Wave and Tidal Energy Generators in Early Warning Buoy Systems

Although higher wave heights exist at higher and lower latitudes, good wave potentials exist around the Indian coast. Potential sites in India have been identified as Gulfs of Cambay, and Kutch in Gujarat with maximum tides in the range of 11 and 8 m and average tidal range of 6.77 and 5.23 m respectively [10]. Power generation potentials are 7000 and 1200 MW. At Durgaduani creek in the Sundarbans, a potential of 100 MW with a maximum tidal range of 5 and 2.97 m has been identified, and in Belladonna creek, mean tidal range of 3.6 m is estimated. All these regions are in gulf and inlet areas which act as natural funnels for the gathering of wave energy. Global maps of wave heights illustrate the typical wave distributions as being highest above and below 40° latitude (Fig. 2).

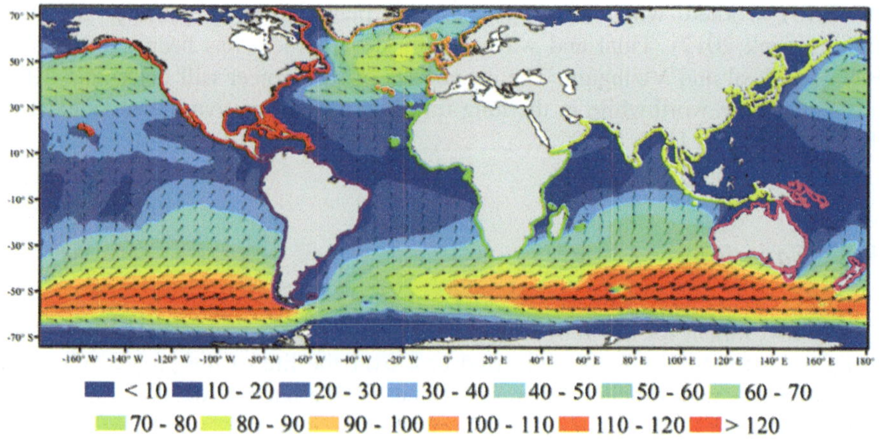

Fig. 2 Global distribution of wave amplitudes [7]

2 Potential for Harvesting Wave Energy

Estimates [2] were that hydroelectric potential was of the order of 10% of the total incident offshore wave energy. A typical wave energy harvesting schematic is given below in Fig. 3

The tabulation shows a cost ranging from 10 cents per kWh to 30 cents per kWh. In Rupee terms, this translates to approx 5 Rs to 20 Rs per kWh. These figures compare favorably with the estimates given in the earlier part of this paper for the Bengal plant, rate of interest 10–15% over 20–25 years (Table 1).

The use of energy transfer from surface waves to the floor has been studied in view of the fact that muddy sea floors can attenuate surface turbulence [1]. Significant surface energy may be extractable using an artificial floor carpet in this manner. A review of the current understanding of tidal energy from stream power generation was given [2]. Around the same time reviews were done, estimating tidal electric power generation and giving the current state of the art and its status [3, 4]. Recently, specific application to Malaysia [14] was done, using a Utility function.

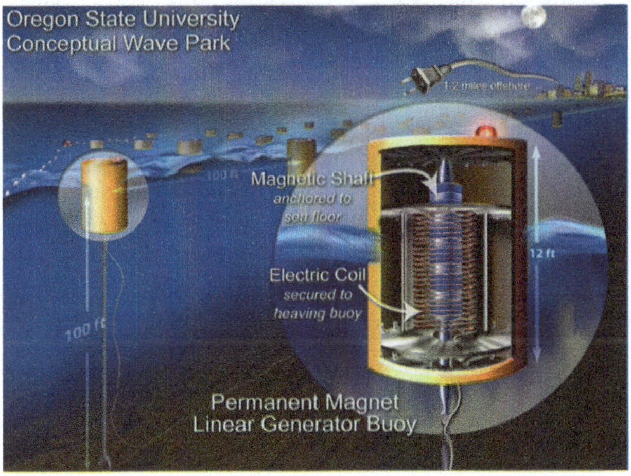

Fig. 3 Wave energy park concept (oregonstate.edu, 2017)

Table 1 Cost per kWh from various power generation units offshore the USA from Bedard [2]

Commercial plant	Hawaii	Oregon	Calif	Calif	Mass	Maine
No of units needed for 300,000 MWh/yr	180	130	215	152	206	615
Total plant investment (m$)	220	235	229	238	273	235
Annual O&M cost (m$)	11	11	13	11	12	33
10 yr. refit cost (m$)	24	23	23	15	26	74
COE (cents/kWh) nominal	12.4	11.6	13.4	11.1	13.4	39.1

A recent study has evaluated the generation of power from ocean-based generators regarding matching it to electric grid systems using network and impedance theory [15].

3 Economic Analysis

Supposing it supplies only 25% of its capacity, then from Eq. (3), the actual output is 3421.9 MWh. At a market rate **i** of 10% over 25 years, the annual capitalized cost is Rs 4.4 crores (one crore is 10^7). Taking an annual maintenance cost of 2% gives 80 lakhs per year (one lakh is 10^5). Hence total annual cost is 5.2 crores. Equating the power output at 25% capacity for 10 h per day gives a cost of 3799 per MWh or Rs 3.8 per kWh. At a rate of 15% over 20 years: annualized cost is 6 crores, with annual maintenance of 80 lakhs per year, give a total of Rs 6.8 crores for 3421.9 MWh, yielding a rate of 19.8 per kWh. These estimates may be compared to the rate in Bangalore of 5–6 Rs per kWh. Appropriate adjustments can be made for inflation and incremental increases in maintenance, depreciation, and other factors. Sufficient sites exist for incorporating wave generators into the information gathering network [7]. It is seen that sufficient potential exists in the Indian Ocean region for energy generation and can be availed of by neighboring countries. As regards the logistics, one must make sense of a large number of devices and agencies that operate data gathering units in the area. A start has been made in the Disaster management regimes initiated amongst the SAARC countries [9, 12] (Wikipedia 2017) and amongst the Asia Pacific nations [6]. The possibility of using these sensors to track aircraft and naval vessels is also possible [13]. A rough calculation of the economics shows that the tide and wave energy production schemes can be self-sustaining at current market rates. However, actual implementation is still far off due to regional differences and bureaucracy.

4 Conclusion

The key to generating power from ocean currents lies in the harvesting of streams situated near the coast lines. Among these are currents such as the Gulf Stream and others in the Indian Ocean. Apart from tidal currents, surface motions (bobbing) can also be harvested based on wave amplitudes and frequencies. Utilization of the density difference to augment energy output results in a diameter reduction of approximately 10, as compared to wind flows. This factor can lead to enormous amounts of energy harvest from ocean currents. One may also consider the economic rate of return (ROI) as an efficiency measure, instead of mechanical efficiency, since the source of power is renewable and sustainable solar and wind energy which impacts and is absorbed by the ocean.

Acknowledgements The author also gratefully acknowledges the helpful comments and suggestions of the reviewers, which have improved the presentation.

References

1. Alam E, Dominey-Howes D, Chagué-Goff C, Goff J (2012) Tsunamis of the Northeast Indian Ocean with a particular focus on the Bay of Bengal region—a synthesis and review. Earth Sci Rev 114:175–193
2. Bedard R et al (2015) Offshore wave power feasibility demonstration project, 2005. E21 EPRI Global WP009-US Rev2, 22 Sept 2005
3. Benelghali S, BenBouzid M, Charpentier JF (2007) Marine tidal current electric power generation technology: state of the art and current status. In: Proceedings of IEEE, IEMDC07, pp 1407–1412 (HAL-00531255)
4. Blumden IS, Bahaj AS (2007) Tidal energy resource assessment for tidal power generators. Proc Inst Mech Eng A J Power Eng. https://doi.org/10.1243/09576509jpe332rences
5. DBEDT (2002) Feasibility of developing wave power as a renewable energy resource for Hawaii. Department of Business, Economic Development, and Tourism, Honolulu
6. Disaster Risk Management in Asia and the Pacific issues paper. A joint study of the Asian development bank and the Asian development bank institute, April 2013
7. Gunn K, Stock-Williams C (2012) Quantifying the global wave power resource. Renew Energy 44:296–304
8. INCOIS. www.incois.gov.in/portal/OON.jsp
9. Kawata Y (2011) Downfall of Tokyo due to devastating compound disaster. J Disaster Res 6(2):176–184
10. Khan BH (2009) Nonconventional energy resources. Tata McGraw Hill, New Delhi
11. Lutjeharms JRE, Bang ND, Duncan CP (1981) Characteristics of the currents east and south of Madagascar. Deep Sea Res A 28(9):879–899
12. Nayak S, SrinivasKumar T (2012) Indian Tsunami warning system. www.isprs.org/proceedings/XXXVII/congress/4_pdf/262.pdf, 2011
13. PRIS "Pacific Risk Information System". pcrafi.spc.int/beta. Accessed 25 July 2017
14. Raja P (2015) Utility function and sustainable development—case study of Malaysia. In: Proceedings of IRES 7th international conference on Kuala Lumpur Malaysia, 18.8.2015. ISBN 978-93-85465-77-2
15. Safayet AM (2015) Distributed voltage regulation and grid connection of renewable energy sources. Dissertation, NCSU
16. Tomczak M, Godfrey JS (2003) Regional oceanography: an introduction, 2nd edn. Daya Publishing House, New Delhi. wavepowerlab.weebly.com/blog. Accessed 24 Sept 2016
17. Yagci B, Wegener P (2009) Determination of sea conditions for wave energy conversion by spectral analysis. In: Proceedings of the 8th European wave and tidal energy conference, Uppsala, Sweden

Web-Based Intelligent RFID Facility Maintenance Systems

Chien-Ho Ko

1 Introduction

The operating condition of any facility directly affects occupant satisfaction. However, a facility's functions deteriorate with time. To extend the service life of constructed facilities, a well-tuned maintenance program is required [2, 10]. Facility functions are provided by the facilities themselves and their equipment. Both depend on continuous maintenance to maintain normal functions. Assigning an ID to each facility and apparatus is a regular way to manage them. Barcodes seem to be the most frequently employed system for acquiring the information. However, barcodes printed on paper are easily broken. In addition, obstacles between the barcode and laser scanner can interrupt optical communications. These problems limit their efficacy in facility maintenance [4, 7].

Radio Frequency Identification (RFID) technology has characteristics of repetitive write and read abilities. It can access tags without physical contact. This technology has become one of the most important management systems in recent decades [9, 14]. Various reports can be found under facility management. For example, [3] surveyed innovations in facility management and claimed RFID a hopeful information technology (IT) in the field. Legner and Thiesse [13] also applied RFID to maintenance at Frankfurt Airport. Wing [16] reported RFID applications in construction site as well as facility management. The authors combined this technology with portable devices for implementing asset management operations. Another study overcame difficulties faced in facilities maintenance

C.-H. Ko (✉)
Department of Architecture, Massachusetts Institute of Technology,
77 Massachusetts Ave, Cambridge, MA 02139, USA
e-mail: ko@mail.npust.edu.tw

C.-H. Ko
Department of Civil Engineering, National Pingtung University of Science
and Technology, 1, Shuefu Road, Neipu, Pingtung 91201, Taiwan

S. Şahin (ed.), *8th International Conference on Engineering, Project, and Product Management (EPPM 2017)*, Lecture Notes in Mechanical Engineering,
https://doi.org/10.1007/978-3-319-74123-9_4

by using the same technology [6]. Ergen et al. [6] improved data transfer between maintenance workers by this technology. Although various research works have investigated advantages of applying RFID technology to facility maintenance, a comprehensive system integrating the required management technologies has rarely been discussed. How does the RFID technology work with the practical maintenance environment? In addition, a tremendous amount of maintenance data are collected and created while implementing maintenance jobs. How the collected records are analyzed through the assistance of RFID technology has rarely been fully addressed [12].

The main purpose of this research work is to improve efficiency of facility management by integrating RFID, web-based, and artificial intelligence. To accomplish the set goal, first, RFID technology is reviewed from published journal and conference papers. Three modules, i.e. prediction, statistical, and data management modules are then developed based on the requirements on facility management. To understand the performance in a practical maintenance environment, an experiment was conducted with RFID devices using environmental challenges. Finally, feasibility of the developed system is tested by a real case. Implementation outcomes are discussed in the paper.

2 RFID Technology

RFID technology is first proposed by Stockman in 1948 [15]. The original purpose of using the method is to fulfill the need to identify useful applications for "reflected-power communication." One of the first large-scale commercial uses was documented in the 1990s in electronic toll collection [1]. After that, RFID has been used in various kinds of fields. These days, RFID is a general technological term that uses radio signal to identify objects.

A classical RFID architecture is demonstrated in Fig. 1. In the figure, radio waves are sent by a transceiver through an antenna. The transponder (noted as RF tag in the figure) is awakened by the radio signal. Data stored on it are read and written by the requested signals emitted from the antenna. RF tag transfers data according to a request sent from the transceiver. Data can thus be transferred for processing accordingly [5, 8].

3 Web-Based RFID Facilities Management System

3.1 Hardware and Software

Facilities operating conditions are referred to functions provided by facilities themselves and their apparatuses. Therefore, facility maintenance as discussed in

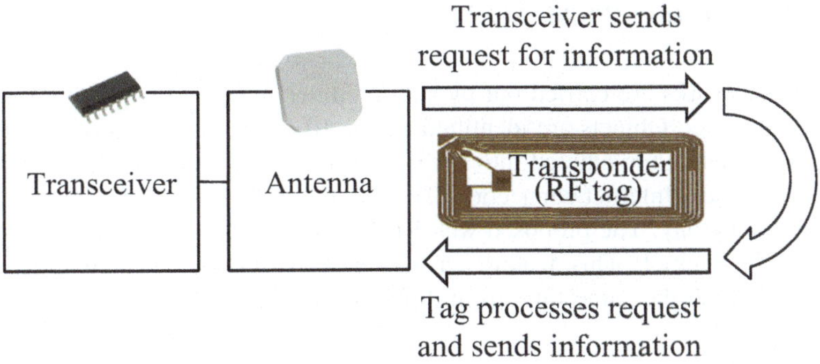

Fig. 1 Schema of typical RFID system

this study includes facilities and any facility apparatus. Portable RFID devices (reader/writer) are adapted to conform to mobile requirements. RFID devices are chosen by considering weight, transmission power, size, interface, price, and frequency range. Ensyc RFID Block shown in Fig. 4 is thus selected. Gen 2 RFID tag displayed in Fig. 2 on Ensyc RFID Block is selected with the corresponding communication protocol.

To allow multiple users to use the system, a web-based application is planned. ASP.NET framework is select as the development platform. Microsoft ASP.NET can integrate the selected RFID hardware with web application. It supports browsers in executing different kinds of platforms such as the PC, mobile devices, Palm, etc. Users can manipulate these systems using all kinds of hardware and software platforms.

(a) RFID Block (http://www.ensyc.com) **(b) Gen 2 RFID tag (http://www.ni.com)**

Fig. 2 Selected RFID hardware and Gen 2 RFID tag

3.2 Data Management Module

Maintenance works are carried out by staff members using portable RFID devices attached on tablet. Objects are identified by the RFID device. As shown in Fig. 3, maintenance data were entered into the system using the web-based RFID system through a wireless Internet connection. The number of facilities and equipment may increase day by day. The proposed web RFID management system connects to the database via internet. This feature enables users to operate maintenance works simultaneously at the same time from different locations.

3.3 Statistical Module

The developed system updates the database in a real time base. The statistical module thus can offer an ability of avoiding data re-typing. In addition, duplicate maintenance and missing maintenance activities can be eliminated from the maintenance works. Maintenance records stored in the database can be analysed and displayed using graphs at any time with up-to-date status from anywhere that has internet connection. A graph summarizing the purchasing cost is shown in Fig. 4.

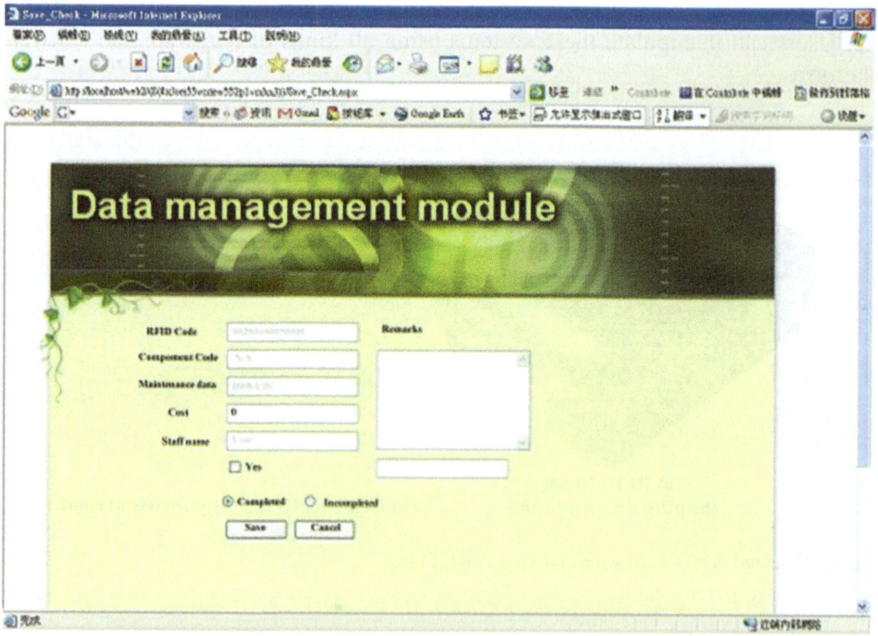

Fig. 3 Data management module

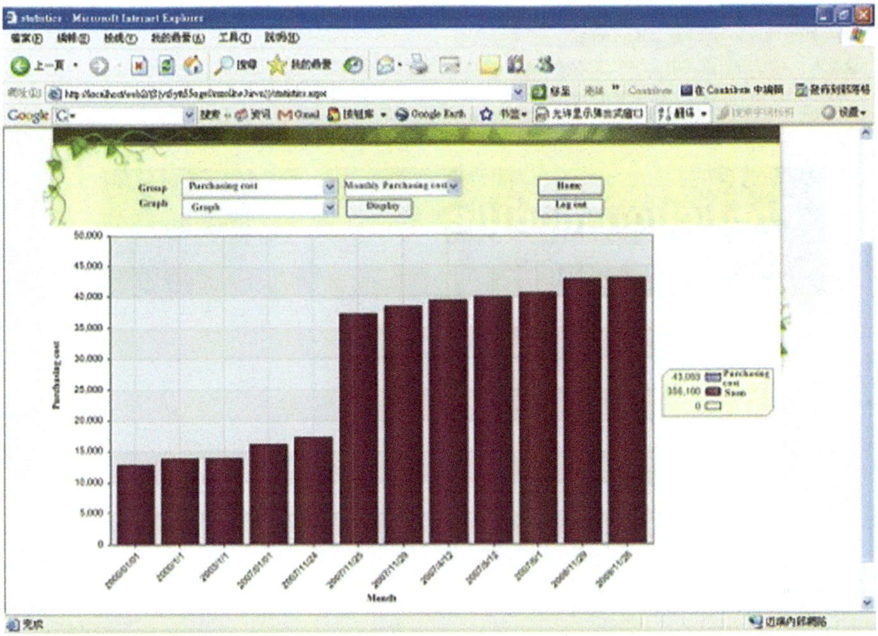

Fig. 4 Statistical module

3.4 *Prediction Module*

Using fuzzy neural networks, the developed prediction module can forecast the possible lifetime of objects through historical maintenance data, as shown in Fig. 5. Users can configure a prediction model for a specific component. Parameters of the prediction model can be modified corresponding to user's requirements. Facility functions can therefore be ensured before the next maintenance.

A fuzzy neural network is employed to develop a prediction model. In this module, uncertain information is processed using fuzzy logic. Fuzzy rules in the fuzzy logic are formulated using artificial neural networks. Mapping relationships of inputs (influencing factors) and output (component lifetime) are allocated using learning algorithms. The fuzzy neural network architecture used in the prediction module is comprised of four layers. First layer (input layer) of the fuzzy neural networks gets the input data and sends them to the backward neurons in the fuzzification layer. The fuzzification layer converts crisp inputs into fuzzy values using membership functions. Intermediate Neurons process the input signals. Finally, output layer processes the fan-in signals and generates output signals. Since this module calculates component lifetime as inference result, the fuzzy neural network only involves single output neuron.

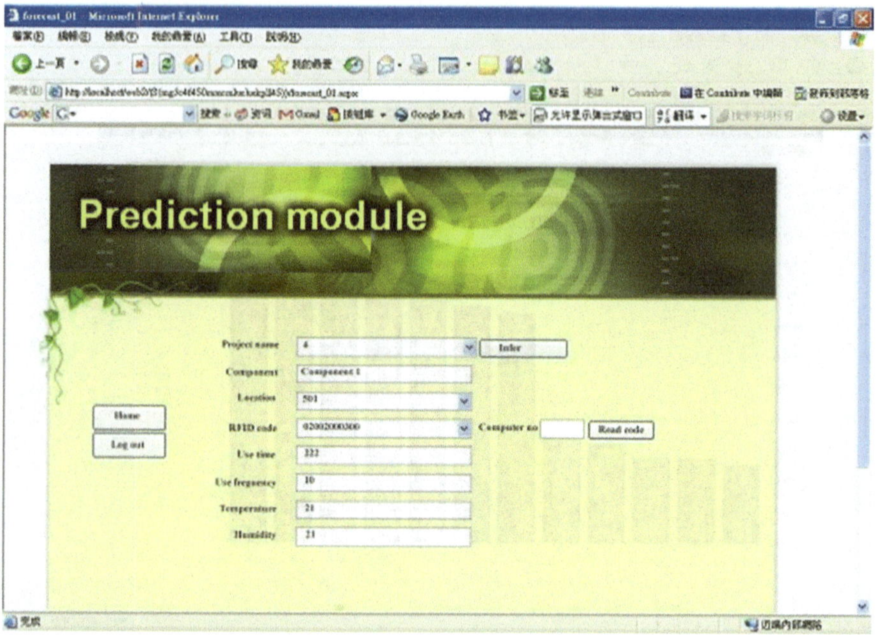

Fig. 5 Prediction module

4 Experiments

Effective ranges of the RFID device are influenced by environmental factors [11, 12]. To analyze the environmental impacts on the used RFID equipment, practical conditions encountered in facilities maintenance were provided. Nearby appliances, metal, and dusty conditions were thus tested. The experimental results are depicted in Table 1. The effective range of the selected RFID device is 10 cm.

RFID tags can become covered with dust over a short time span. To understand the influence of dust, grime conditions were conducted. Table 2 demonstrates the experiment results. The results show that grime and dust have limited influence on the effective RFID reading range.

Facilities and their equipment frequently consist of metal parts. Because metal reflects radio waves, metal parts have been regarded as the most challenging issue for effective RFID reading ranges. Table 3 shows the experimental results. From

Table 1 Effective ranges with normal conditions

Distance (cm)	Successful times	Failure times
8	5	0
9	5	0
10	5	0
11	0	5
12	0	5

Table 2 Effective ranges with grime conditions

Distance (cm)	Successful times	Failure times
7	5	0
8	5	0
9	5	0
10	0	5

Table 3 Effective ranges with metal conditions

Distance (cm)	Successful times	Failure times
0	0	5
1	3	2
2	5	0
3	5	0

Table 4 Effective ranges nearby running appliance

Distance (cm)	Successful times	Failure times
0	0	5
1	0	5
2	0	5
3	5	0
4	5	0
5	5	0

the table, in RFID applications, it is inappropriate to Placing an RFID tag directly onto a metal surface is.

Radio signals may collide with surrounding electric waves. Electrical equipment emits electromagnetic waves that affect radio communications. Consequently, the effect of electrical appliances on RFID signals was examined. A personal computer, regarded as one of the most popular electrical devices, was used in this experiment. The experimental results summarized in Table 4 show that a minimum distance of 3 cm is required to avoid the impact of the electromagnetic waves emitted by electrical appliances like PCs.

A real case of building maintenance was used to validate the feasibility of the developed prediction module. More than 72 historical fluorescent maintenance records were used to train the module. The derived prediction module could obtain 79% accuracy.

5 Conclusions

The development of a web-based RFID facility management system was presented in this study. The system consists of a prediction module, a data management module, and a statistical module. Maintenance jobs are carried out using a tablet PC attached with a portable RFID reading/writing device. RFID performance was examined to verify its applicability in a practical environment.

The proposed application software combined internet with RFID technology. The system could identify object IDs through electronic tags, thus prevents typing errors and reduce maintenance time. Unlike conventional barcode systems, information stored in RFID tags can be re-written conveniently. In addition, RFID tags have been proven practical under challenging conditions.

The developed web facility maintenance system can be operated with wire or wireless Internet connections. This characteristic improves facility management efficiency. Regarding statistical module, it provides functions to display historical maintenance records using graphical charts. Maintenance data could be visualized accordingly. This research forecasts possible lifetime of facility components using a fuzzy neural network. Using the predicted results, maintenance staff could arrange maintenance schedules. Therefore, malfunctions and unexpected breakdowns could be reduced.

References

1. Association for Automation Identification and Data Capture Technologies (AIM) (2002) Shrouds of time: the history of RFID. http://www.aimglobal.org. Accessed 30 March 2017
2. Barco AL (1994) Budgeting for facility repair and maintenance. J Manag Eng 10(4):28–34
3. Cardellino P, Finch E (2006) Mapping IT innovation in facilities management. Electron J Inf Technol Constr 11:673–684
4. Djerdjouri M (2005) Assessing and benchmarking maintenance performance in a manufacturing facility: a data envelopment analysis approach. INFOR 43(2):121–133
5. Domdouzis K, Kumar B, Anumba C (2007) Radio-frequency identification (RFID) applications: a brief introduction. Adv Eng Inform 21(4):350–355
6. Ergen E, Akinci B, East B, Kirby F (2007) Tracking components and maintenance history within a facility utilizing radio frequency identification technology. J Comput Civil Eng ASCE 21(1):11–20
7. Ko CH (2009) RFID-based building maintenance system. Autom Constr 18(3):275–284
8. Ko CH (2010) RFID 3D location sensing algorithms. Autom Constr 19(5):588–595
9. Ko CH (2013) 3D-Web-GIS RFID location sensing system for construction objects. Sci World J Article ID 217972, p 8
10. Ko CH, Pan NF, Chiou CC (2013) Web-based radio frequency identification facility management systems. Struct Infrastruct Eng 9(5):465–480
11. Ko CH (2015) Integrating RFID, web-based technology, and artificial intelligence in engineering management. Sci Iran 22(2):299–312
12. Ko CH (2017) Accessibility of radio frequency identification technology in facilities maintenance. J Eng Proj Prod Manag 7(1):45–53
13. Legner C, Thiesse F (2006) RFID-based maintenance at Frankfurt airport. IEEE Pervasive Comput 5(1):34–39
14. Ngai E, Riggins F (2008) RFID: technology, applications, and impact on business operations. Int J Prod Econ 112(2):507–509
15. Stockman H (1948) Communication by means of reflected power. Proc Inst Radio Eng 36(10):1196–1204
16. Wing R (2006) RFID applications in construction and facilities management. Electron J Inf Technol Constr 11711–11721

Causal Relationships of Construction Performance Using the Balanced Scorecard

Jainnarong Jantan, Veedard Tesan, Pitchayanan Purirodbhokhin, Sasawat Aree, Jean Meenchainant, Katawut Noinonthong and Thanwadee Chinda

1 Introduction

Construction industry is one of the important industries, as it has a great impact on the economy of the nation. It makes a vital contribution to the competitiveness and prosperity of the economy. It has a major role in delivering the built infrastructure in an innovative and cost effective way. Firms throughout the economy are dependent on the performance of built infrastructure, such as roads, rail, power stations, and telecoms networks to remain competitive, and make location decisions. In Thailand, the growth of construction spending in 2014–2019 is around 4.5%, which is the second rank in Southeast Asian region following Vietnam (see Fig. 1) [3].

Based on SAP [20], construction companies now face four major challenges to survive and improve its growth, including (1) poor productivity and profitability, (2) project performance, (3) skilled labor shortages, and (4) sustainability concerns. These main outputs will indicate the performance of the organization and quality. However, the productivity also depends on the efficiency and nature of the built environment. The flexibility, mobility and effectiveness of the workforce and the productivity of firms depend on the availability of properly configured and located houses and premises [20].

Many research studies attempt to improve project performance in the construction industry utilizing a number of international performance measurement

J. Jantan (✉) · V. Tesan · P. Purirodbhokhin · S. Aree
J. Meenchainant · T. Chinda
School of Management Technology, Sirindhorn International Institute of Technology, Thammasat University, 131 Tiwanont Rd. Bangkadi, Muang, Pathumthani 12000, Thailand
e-mail: jainnarongjane@gmail.com

K. Noinonthong
J.A.T. Ground Expert Co. Ltd, 8/2 Buengkhamphroi, Lumlukka, Pathumthani 12150, Thailand

S. Şahin (ed.), *8th International Conference on Engineering, Project, and Product Management (EPPM 2017)*, Lecture Notes in Mechanical Engineering,
https://doi.org/10.1007/978-3-319-74123-9_5

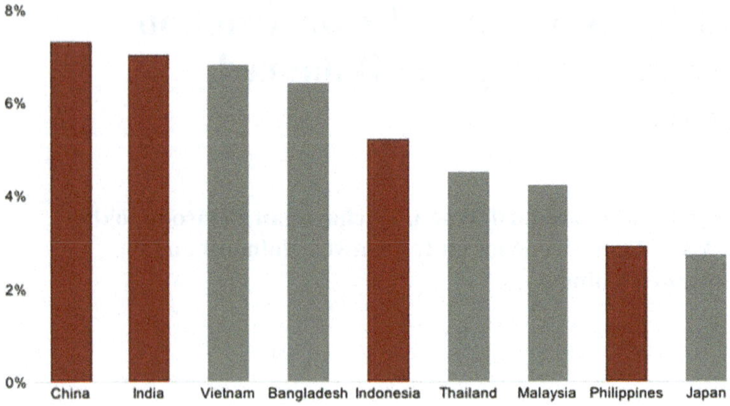

Fig. 1 Growth of construction spending in 2014–2019 [3]

models. Enshassi et al. [8], for example, examined performance of a construction project in Gaza Strip, and summarized six key factors, including (1) delays because of borders/roads closure leading to materials shortage, (2) unavailability of resources, (3) low level of project leadership skills, (4) escalation of material prices, (5) unavailability of highly experienced and qualified personnel and (6) poor quality of available equipment and raw materials. Al-Otaibi et al. [1] adopted a new standard design model (SDM) approach to improve a construction project performance in Saudi Arabia in the design modification, pre-construction, construction, and post-construction phases. Davis and Sammy [7] utilized a balanced scorecard (BSC) to achieve a quality excellence in the construction industry, and mentioned four main perspectives, namely financial, internal business process, learning and growth, and customer perspectives. Spender [21] stated that the BSC has been adopted in more than half of all major firms to improve their performance.

Despite the above researches, there is a need to examine causal relationships among key factors affecting a construction performance to better plan for construction improvement. This paper, therefore, develops a causal loop diagram, based on the BSC, to understand the relationships and feedback of factors affecting construction performance. It is expected that the study results assist a construction company to better plan for its performance.

2 Balanced Scorecard: Key Factors and Items Affecting Construction Performance

Balanced scorecard is a tool for performance management that an organization uses to measure the performance. It has been used in many construction-related studies. Oyewobi et al. [16], for example, examined and compared a performance measurement system and performance frameworks commonly used within the construction industry, and concluded that the BSC can be used to help organizations achieve performance excellence, financial integrity, and continuous improvement in business results to sustain competitive advantage. Maya [13] examined the application of balanced, integrated, and recent performance management system to be used as a tool to measure and manage Syrian construction projects performance, and concluded that the BSC performance management framework proved to be an excellent tool for performance measurement and management.

The BSC framework is not only used to measure financial perspective, but also customer, internal business, and learning and growth perspectives (see Fig. 2) [12]. Financial perspective concerns about the financial performance in company in relation with profitability, growth, and shareholders' value. Customer perspective, on the other hand, concerns about customer satisfaction and customer focus. Internal Process perspective assists companies in deciding processes to be implemented, such as, training program, communication tools, and green image.

Learning and Growth perspective examines three main parts, including leadership, employee, and innovation. Leadership considers leaders' roles in motivating their employees to work more effective. Employees, on the other hand, should have skills, abilities, and knowledge to improve performance of the organization. Various innovative and information technology should also be provided to enhance work performance and reduce rework.

In this study, a total of six key factors, including (1) Financial, (2) Customer, (3) Internal Process, (4) Leadership, (5) Employee, and (6) Innovation perspectives are used, based on the BSC, to improve a construction performance. Based on a number of construction-related literatures, each factor consists of a number of items to explain its construct.

- Financial factor: This factor consists of three items, including (1) cost, (2) available budget, and (3) market expansion. Tenant and Langford [22], for example, mentioned construction cost as a key criterion of financial perspective.
- Customer factor: This factor is associated with five items, namely (1) customer satisfaction, (2) customer focus, (3) customer relationship, (4) competitiveness, and (5) time. Davis and Sammy [7], for example, mentioned customer satisfaction as an indicator in the customer perspective. BSC Designer [2], on the

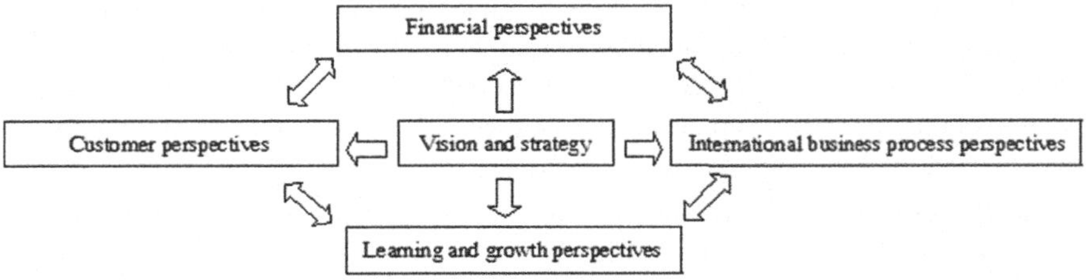

Fig. 2 Balanced scorecard tool [7]

other hand, stated three objectives of customer perspective, including product quality, customer experience, and time.

- Internal Process factor: This factor consists of five associated items, namely quality management checklist, green image, specific training program, quality policy, and rework. Environmental impact must be considered when improving the internal process perspective [7]. Mistakes and rework should also be minimized to enhance work performance [2].
- Leadership factor: Role model, management review, management commitment, and top-down communication are four items explaining the Leadership factor [2, 4].
- Employee factor: This factor consists of seven items, including incentive, employee involvement, employee attitude, worker's skill, motivation, teamwork, and turnover rate [4, 12].
- Innovation factor: This factor is associated with three items, namely communication tool, research and development, and continuous improvement [2, 7].

In summary, six key factors, including (1) financial, (2) customer, (3) internal process, (4) leadership, (5) employee, and (6) innovation perspectives are used, together with a total of 27 items, to investigate causal relationships to enhance performance of the construction industry.

3 Causal Relationships Among Key Factors and Items Affecting Construction Performance

Causal loop diagram is a diagram used in visualizing how interrelated items or factors affect one another [18]. It identifies processes and root causes. It also brings out the systematic feedback in processes. The diagram consists of a set of nodes representing the items connected together. The relationships between the items and factors are represented by arrows, and can be considered as positive or negative causal links [19]. Positive causal link ($\xrightarrow{+}$) explains that the two nodes change in the same direction, while in negative causal link ($\xrightarrow{-}$), a change in a node changes the other node in opposite direction.

The relationships between the items or factors can also be defined in terms of positive or negative feedback loops. A positive feedback loop ((+) or (+)) explains that if a particular element starts the loop by changing its value in one direction, and closes the loop with the value changed in the same direction, it is a positive feedback loop. If a particular element, on the other hand, starts the loop by changing its value in one direction, and closes the loop with the value changed in the opposite direction, it is then a negative feedback loop ((−) or (−)).

The 27 items under six key factors affecting a construction performance form a causal loop diagram, as shown in Fig. 3.

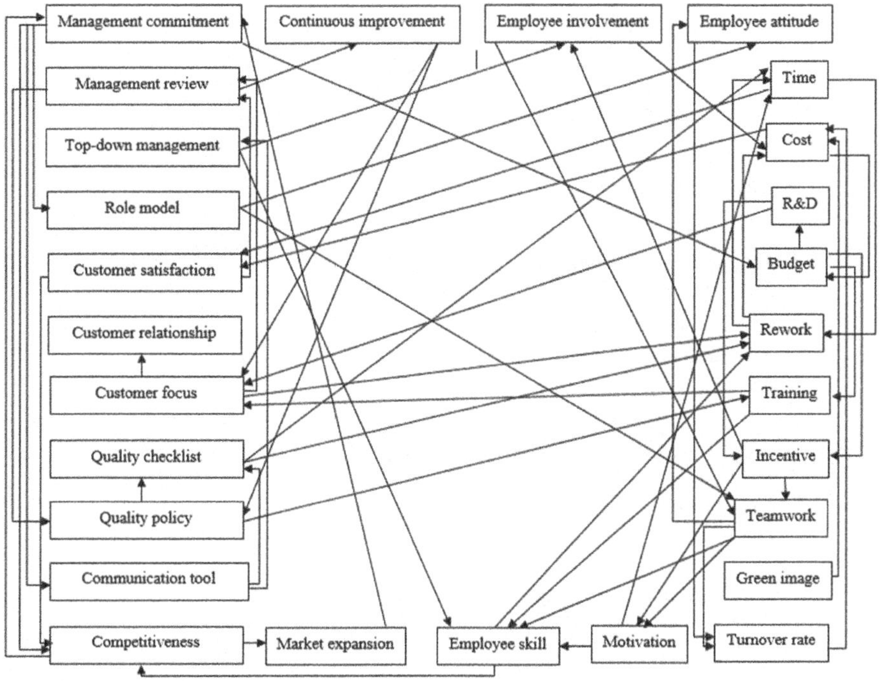

Fig. 3 A causal loop diagram of construction performance using the balanced scorecard

The diagram explains a number of positive and negative causal loops. With management commitment in improving performance, for example, a number of communication tools are used to manage top-down communication (positive links) [5]. This brings more employees' involvement and teamwork in performance program implementation (positive links) [6]. With good cooperation in the team, work skill is enhanced (a positive link), resulting in less rework (a negative link) and cost (a positive link), and high customer satisfaction (a positive link) [2, 12]. This raises company's competitiveness, and expands the market, resulting in more commitment from management (positive links) [5, 14, 17]. This thus closes a positive loop of the 11 items, and the six key factors affecting construction performance (see Figs. 4 and 5).

Good management review leads to a better and realistic quality policy, in which it could result in proper training for employees [10, 11]. This represents a positive link from the management review to quality policy items, and from the quality policy to training items. Employees have higher working skill when received the training. This thus reduces rework, representing a negative causal link from the employees' skill to rework items [12]. Less rework reduces cost, and enhances customer satisfaction, which, in turn, brings a better management review [17]. This closes a positive causal loop of the seven items, as shown in Fig. 6.

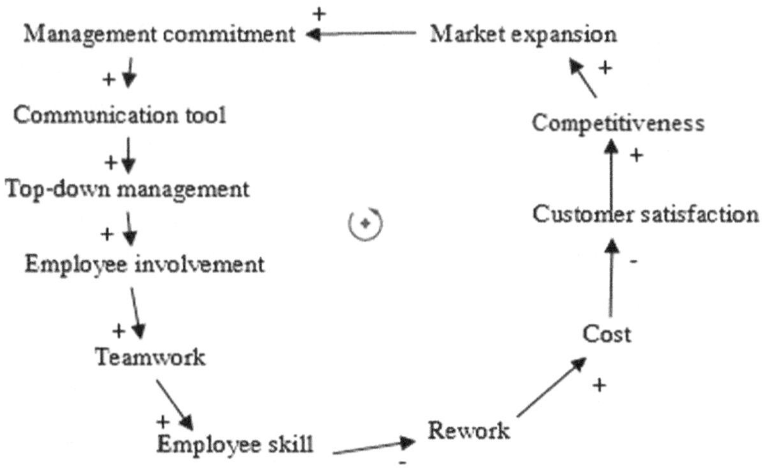

Fig. 4 A positive causal loop of the 11 items

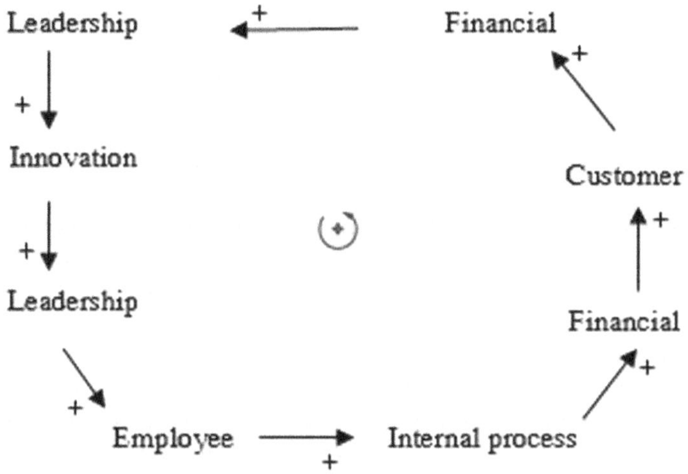

Fig. 5 A positive causal loop of the six key factors

When budget is available, incentive program can be used to motivate workers to achieve a better performance by reducing working time [12, 15]. This, however, can lead to more rework and cost, resulting in fewer budgets provided for the performance improvement program [12]. This closes a negative causal loop of the six items, as shown in Fig. 7.

Fig. 6 A positive causal loop of the seven items

Fig. 7 A negative causal loop of the six items

4 Continuing Study

The causal loop diagram of construction performance using the balanced scorecard will further be used to develop a dynamic model of construction performance assessment utilizing a system dynamics modeling. Causal relationships between the six key factors, together with their 27 items, are used to set equations for the dynamic model. Applications of the BSC framework will further be investigated through the developed dynamic model.

5 Conclusion

The balanced scorecard is an effective tool for improving construction performance. In this study, six key factors, including (1) financial, (2) customer, (3) internal process, (4) leadership, (5) employee, and (6) innovation factors are concluded based on the balanced scorecard. They are associated with a total of 27 items, to form a causal loop diagram to explain a number of positive and negative links, as well as positive and negative causal loops of construction performance. The construction company can use the study results to better understand the relationships between factors and items affecting construction performance, and plan for a better improvement.

References

1. Al-Otaibi S, Osmani M, Price ADF (2013) A framework for improving project performance of standard design models in Saudi Arabia. J Eng Proj Prod Manag 3(2):85–98
2. BSC Designer (2017) The customer's perspective of the balanced scorecard. http://www.bscdesigner.com/customer-perspective.htm. Accessed 13 June 2017
3. Building Radar (2016) Asian construction market forecast from 2015–2020. https://buildingradar.com/construction-blog/asian-construction-market-forecast-from-2015-2020/. Accessed 13 June 2017
4. Chinda T (2008) Mapping the European foundation for quality management excellence model with the construction safety management. In: Proceedings of the 4th international global academy of business and economic research conference, Bangkok, Thailand, pp 418–437, 28–30 Dec 2008
5. Chinda T (2012) A safety assessment approach using safety enablers and results. Int J Occup Saf Ergon 18(3):343–361
6. Chinda T (2014) Organizational factors affecting safety implementation in food companies in Thailand. Int J Occup Saf Ergon 20(2):213–225
7. Davis LTC, Sammy WKM (2017) Balanced scorecard for quality excellence in the construction industry: a success story. http://www.hksq.org/bsc_atal.pdf. Accessed 13 June 2017
8. Enshassi A, Mohamed S, Abushaban S (2009) Factors affecting the performance of construction projects in the Gaza Strip. J Civ Eng Manag 15(3):269–280
9. Fayek R, Hafez SM (2013) Applying lean thinking in construction and performance improvement. Alexandria Eng J 52(4):679–695
10. ISO9001 (2008a) Management review. https://www.iso-9001-checklist.co.uk/tutorial/5.6-management-review.htm. Accessed 13 June 2017
11. ISO9001 (2008b) Quality policy. https://www.iso-9001-checklist.co.uk/tutorial/5.3-quality-policy.htm. Accessed 13 June 2017
12. Kaplan RS, Norton DP (1992) The balanced scorecard—measures that drive performance. Havard Bus Rev. https://hbr.org/1992/01/the-balanced-scorecard-measures-that-drive-performance-2. Accessed 10 June 2017
13. Maya RA (2016) Performance management for Syrian construction projects. Int J Constr Eng Manag 5(3):65–78
14. Milekić MJ (2013) Balanced scorecard concept of balanced measurement of performances. In: Proceedings of the 17th international research/expert conference: trends in the development of machinery and associated technology, Istanbul, Turkey, 10–11 Sept 2013

15. Niven PR (2005) Balanced scorecard diagnostics. John Wiley and Sons, New Jersey
16. Oyewobi LO, Windapo AO, Rotimi JOB (2015) Measuring strategic performance in construction companies: a proposed integrated model. J Facil Manag 13(2):109–132
17. Poveda-Bautista R, Baptista DC, Garcia-Melon M (2012) Setting competitiveness indicators using BSC and ANP. Int J Prod Res 50(17):4738–4752
18. Rasmussen A, Muller G, Pennotti M (2010) Causal loop based change propagation and risk assessment. http://www.gaudisite.nl/EUSEC2010_RasmussenEtAl_CausalLoop.pdf. Accessed 27 Dec 2010
19. Richardson GP (1986) Problems with causal-loop diagrams. Syst Dyn Soc 2(2):158–170
20. SAP (2017) Top 4 challenges facing the construction industry. http://www.digitalistmag.com/future-of-work/2016/08/15/top-4-challenges-facing-construction-industry-04388065. Accessed 13 June 2017
21. Spender JC (2014) Business strategy: managing uncertainty, opportunity, and enterprise. Oxford University Press, Oxford
22. Tenant S, Langford D (2017) The construction project balanced scorecard. http://www.arcom.ac.uk/-docs/proceedings/ar2008-361-370_Tennant_and_Langford.pdf. Accessed 12 June 2017

Understanding Associations Between Project Team Involvement, Project Design and Project Outcomes: A Case Study of Health Development Projects in Thailand

Jantanee Dumrak, Nick Hadjinicolaou, Bassam Baroudi and Sherif Mostafa

1 Introduction

How a project is designed generally impacts on project implementation and outcome delivery. Judicious project design will hopefully produce successful and satisfactory results. Thus, including the project team in the design process at an early stage is deemed necessary. Like many development projects, reproductive health development (RHD) projects in Thailand are commonly funded by international donors through international development agencies. These projects are implemented by governmental recipients and/or non-governmental organizations under agreements between the participating parties. According to Golini et al. [1], international projects are defined and designed in develop countries but executed in different context. Project donors and sponsors are likely to influence on the initial design of projects at the high level when projects are initiated and not fully developed [2]. At this stage, project strategies are aligned with the strategies of the program before a detailed design phase is proceeded whereas project plans, time and cost estimation as well as monitoring and control processes are developed by project designers assigned by project donors and sponsors [3]. Relying solely on

J. Dumrak (✉) · N. Hadjinicolaou
Global Project Management, Torrens University Australia,
88 Wakefield Street, Adelaide, SA 5000, Australia
e-mail: jdumrak@laureate.net.au

B. Baroudi
Entrepreneurship, Commercialisation and Innovation Centre,
University of Adelaide, Nexus 10 Building, North Terrace,
Adelaide, SA 5000, Australia

S. Mostafa
Griffith School of Engineering, Griffith University,
Gold Coast Campus, Gold Coast, QLD 4222, Australia

S. Şahin (ed.), *8th International Conference on Engineering, Project, and Product Management (EPPM 2017)*, Lecture Notes in Mechanical Engineering,
https://doi.org/10.1007/978-3-319-74123-9_6

project designers for detailed design may misread micro politics, cultural sensitivity of implementing organizations, organizational and team dynamics, needs and capabilities of project teams etc. that result in weak design and project failure. Although project design and project team performance are proven as key critical success factors in many development projects, an indication of early project team involvement in project design is not clearly presented. This research paper, therefore, aims to examine the associations between the involvement in project design, project management (PM) application and implement of PM tools as well as project outcomes.

2 Literature Review

Project design is a process which a set of activities is established to satisfy project objectives [4]. Many studies [5–8] agree that effective project design results in project desirable outcomes and that an effective project design should include the expected goals and objectives to achieve the planned results [5, 7]. Project design that fails to deliver expected and satisfied outcomes can impede the project achievement.

Toakley and Marosszeky [9] identify the association between project design and project quality management stating that key quality decisions are generally made during the design stage prior to quality management implementation and assessment. Imran and Zaki [10] mention that project and organizational risks as well as logistic problems The significance of project design is to not only PM knowledge and implementation but also to overall success of a project. According to Ika et al. [11], project design is among other factors that are defined as critical success factors for projects initiated by the World Bank.

Early involvement of project teams and other stakeholders in project design can result in prevention of project failure. According to Ika [3], fatal errors in the execution of projects are caused by exclusion of key stakeholders during the project design stage. Environmental factors e.g. socio-cultural and political factors can influence successful design of a project [12]. In development projects such as RHD projects, cultural perspectives, knowledge and expectations of the project donors (sponsors) and target recipients (customers) are not always well-aligned [13] which can lead to a poorly designed project. Therefore, the role of a project team in this situation is to bring collaboration, mutual understanding and alignment of expectations during the design stage while maintaining the compliance with the funding policies. Despite unclear classification of project stakeholders in the studied projects, Mathur et al. [14] affirm stakeholder involvement is the center of project design and as a vehicle to project-related dialogues.

It can be said that to accomplish in delivering desirable project outcomes, the aspects contributing to project performance should be given attention at all stages of project management including project planning, design, implementation and operation [15]. Achieving targeted results of a project requires strong integration and penetration of project management knowledge, techniques and tools from the beginning of the project implementation. To enhance the ability in maximizing project performance, [16] argue that project design which can become incorporated into the day-to-day operations will create greater results. Project design also covers how the projects are structured as well as other important project components in respect to achieving project objectives.

3 Research Methodology

This empirical research was conducted using a questionnaire survey to obtain data from the Thai RHD projects. The distributed questionnaires targeted at the project personnel to identify the relationships between the project team involvement in project design and effectiveness in project management. Data collection was conducted from 75 participants from four RHD projects. The construction of the research questionnaire was based on *A Guide to the Project Management Body of Knowledge (5th Edition)* published by Project Management Institute [17]. Additional questions in the research questionnaire were developed using a group of experts recruited from provincial public health offices (PPHO) and a regional health promotion center (HPCR) in Thailand to ensure that this study incorporated other criteria that could result from project design. The Statistical Product and Service Solutions (SPSS) was employed to obtain statistical results mainly from the questionnaire survey. The descriptive, correlation, and inferential statistics were included in data analysis.

To achieve the aim of this research, the research analysis employed descriptive statistics, Spearman's Rho Correlation and Kruskal-Wallis Test. The descriptive statistics was mainly to present demographic information while Spearman's Rho was used to investigate the directional relationship between project design involvement of the studied project teams, effectiveness application of PM knowledge and implementation of PM tools, and other managerial criteria of the projects that could be impacted by project team involvement in project design. According to Jackson [18], when correlation coefficient (r) values are between ± 0.00 to 0.29, ± 0.3 to 0.69 and ± 0.7 to 1.00, they indicate no relationship to a weak relationship, a moderate level of relationship and strong relationship respectively. Kuskal-Wallis Test was performed to produce the results on differences of opinions between three or more groups of the respondents. The test was to examine if project team involvement was the same or different in funding sizes, project durations, number of sites and sizes of project teams.

4 Data Analysis and Findings

The research data was obtained from 75 public health and project management practitioners who were recruited to undertake four RHD projects in Thailand. Experience in managing RHD projects and roles of the respondents are illustrated in Fig. 1. It can be seen that 'Project team' was the largest group among others (74.7%) that undertook the RHD projects. Project experience of the project teams was found diverse ranging from less than 3.0 years (34.7%), 3.0 to 4.99 years (28.0%), 5.0 to 6.99 years (1.3%), 7.0 to 9.99 years (4.0%) and 10 years or more (6.7%) respectively.

An overview of RHD project team involvement in project design was taken into account using the quantitative data obtained from the questionnaire survey which is presented in Table 1. The results below responded to a question asking if the respondents were ever involved or included in the RHD project designing process. It was found that the majority of respondents (46.7% of 75 respondents) highly participated in the RHD project design process. A large percentage (34.7% of 75 respondents) responded that they moderately involved in the design process. However, the number of respondents who rarely participated in the process was accumulatively 16% of overall respondents.

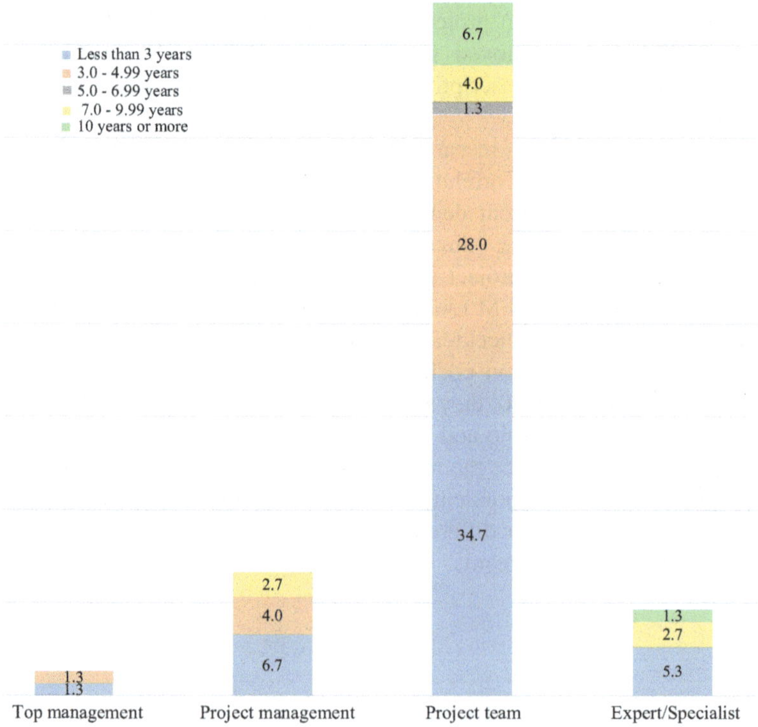

Fig. 1 Project experience roles and experience (in % of respondents)

Table 1 Involvement of RHD project team in project design

Involvement in project design	% of respondents ($N = 75$)
Very high	2.7
High	46.7
Moderate	34.7
Rare	13.3
Very rare	2.7

A further examination of project design involvement was conducted and the results are revealed in Table 2 in the context of the project phases. It was discovered that there was a very low level of involvement for designing most project phases namely *Initiation*, *Planning*, *Monitoring and Control*, and *Closing* of the RHD projects. It can be said that a large number of respondents had *Not at all* involved on how each phase should be structured and managed. Nevertheless, 56% of the respondents reported that they were very highly involved in the design of the *Execution* phase.

To confirm the level of team involvement allowed in project design as shown in Table 2, Kruskal-Wallis Test was conducted across different funding size, project duration, number of sites and size of project team. The statistical significance (Asymp. Sig.) presented in Table 3 for all tested data is greater than 0.05. This indicates that all respondents from different grouping agreed to the same level of the studied variable namely the involvement level in RHD project design.

To understand the relationship between project team involvement and project design, the study investigated further into the involvement and effectiveness of PM knowledge application and implementation of PM tools in the projects using

Table 2 Involvement of RHD project team in project phase design

Involvement in phase-based project design	(% of respondents, $N = 75$)				
	Initiation	Planning	Execution	Monitoring and control	Closing
Very highly	14.7	14.7	56.0	12.0	6.7
Highly	1.3	1.3	1.3	2.7	0.0
Moderately	0.0	1.3	1.3	2.7	1.3
Somehow involved	4.0	4.0	25.3	30.7	25.3
Rarely	16.0	26.7	8.0	2.7	1.3
Not at all	64.0	52.0	8.0	49.3	65.3

Table 3 Kruskal-Wallis test of project team involvement in project design

	Funding size	Project duration	Number of sites	Size of project team
Chi-square	.307	3.311	1.211	3.860
df	2	3	4	3
Asymp. Sig.	.858	.346	.876	.277

Table 4 Spearman's rho correlations between project design and effectiveness of PM knowledge application and implementation of PM tools

Involvement in project design	PM knowledge	PM tools
Integration management	.276[b]	.265[b]
Scope management	.391[a]	–
Time management	.335[a]	.378[a]
Cost management	.316[b]	.417[a]
Quality management	.504[a]	.292[b]
Human resource management	.297[b]	.289[b]
Communication management	.308[b]	.350[a]
Risk management	.316[b]	.380[a]
Procurement management	–	.256[b]
Stakeholder management	.254[b]	.791[a]

[a]Correlation is significant at the 0.01 level (2-tailed)
[b]Correlation is significant at the 0.05 level (2-tailed)

Spearman's Rho Correlation. The term 'effectiveness' in this research was referred to the objectively verifiable indicator (OVI) that the respondents would be at least 80% satisfied with the PM knowledge and tools designed for implementation in the studied projects. The results are illustrated in Table 4.

The analysis shows that the level of involvement in project design mostly contributes to the effectiveness of PM knowledge application and implementation of PM tools of the studied projects. In other words, more involvement in project designed can lead to more effectiveness of PM knowledge and tools applied to the projects. According to the result, quality management knowledge ($r = .504$, $p < 0.001$) and stakeholder management tools ($r = .791$, $p < 0.001$) contain the strongest relationships to project design. This is interpreted that the more involvement the project teams contributes to project design, the higher level of effectiveness in PM knowledge and PM tools utilized in the projects can be expected. The study extended the investigation into the correlations between other managerial criteria that were associated to team involvement in project design. These aspects were developed from the experts' recommendations specifically for these studied RHD projects.

The results from Table 5 reveal that the involvement in project design in the RHD projects did not only contribute to PM knowledge application and PM tools implementation but also to other criteria that the projects were measured against. These criteria were generally to ensure that the projects worked in alignment with the project goals and objectives in addition to effectively apply PM knowledge and tools. It was found that the level of involvement in project design is strongly associated to the ability of the projects to conduct environmental analysis ($r = .791$, $p < 0.001$) and attain project success ($r = .672$, $p < 0.001$).

Table 5 Spearman's Rho Correlations between project design and effectiveness of PM knowledge application and implementation of PM tools

Recommended managerial criteria	Spearman's Rho (*r*)
Project management training	.617[b]
Project team competency development	.652[a]
Project environment analysis	.791[a]
Customer engagement	.266[b]
Customer satisfaction	.425[a]
Operations management	.400[a]
Shared lessons learned	.283[b]
Project success	.672[a]

[a]Correlation is significant at the 0.01 level (2-tailed)
[b]Correlation is significant at the 0.05 level (2-tailed)

5 Discussion and Conclusion

According to the results shown in the Data Analysis and Findings section, the study found that the studied RHD project teams consisted of project personnel with diverse roles and experience. Where percentage of project staff with less than three years of experience was reported as 48%, the remaining 52% of the project teams had project experience from three years and more. Moreover, approximately 10% of the respondents possessed rich experience in working with projects for 10 years or more. This enriched experience is considered as valuable resources to RHD projects if opportunities for project staff early involvement in project design of all project phases are provided. It was discovered that only 58.6% of project teams were significantly involved only in the design of project execution phase while they were mostly excluded in the design of other phases including monitoring and control where project performance assessments and change management are generally the key activities undertaken in any projects. The impact of not having project staff involvement in the project phase design can lead to under performance of the project teams as the research results show that involvement in project design is associated to project management application and implementation of project management tools in the RHD projects, especially on quality management and stakeholder management. Furthermore, project design also correlates to the ability of the project teams to conduct effective project environmental analysis and accurately understand both internal and external forces that can influence on project implementation and project success. Allowing early involvement of project teams during the project design can overcome the issues of delayed project appraisal, poor impact assessment and ineffective project management application and tools stated in Mosley [19], Hekala [20] and Ika et al. [11].

Acknowledgements The authors also gratefully acknowledge the helpful comments and suggestions of the reviewers, which have improved the presentation.

References

1. Golini R, Corti B, Landoni P (2017) More efficient project execution and evaluation with logical framework and project cycle management: evidence from international development projects. Impact Assess Proj Apprais 35(2):128–138
2. Ahsan K, Gunawan I (2010) Analysis of cost and schedule performance of international development projects. Int J Proj Manag 28(1):68–78
3. Ika LA (2012) Project management for development in Africa: why projects are failing and what can be done about it. Proj Manag J 43(4):27–41
4. Girard P, Robin V (2006) Analysis of collaboration for project design management. Comput Ind 57(8):817–826
5. Bossert TJ (1990) Can they get along without us? sustainability of donor-supported health projects in Central America and Africa. Soc Sci Med 30(9):1015–1023
6. Gow DD, Morss ER (1988) The notorious nine: critical problems in project implementation. World Dev 16(12):1399–1418
7. Shediac-Rizkallah MC, Bone LR (1998) Planning for the sustainability of community-based health programs: conceptual frameworks and future directions for research, practice and policy. Health Educ Res 13(1):87–108
8. Savaya R, Spiro S, Elran-Barak R (2008) Sustainability of social programs: a comparative case study analysis. Am J Eval 29(4):478–493
9. Toakley AR, Marosszeky M (2003) Towards total project quality—a review of research needs. Eng Constr Architect Manag 10(3):219–228
10. Imran A, Zaki A (2016) Impact of human capital practices on project success. Kuwait Chapter of AJBMR 5(6):1–16
11. Ika LA, Diallo A, Thuillier D (2012) Critical success factors for World Bank projects: an empirical investigation. Int J Proj Manag 30(1):105–116
12. Ofori DF (2013) Project management practices and critical success factors—a developing country perspective. Int J Bus Manag 8(21):14
13. Khang DB, Moe TL (2008) Success criteria and factors for international development projects: a life-cycle-based framework. Proj Manag J 39(1):72–84
14. Mathur VN, Price AD, Austin S (2008) Conceptualizing stakeholder engagement in the context of sustainability and its assessment. Constr Manag Econ 26(6):601–609
15. Bamberger M, Cheema S (1990) Case studies of project sustainability: implications for policy and operations from Asian experience. The World Bank
16. Stange KC, Goodwin MA, Zyzanski SJ, Dietrich AJ (2003) Sustainability of a practice-individualized preventive service delivery intervention. Am J Prev Med 25(4):296–300
17. PMI (2013) A guide to the project management body of knowledge. PMI
18. Jackson SL (2015) Research methods and statistics: a critical thinking approach. Cengage Learning
19. Mosley P (2001) A simple technology for poverty-oriented project assessment. Impact Assess Proj Apprais 19(1):53–67
20. Hekala W (2012) Why donors should care more about project management. Devex. Retrieved from http://www.devex.com/en/news/why-donors-shouldcare-more-about-project/77595

Improving Project Success with Project Portfolio Management Practices

Nick Hadjinicolaou, Jantanee Dumrak and Sherif Mostafa

1 Introduction

According to Turner [1], it has been estimated that one third of the world's economy is generated through projects. The tangible benefits of projects may include increasing sales, improved profit margins and cash flows through increased revenue or reduced costs are amongst these. Intangible benefits might include areas of safety, improving customer service, relationships with stakeholders, and organisational capability [2]. Despite the benefits gained from projects, the Standish Groups' 2015 Chaos Report [3] measured project success for information technology project sand found that success rates of these projects are below 50% and continue to be a problem. As a result, the group's definition of a successful project was redefined to include a measure of perceived value by the customer in addition to the triple constraint of delivering on time, within budget and to the required scope. By adding perceived value, project success rates dropped further by 7%.

The need to align project delivery capability with corporate strategy is well-recognized [4]. Organisations increasingly realise that corporate strategy is delivered through projects, and selecting the right projects is key to their ability to deliver their strategic intent is required for strategic alignment [5].

N. Hadjinicolaou (✉) · J. Dumrak
Global Project Management, Torrens University Australia,
88 Wakefield Street, Adelaide SA 5000, Australia
e-mail: nhadjinicolaou@laureate.net.au

S. Mostafa
Griffith School of Engineering, Griffith University, Gold Coast Campus,
Gold Coast QLD 4222, Australia

S. Şahin (ed.), *8th International Conference on Engineering, Project, and Product Management (EPPM 2017)*, Lecture Notes in Mechanical Engineering,
https://doi.org/10.1007/978-3-319-74123-9_7

2 Project Portfolio Management

The *Standard for Portfolio Management* by PMI [6] defines project portfolio management (PPM) as the management that coordinates one or more organisational portfolios i.e. component collections of programs, projects, or operations, to achieve strategic objectives. One of the benefits gained from PPM is the right projects are chosen and implemented for organisational success. All types of organisations can benefit from implementing PPM regarding monetary return and/ or strategic achievement. Projects and project management practices have been employed as key enablers to serve success in strategy implementation. In the early days, PPM was executed mainly to make rational investment decisions for organizations [7]. Decades later, the utilisations of PPM have been extended to project prioritization, selection of projects [8], project evaluation and control [9–11], optimisation of decision, processes and resources [12, 13] and knowledge sharing between projects [14].

Portfolio managers are responsible for executing the PPM process. Responsibilities of portfolio managers include, but not limited to, establishing and maintaining PPM framework, methodology and processes; selecting, prioritizing and managing portfolio components; establishing and maintaining portfolio infrastructure and systems; reviewing, reprioritizing and optimizing portfolio; measuring and monitoring portfolio performance and value; supporting management decision making; and influencing sponsorship engagement [6]. Despite the availability of PPM standards and decades of practices, Patanakul [15] points out that understanding of research and practitioners on constituents of PPM effectiveness remains insufficient and that affects business outcomes, financial performance, productivity as well as morale of project stakeholders. The conceptual problems in the existing body of PPM knowledge highlighted in the study of Young and Conboy [11] could explain the lack of understanding in PPM effectiveness. These are PPM competency-related problems caused by the lack of: cumulative tradition, clarity, theoretical glue, parsimony, and applicability.

The role of PPM is not crucial at only the organisational level but also at the project level. According to Killen and Hunt [16], the processes of PPM aim to improve project success rate through aholistic and responsive decision-making environment to maximise the long-term value of the project portfolio. Pajares and López [17] state forward and backward interrelations between projects and PPM especially on the aspects of cost and risk management. The study urges further research to investigate the role of interactions between projects and PPM. Regardless the growing studies on PPM, existing research works on the relationships between PPM practices and project success appear limited. This paper, therefore, aims to contribute to broader understanding of PPM practices about project success, particularly on applicability and interconnections between these two variables.

3 Project Success Criteria

Project success has been measured in many different ways. Traditionally, the focus of success was on the iron triangle and tangibles such as meeting scope, time and budget goals [18]. It is evident in that the three traditional dimensions of project efficiency were time, budget and scope in which scope had the largest role of impact on the customer and their satisfaction as well as on the business [19]. Munns and Bjeirmi [20] note that much of the previous project management literature considered measuring success at the end of projects when the projects are delivered to the sponsor, project management is terminated, and the project managers move on to other projects.

Years later, researchers increasingly measured success by examining the impact on the organisation rather than success at only meeting the triple constraint. Cooke-Davies [21] differentiates between project management success, where the project is well-managed to finish the desired scope within time and cost, and project success, where the project achieves its business objectives. Jugdev and Moller [22] review the project success literature over the past 40 years and report that a more holistic approach incorporating several dimensions to measuring success was becoming more evident. Thomas et al. [4] emphasise that measuring project success is not straightforward. Shenhar and Dvir [23] suggest a model of success based on five dimensions, judged over different timescales. These five dimensions of project success are as follows:

1. Project efficiency (end of project)
2. Team satisfaction (end of project)
3. Impact on the customer (months following the project)
4. Business success (years following the project)
5. Preparing for the future (years following the project).

Turner [1] also states that the reward structure in many organisations encourages the project manager to finish the project on cost and time and nothing else. Moreover, it is suggested in the current thinking that stakeholders' satisfaction is a primary measure, especially the primary sponsor [24]. At the end of the projects, project success is judged by whether the scope is completed on time and budget, and the project outputs are delivered to the specification as well as whether benefits are delivered.

One can argue time, budget and scope are an important part of project success. However, they are only necessary conditions, but not sufficient conditions [24]. The importance of broader success measures for projects is now the norm. The most recent version of PMBOK [6] as an example, no longer mentions the triple constraint. In addition to scope, time and cost, stakeholder satisfaction has been incorporated. The future research extended from this proposal will align with PMBOK and focus on the modern concepts of the project success and its factors including scope, time, budget, team satisfaction, customer satisfaction, business success, preparing for the future.

4 Research Methodology

In 1736, Leonhard Euler, the Swiss mathematician, introduced the application of Graph Theoretic Approach (GTA) through his work on the Königsberg Bridge Problem to list all of the possible routes among the seven bridges and find any route that satisfies the conditions of the problem [25]. Today, the application of GTA is widely utilised in various field of science and technology to deal with problems of structural relationship and decision-making [26]. In this research, GTA is employed to quantify and prioritise project success resulted from project portfolio management (PPM) practices. GTA presents an analysis in three representations, the matrix presentation, the directed graph (digraph) representation and the permanent representation. All representations were constructed to demonstrate the research results.

In this research, the main objective is to determine the intensity of PPM practices (*P*) about project success (*S*) by quantifying them. A questionnaire survey was utilised as a data-collection method. The data was obtained from 64 respondents in differing Australian sectors including (1) Transport and Logistics, (2) Banking and Insurance, (3) Construction and Engineering, (4) Consulting, (5) Defence, (6) Education, (7) Energy and Utilities, (8) Government, (9) Healthcare and Pharmaceutical, (10) Information Technology, and (11) Telecommunications to identify the interrelationships between PPM practices and project success as illustrated in Fig. 1. The development of the research variables and the representation in Fig. 1 derived from the existing related scholarly publications (as discussed in Project Portfolio Management and Project Success Criteria). The selection criteria of the participants included current positions in portfolio management level and minimum of 2-years involvement in PPM practices.

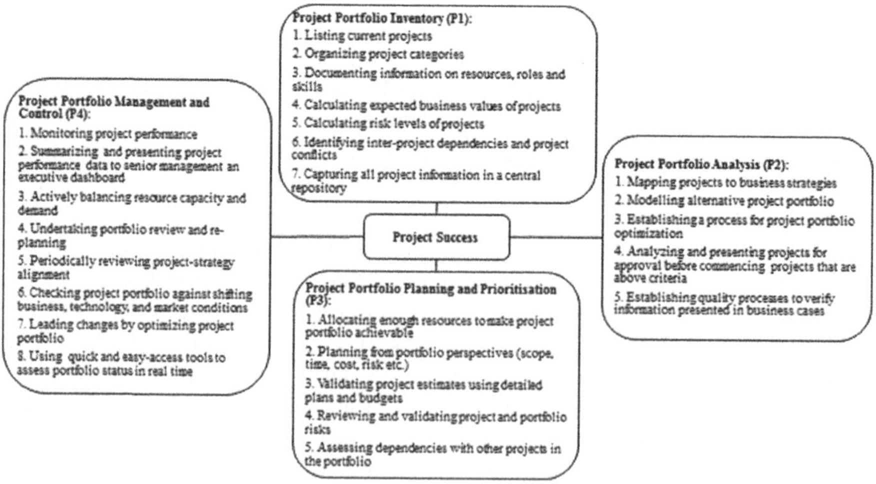

Fig. 1 Representation of PPM practices for project success

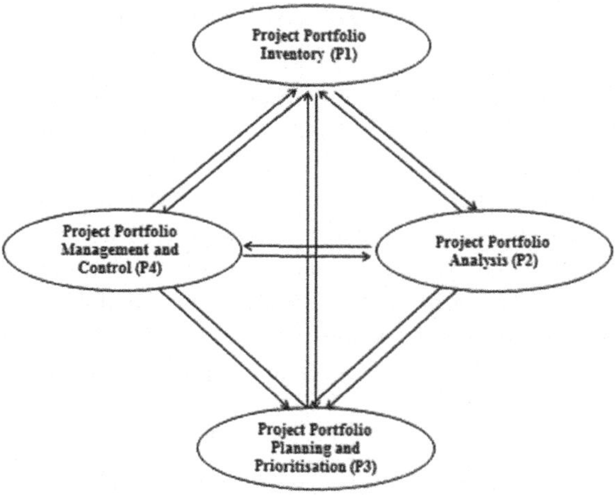

Fig. 2 Interactions of PPM practices

The intensity of these relationships indicates the deterring strength in the implementation of PPM practices to attain project success and depends upon their inheritance and the extent of the interaction among the studied variables (Fig. 2). A panel of experts was invited to determine the interconnections amongst PPM practices and the sub-factors nominated under each PPM practice. The permanent function, a mathematical expression of combinatorial mathematics based on the P_i and p_{ij} values, is used to construct an index of project success. The four categories of PPM practices and their sub-elements are used to evaluate project success by (1):

$$\text{PPM for "Project Success"} = f(\text{Practices}) \tag{1}$$

4.1 Matrix Representations

The relative importance matrix (RIM) representation of the digraph presented in Fig. 1 demonstrates its one to one representation. The matrix consists of two key elements, the off-diagonal and diagonal elements. This is an $N \times N$ matrix, which represents all project success about PPM practices and relative importance of the practices. In this research, there are four different categories of PPM practices in relation to project success (S). The matrix is represented as:

$$S = \begin{bmatrix} S_1 & r_{12} & r_{13} & r_{14} \\ r_{21} & S_2 & r_{23} & r_{24} \\ r_{31} & r_{32} & S_3 & r_{34} \\ r_{41} & r_{42} & r_{43} & S_4 \end{bmatrix} \tag{2}$$

where, S_i is the value of the factor represented by node and r_{ij} is the relative importance of the ith factor over jth factor represented by edge r_{ij} (as in Fig. 1). Where, $S_1^1, S_2^1, S_3^1, S_4^1$ represent P_1, P_2, P_3, and P_4 respectively.

5 Research Results and Discussion

This section presents the research data analysis and its results of relationships between PPM practices and project success. Graph Theory Approach (GTA) was employed to represent relationships between the studied variables. The inheritance and interdependencies of the four PPM practices categories are presented in the permanent function matrices. The inheritance and interdependencies were determined from the questionnaire survey with the industry practitioners using two proposed scales, i.e. 1–9 for inheritance and 1–5 for interdependencies. Then, the permanent function values were obtained from the computation using MatLab. The values of all PPM practices about project success are indexed as below.

$$Per(p_1) = \begin{bmatrix} P_1^1 & r_{12}^1 & r_{13}^1 & r_{14}^1 & r_{15}^1 & r_{16}^1 & r_{17}^1 \\ r_{21}^1 & r_2^1 & r_{23}^1 & r_{24}^1 & r_{25}^1 & r_{26}^1 & r_{27}^1 \\ r_{31}^1 & r_{32}^1 & r_3^1 & r_{34}^1 & r_{35}^1 & r_{36}^1 & r_{37}^1 \\ r_{41}^1 & r_{42}^1 & r_{43}^1 & r_4^1 & r_{45}^1 & r_{46}^1 & r_{47}^1 \\ r_{51}^1 & r_{52}^1 & r_{53}^1 & r_{54}^1 & r_5^1 & r_{56}^1 & r_{57}^1 \\ r_{61}^1 & r_{62}^1 & r_{63}^1 & r_{64}^1 & r_{65}^1 & r_6^1 & r_{67}^1 \\ r_{71}^1 & r_{72}^1 & r_{73}^1 & r_{74}^1 & r_{75}^1 & r_{76}^1 & r_7^1 \end{bmatrix}$$

$$= \begin{bmatrix} 0.7 & 0.6 & 0.8 & 0.7 & 0.7 & 0.7 & 0.7 \\ 0.4 & 0.6 & 0.7 & 0.6 & 0.5 & 0.5 & 0.6 \\ 0.2 & 0.3 & 0.6 & 0.4 & 0.4 & 0.4 & 0.4 \\ 0.3 & 0.4 & 0.6 & 0.6 & 0.5 & 0.5 & 0.5 \\ 0.3 & 0.5 & 0.6 & 0.5 & 0.6 & 0.5 & 0.5 \\ 0.3 & 0.5 & 0.6 & 0.5 & 0.5 & 0.6 & 0.5 \\ 0.3 & 0.4 & 0.6 & 0.5 & 0.5 & 0.5 & 0.7 \end{bmatrix} = 40.24$$

$$Per(P_2) = \begin{bmatrix} P_1^2 & r_{12}^2 & r_{13}^2 & r_{14}^2 & r_{15}^2 \\ r_{21}^2 & P_2^2 & r_{23}^2 & r_{24}^2 & r_{25}^2 \\ r_{31}^2 & r_{32}^2 & P_3^2 & r_{34}^2 & r_{35}^2 \\ r_{41}^2 & r_{42}^2 & r_{43}^2 & P_4^2 & r_{45}^2 \\ r_{51}^2 & r_{52}^2 & r_{53}^2 & r_{54}^2 & P_5^2 \end{bmatrix} = \begin{bmatrix} 0.7 & 1.0 & 0.9 & 0.5 & 0.7 \\ 0.0 & 0.6 & 0.5 & 0.0 & 0.3 \\ 0.1 & 0.5 & 0.5 & 0.1 & 0.3 \\ 0.5 & 1.0 & 0.9 & 0.6 & 0.7 \\ 0.3 & 0.7 & 0.7 & 0.3 & 0.6 \end{bmatrix} = 1.67$$

$$Per(P_3) = \begin{bmatrix} P_1^3 & r_{12}^3 & r_{13}^3 & r_{14}^3 & r_{15}^3 \\ r_{21}^3 & P_2^3 & r_{23}^3 & r_{24}^3 & r_{25}^3 \\ r_{31}^3 & r_{32}^3 & P_3^3 & r_{34}^3 & r_{35}^3 \\ r_{41}^3 & r_{42}^3 & r_{43}^3 & P_4^3 & r_{45}^3 \\ r_{51}^3 & r_{52}^3 & r_{53}^3 & r_{54}^3 & P_5^3 \end{bmatrix} = \begin{bmatrix} 0.5 & 0.5 & 0.4 & 0.4 & 0.4 \\ 0.5 & 0.5 & 0.4 & 0.4 & 0.4 \\ 0.6 & 0.6 & 0.6 & 0.5 & 0.5 \\ 0.6 & 0.6 & 0.5 & 0.5 & 0.5 \\ 0.6 & 0.6 & 0.5 & 0.5 & 0.5 \end{bmatrix} = 3.72$$

$$Per(P_4) = \begin{bmatrix} P_1^4 & r_{12}^4 & r_{13}^4 & r_{14}^4 & r_{15}^4 & r_{16}^4 & r_{17}^4 & r_{18}^4 \\ r_{21}^4 & r_2^4 & r_{23}^4 & r_{24}^4 & r_{25}^4 & r_{26}^4 & r_{27}^4 & r_{18}^4 \\ r_{31}^4 & r_{32}^4 & r_3^4 & r_{34}^4 & r_{35}^4 & r_{36}^4 & r_{37}^4 & r_{18}^4 \\ r_{41}^4 & r_{42}^4 & r_{43}^4 & r_4^4 & r_{45}^4 & r_{46}^4 & r_{47}^4 & r_{18}^4 \\ r_{51}^4 & r_{52}^4 & r_{53}^4 & r_{54}^4 & r_5^4 & r_{56}^4 & r_{57}^4 & r_{18}^4 \\ r_{61}^4 & r_{62}^4 & r_{63}^4 & r_{64}^4 & r_{65}^4 & r_6^4 & r_{67}^4 & r_{18}^4 \\ r_{71}^4 & r_{72}^4 & r_{73}^4 & r_{74}^4 & r_{75}^4 & r_{76}^4 & r_7^4 & r_{18}^4 \\ r_{81}^4 & r_{82}^4 & r_{83}^4 & r_{84}^4 & r_{85}^4 & r_{86}^4 & r_{87}^4 & r_{88}^4 \end{bmatrix}$$

$$= \begin{bmatrix} 0.6 & 0.5 & 0.8 & 0.8 & 0.8 & 1.0 & 0.9 & 1.0 \\ 0.5 & 0.6 & 0.8 & 0.8 & 0.8 & 1.0 & 0.9 & 1.0 \\ 0.2 & 0.2 & 0.6 & 0.5 & 0.4 & 0.6 & 0.4 & 0.6 \\ 0.2 & 0.2 & 0.5 & 0.6 & 0.5 & 0.6 & 0.5 & 0.6 \\ 0.2 & 0.2 & 0.5 & 0.5 & 0.6 & 0.6 & 0.5 & 0.6 \\ 0.0 & 0.0 & 0.4 & 0.4 & 0.4 & 0.6 & 0.5 & 0.5 \\ 0.1 & 0.1 & 0.5 & 0.5 & 0.5 & 0.5 & 0.6 & 0.5 \\ 0.0 & 0.0 & 0.4 & 0.4 & 0.4 & 0.5 & 0.5 & 0.6 \end{bmatrix} = 53.66$$

$$Per(S_{PPM}) = \begin{bmatrix} 40.24 & r_{12} & r_{13} & r_{14} \\ r_{21} & 1.67 & r_{23} & r_{24} \\ r_{31} & r_{32} & 3.72 & r_{34} \\ r_{41} & r_{42} & r_{43} & 53.66 \end{bmatrix} = \begin{bmatrix} 40.24 & 0.5 & 0.5 & 0.5 \\ 0.5 & 1.67 & 0.5 & 0.5 \\ 0.5 & 0.5 & 3.72 & 0.5 \\ 0.5 & 0.5 & 0.5 & 53.66 \end{bmatrix}$$

$$= 14107.59$$

The value of the permanent function is computed as demonstrated in the above matrices for each category of PPM practices and the overall project success (S_{PPM}) The matrices of the permanent value indicate the values of each PPM category in project performance in overall studied industrial sectors. Moreover, it represents the inherited power of each category mathematically with the project success. Computing the hypothetical lowest and highest values of S_{PPM} were proposed in this paper to calculate the range within which the values of S_{PPM} can vary. These values are the maximum and minimum values of S_{PPM}. The S_{PPM} is maximum or minimum when the inheritance of all PPM practices is maximum or minimum. The inheritance of each PPM category is calculated at the sub-system level by applying GTA. Therefore, the inheritance of each practices category depends upon its sub-practices. S_{PPM} is at the lowest value when the inheritance of all practices is at lowest level value i.e. 0 and highest value i.e. 1 (from scale of 0 to 1 for inherence).

For instance, permanent function value for the first category of PPM practices, project portfolio inventory, is minimum when the inheritance of all its sub-practices is minimum, i.e. 0 (from a scale of 0 to 1 inheritance). Hence, the matrix of project portfolio inventory category for the minimum value of S_{PPM} is written as:

$$(P_1)_{min} = \begin{bmatrix} \mathbf{0} & 0.6 & 0.8 & 0.7 & 0.7 & 0.7 & 0.7 \\ 0.4 & \mathbf{0} & 0.7 & 0.6 & 0.5 & 0.5 & 0.6 \\ 0.2 & 0.3 & \mathbf{0} & 0.4 & 0.4 & 0.4 & 0.4 \\ 0.3 & 0.4 & 0.6 & \mathbf{0} & 0.5 & 0.5 & 0.5 \\ 0.3 & 0.5 & 0.6 & 0.5 & \mathbf{0} & 0.5 & 0.5 \\ 0.3 & 0.5 & 0.6 & 0.5 & 0.5 & \mathbf{0} & 0.5 \\ 0.3 & 0.4 & 0.6 & 0.5 & 0.5 & 0.5 & \mathbf{0} \end{bmatrix}$$

On calculating the permanent value of the above matrix, the value of $P1_{min}$ is 9.5. Likewise, S_{PPM} is at its highest value when the inheritance of all its practices at its highest value, i.e. 1. Therefore, the matrix for project portfolio inventory category for the maximum value of S_{PPM} is written as:

$$(P_1)_{max} = \begin{bmatrix} \mathbf{1} & 0.6 & 0.8 & 0.7 & 0.7 & 0.7 & 0.7 \\ 0.4 & \mathbf{1} & 0.7 & 0.6 & 0.5 & 0.5 & 0.6 \\ 0.2 & 0.3 & \mathbf{1} & 0.4 & 0.4 & 0.4 & 0.4 \\ 0.3 & 0.4 & 0.6 & \mathbf{1} & 0.5 & 0.5 & 0.5 \\ 0.3 & 0.5 & 0.6 & 0.5 & \mathbf{1} & 0.5 & 0.5 \\ 0.3 & 0.5 & 0.6 & 0.5 & 0.5 & \mathbf{1} & 0.5 \\ 0.3 & 0.4 & 0.6 & 0.5 & 0.5 & 0.5 & \mathbf{1} \end{bmatrix}$$

On calculating the permanent function of the above matrix, the value of $P1_{max}$ is 79.64. Consistently, the maximum and minimum values of each S_{PPM} category are calculated as presented in Table 1.

Table 1 displays the maximum, minimum, and current value for each PPM category at system (S_{PPM}) and sub-system (P_1, P_2, P_3, P_4) level. Therefore, the maximum and minimum value of the S_{PPM} indicates the range within which the PPM practices groups can diverge. The analysis of the GTA revealed that the practices of PPM benefited competency improvement in projects to the greatest degree when compared to other studied categories whereas project management improvement showed the least affected results from PPM implementation. It should be noted that these values may be used to develop self-assessment baselines for organisations that wish to pursue the PPM implementation and maturity.

Table 1 Maximum, min and current values of the permanent functions

Permanent function	Max value	Min value	Current value
Per P_1	79.64	9.50	40.24
Per P_2	4.82	0.33	1.67
Per P_3	9.82	1.29	3.72
Per P_4	144.83	11.90	53.66
Per S_{PPM}	546,963.46	91.48	14,107.59

6 Conclusion

Projects are means serving organizational objectives and goals. Project portfolio management is to ensure that only the right projects are selected and resources supporting the projects are optimised and effectively managed. As a result, a practice of project portfolio management (PPM) generally contributes to success of a project. This research demonstrates the interrelations between PPM and project success by developing indexes from the determined values of PPM practices and project success obtained from differing Australian sectors. The research results include the maximum and minimum values of PPM practices and overall project success using permanent functions which allow a benchmarking range to be established. Moreover, the presented values indicate the current performance of the 11 Australian sectors as indicated in the research methodology section. Project and portfolio management practitioners and executives can gain benefits from this research by conducting self-assessment and mapping their PPM performance to improve future project success.

References

1. Turner JR (2008) Handbook of project-based management. McGraw-Hill Professional Publishing
2. Archibald R, Prado D (2014) The importance of knowing your project, program and portfolio management maturity. PM World J 3(2):1–8
3. The Standish Group International (2015) CHAOS report 2015. The Standish Group International. Retrieved from https://www.standishgroup.com/
4. Thomas J, Delisle C, Jugdev K, Buckle P (2002) Selling project management to senior executives: the case for avoiding crisis sales? Project Management Institute
5. Meskendahl S (2010) The influence of business strategy on project portfolio management and its success—a conceptual framework. Int J Project Manage 28(8):807–817
6. PMI (2013) The standard for portfolio management, 3rd ed. Project Management Institute
7. Markowitz H (1952) Portfolio selection. J Finan 7(1):77–91
8. Thiry M, Deguire M (2007) Recent developments in project-based organisations. Int J Project Manage 25(7):649–658
9. Jeffery M, Leliveld I (2004) Best practices in IT portfolio management. MIT Sloan Manage Rev 45(3):41
10. Müller R, Martinsuo M, Blomquist T (2008) Project portfolio control and portfolio management performance in different contexts. Project manage J 39(3):28–42
11. Young M, Conboy K (2013) Contemporary project portfolio management: reflections on the development of an Australian competency standard for project portfolio management. Int J Project Manage 31(8):1089–1100
12. Martinsuo M, Lehtonen P (2009) Project autonomy in complex service development networks. Int J Managing Projects Bus 2(2):261–281
13. Koh A, Crawford L (2012) Portfolio management: the Australian experience. Project Manage J 43(6):33–42
14. Martinsuo M, Lehtonen P (2007) Role of single-project management in achieving portfolio management efficiency. Int J Project Manage 25(1):56–65

15. Patanakul P (2015) Key attributes of effectiveness in managing project portfolio. Int J Project Manage 33(5):1084–1097
16. Killen CP, Hunt RA (2013) Robust project portfolio management: capability evolution and maturity. Int J Managing Projects Bus 6(1):131–151
17. Pajares J, López A (2014) New methodological approaches to project portfolio management: the role of interactions within projects and portfolios. Procedia Soc Behav Sci 119:645–652
18. Atkinson R (1999) Project management: cost, time and quality, two best guesses and a phenomenon, its time to accept other success criteria. Int J Project Manage 17(6):337–342
19. Shrnhur AJ, Levy O, Dvir D (1997) Mapping the dimensions of project success. Project manage J 28(2):5–13
20. Munns AK, Bjeirmi BF (1996) The role of project management in achieving project success. Int J Project Manage 14(2):81–87
21. Cooke-Davies T (2002) The "real" success factors on projects. Int J Project Manage 20(3):185–190
22. Jugdev K, Moller R (2006) A retrospective look at our evolving understanding of project success. IEEE Eng Manage Rev 3(34):110
23. Shenhar AJ, Dvir D (2007) Reinventing project management: the diamond approach to successful growth and innovation. Harvard Business Review Press
24. Turner R, Zolin R (2012) Forecasting success on large projects: developing reliable scales to predict multiple perspectives by multiple stakeholders over multiple time frames. Project Manage J 43(5):87–99
25. Harris JM, Hirst JL, Mossinghoff MJ (2008) Combinatorics and graph theory, vol 2. New York: Springer, p 139
26. Rao RV (2007) Decision making in the manufacturing environment: using graph theory and fuzzy multiple attribute decision making methods. Springer Science & Business Media

The Model of Assessment for Flexographic Printing Technology

Krzysztof Ejsmont and Jan Lipiak

1 Introduction

In the last decades the standard of life of the population has significantly improved. That brought about growing production of consumer goods, and consequently, the development of packaging and label industry. Companies operating in the FCMG sector must offer its clients fast, safe, effective and sustainable solutions. An answer to that issue can be provided by flexographic printing that has appeared a rational alternative to heat-set offset or screen printing [11]. Flexography offers faster, more efficient and frequently more profitable production of packages, but it is also effective in low and specialised stocks as well as large, but highly diversified orders, e.g. in the pharmaceutical branch.

Flexography has evolved from low to high-quality printing and at present is widely used in package and label production with its diversified applications, e.g.: flexible packages, plastic bags, carton and cardboard packages, shrink-sleeve and shrink-wrap packages [10]. Flexographic technology is a standard applied in the packaging sector all over the world. The lifecycle of technology is a developmental phase, thanks to which it is subject to constant modifications and modernisations, which in turn allows for strengthening its leading position on package printing market [18]. The development concerns each stage of technology: from the package's design, through the preparation and creation of the printing plate, until the final product is obtained. It seems that further expansion of the flexographic technology is eminent and constitutes an interesting research area, particularly in terms of new technological opportunities and their rational assessment.

Despite numerous publications on flexographic printing technology [8, 14], the authors failed to identify a complex assessment model dedicated to this printing

K. Ejsmont (✉) · J. Lipiak
Faculty of Production Engineering, Warsaw University of Technology, Warsaw, Poland
e-mail: krzysztof.ejsmont@wp.pl

S. Şahin (ed.), *8th International Conference on Engineering, Project, and Product Management (EPPM 2017)*, Lecture Notes in Mechanical Engineering,
https://doi.org/10.1007/978-3-319-74123-9_8

technology. Scientific literature has dealt with the problem of quality (TPM, SMED, standardisation) and effectiveness (OEE) of flexographic printing [4], and described key assessment measures for this printing technology [5, 13]. Still, there is no assessment model that combines the measures in a coherent structure, at the same time observing the principles of system and holistic approach [1].

2 Flexographic Printing Technology

Flexographic printing technology, apart from the printing process (production) itself, involves many other elements (Fig. 1). There are three most general stages before the product reaches the final client: printing preparation, the actual printing and finishing processes [12].

To facilitate a better understanding of the flexographic printing technology, the authors present the characteristics of its components as follows:

- Marketing—prepares and supervises the graphic design of the flexographic product as well as acquiring clients (e.g. with the use of advertising).
- Development department—in cooperation with the printing company it studies and creates a package suited to the product's needs. It verifies the profitability of the printing process, types of materials and difficulties in design reproduction. While cooperating with the development division, the management defines the production schedule and initial product distribution.
- Production—evaluates the design's prototype on the production line and counteracts potential problems in the product's exploitation.
- Suppliers—provide materials necessary for the execution of the production and ensure a reliable supply process.

Fig. 1 Elements comprising flexographic printing technology

Flexographic overprint is a complex production process exposed to many variable parameters (Fig. 2) connected with such activities as: manual print adding, matching colours, adjusting shades, washing of ink units. These are operations consuming a significant percentage of the total production time and generating various waste—frequently quantities difficult to estimate and calculate in the final price of the product. Controlling variables means precise familiarisation with all stages of the image's reproduction in the printing company, not only variables directly connected with the printing machine. It is crucial that various divisions of the printing company are accordingly responsible for: accepting the order, its graphic processing, mixing inks, controlling supplies of raw materials, printing and eventually quality control—interact within the same plane that should be spectral data.

There are many variable parameters affecting the result of flexographic printing, among which the most significant ones are:

- Printing machine—each machine is different, it has its advantages and disadvantages. It may exhibit various inaccuracies, clearances, construction faults. That situation makes it an important variable in the printing process.
- Print form—while printing, the printing plate directly contacts the printing substrate, and thus is prone to deformations. Upon its mounting on the printing cylinder, the image that is reflected on the convex plate is subject to deformation, leading to the raster point gain and the resulting tonal change of the overprint.
- Ink—has a liquid form, therefore it is vital to control its dosage so that the image made on the plate is not overflown. Its viscosity, fixation speed and shade should be controlled. Also the anilox cylinder has an important function, being the main tool for dosing ink.

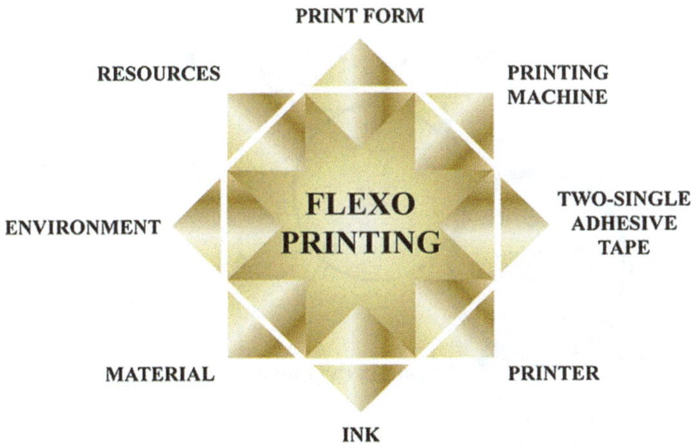

Fig. 2 Variables affecting flexographic printing (production)

- Printer—the human factor that primarily influences the quality of the printed copy. This is affected by human personal features, e.g.: experience, precision, diligence. The printer decides on the plate coating, its proper pressure against the substrate, the choice of the anilox cylinder, a standardised choice of ink colours, the choice of the best double-sided adhesive tape.

3 Characteristics of Flexographic Printing

The process of flexographic printing is unique in its use of convex printing plates that function as stamps pressing a pattern on the printing substrate. These plates are most frequently made of polymer or rubber, which allows for printing on substrates that are not ideally even. The printing unit is made of a steel printing cylinder that is free of dressing and a plate cylinder onto which the plate is mounted (Fig. 3). The ink is applied on the plate by e.g. the anilox. The ink is directly transferred from the elastic plate onto the substrate by being pressed against with the printing cylinder [9]. The ink unit of the flexographic machine is composed of: the ink roll, the anilox roll, the ink tray and the tank. The ink roll is partly infused in the ink and with its rotations, it collects the ink [19]. It contacts the anilox, whose surface has identical hollows. The doctor blade is used for removing the excessive ink from the anilox that directly touches the printing plate. Next, the ink from the plate is transferred onto the substrate, which is supported by the pressure exerted by the impression

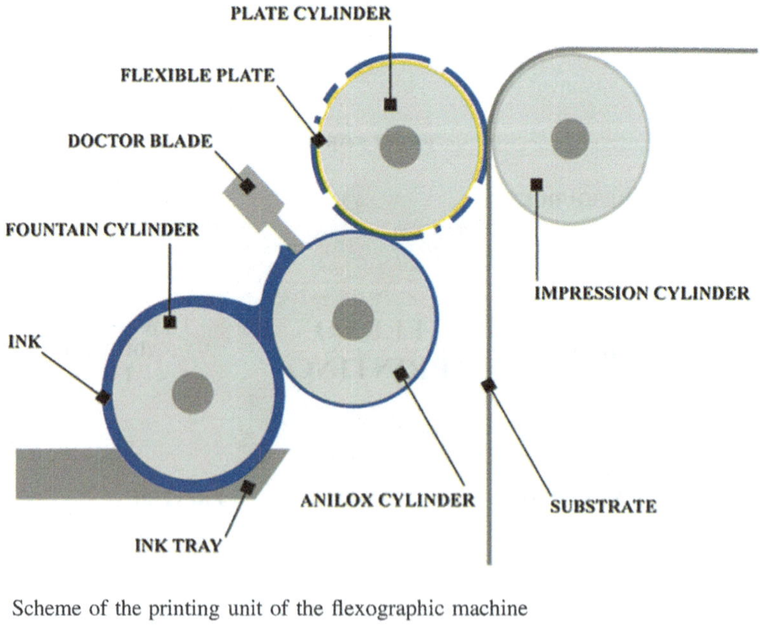

Fig. 3 Scheme of the printing unit of the flexographic machine

cylinder [2]. Applying well-adhesive inks with their possible fast vaporisation on non-absorbent substrates allows flexography to take all the greater share in the packaging sector [15].

4 Model of Assessing Flexographic Printing Technology

Upon characterising key elements comprising flexographic printing technology, the article can present a suggested model of assessment.

The authors of the model's construction have applied a integrated method, created by Prof. S. Marciniak in 1989 [16]. The integrated method is characterised by three main properties: complexity, coherence and a well-organised choice of measures, according to the principles of multi-dimensional structure. Thanks to the modification of the method as well as accounting for a paradigm of new economy—the authors propose the structure of the assessment process that is based on a two-stage approach: dimension—measure (Fig. 4).

The first stage of this structure was described in the earlier part of this article. The second stage focuses on characterising the properties that, according to the authors, will allow for reaching the possibly fullest assessment of flexographic printing technology. Such assessment should be holistic, complex and coherent. The holistic approach refers to Aristoteles's paradigm claiming that the entirety precedes components, and thus the holistic approach dominates over the piecemeal approach. Complexity means that the measures of the assessed undertaking are

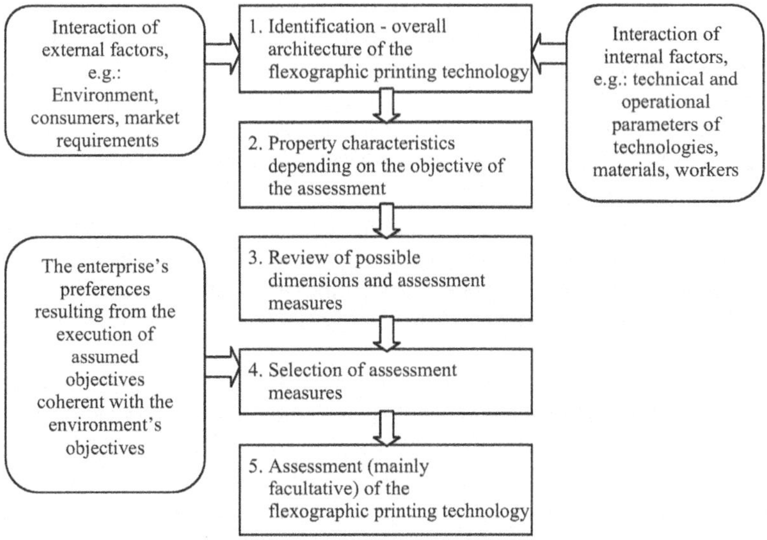

Fig. 4 Proposed structure of the process of assessing the technology of flexographic printing

addressed and interconnected. Coherence is expressed through the connections between the assessment criteria and the elements of the system, as well as their direction (e.g. through the assessment system).

At the third stage the authors decided to account for the economic, technological and environmental dimensions.

The role of the economic dimension in the assessment of the flexographic printing technology is unquestionable. Almost all manufacturing and servicing companies that execute orders primarily rely on the calculation of the profitability of the accepted order in their short-term planning. Before the contract on manufacturing the stock is signed, the company must know the cost of its execution. Then it is possible to determine the profitability of the order and define the profit, e.g. per 1 m^2 of the printed sheet. These figures are most frequently estimated on the basis of currently executed orders (historical data) although it is difficult for many companies to properly calculate indirect costs and their connection with the executed orders for a given recipient. The assessment model should account for the largest possible number of economic measures that will allow for precise determination of the profitability of a given order.

The technological dimension is also highly significant in the assessment of the analysed technology. This stems from the fact that many technical parameters of flexographic machines considerably and directly affect the final product and its ultimate quality. Any technological advancement in the technological aspect may bring lower manufacturing costs, lower energy consumption, higher efficiency etc.

The significance of the environmental dimension in the developed model is extensively described in the further part of this article.

Stage four includes assessment measures grouped in accordance with the adopted research approach as well as the units for which their values should be defined (Table 1). Before proceeding to a detailed assessment of the technology, it is crucial to determine the set of clients' requirements (acceptable and unacceptable) that concern the print-out of a given order. For this reason, the norm ISO 2859 AQL can appear helpful. It is important to determine the possibility of order execution (while preserving the parameters demanded by the client) with the use of this printing technology. If there exists such a possibility, one should proceed to further analyses and assessment.

Such a set of measures allows for entering the obtained figures for any order and comparing them with the planning figures (e.g. determined by the CEO, planner, and technologist). On the one hand, this will allow for organising orders from the most to the least profitable, on the other—it will indicate areas for improving the functioning technology.

It is also crucial to pay attention to the fact that the presented measures are interconnected. For instance, the type of material or the printing speed affect the quality of the overprint, and energy consumption or the volume of the resulting waste is reflected in the order's profitability. Table 1 also presents conditions that determine the value of a given measure (e.g. it is the client who chooses the type of

Table 1 Measures used for assessing flexographic printing technology

Dimension	Measure	Unit	Condition/ Improvement
Economic	Cost of resetting the machine	€	5S/TPM/SMED
	Cost of materials	€	Negotiations
	Machine resetting	h	Better planning
	Gross order stock	mb	Customer
	Added production per man-hour	€	>0
	Order profitability	€/m^2	>0
Technological	Type of material	[name]	Customer
	Thickness of material	mm	Machine
	Width of material	mm	Machine
	Manufacturer of material	[name]	Customer
	Quality of used material	pattern	Control
	Number of rolls used for printing the order	pcs.	Order
	Type of ink	[name]	Customer
	Number of ink overlays	pcs.	Order
	Number of colours	pcs.	Order
	Lacquer layer	[0; 1]	Order
	Upgrading in the line	[0; 1]	Order
	Level of printing difficulty	[1–3]	Machine
	Printing speed	m/min	Machine
	Quality of tonal transitions	[0; 1]	Control
	Colour compatibility	ΔE	Control
	Blanking quality	% of cracks	Control
	Overprint quality	pattern	Control
Environmental	Energy consumption	kWh	Reduction
	Volume of waste	m^2	Better planning
	Material and ink waste	kg	Reduction
	UV or LED fixation	W/m	Minimisation
	CO_2 emission	g/Mg	Reduction
	Adhesion strength	J/m^2	Control

material, whereas its width and thickness depend on the machine capability) or improvement that indicate directions of the activities that should be undertaken in more effective use of flexographic printing technology.

5 Significance of the Environmental Dimension in the Developed Model

The environmental dimension plays a growing role in technology assessment models. It is compatible with the paradigm assuming a balance between the economy, environment and society as well as the idea of sustainable growth. These days, package production of minor environmental impact is a necessity and at the same times a technological challenge. On the one hand, they should be a product that is attractive for the consumer and consequently create a positive image of both the manufacturer and the distributor; on the other—they should protect the natural environment by e.g. the possibility of their recycling [6]. Some market sectors are growingly interested in flexography that gains popularity in, to start with, the production of food packages [7].

The progress that flexography has made in recent years in terms of ecological production is also visible in the used substrates [3]. Electron beam inks (EB) gain popularity since they are characterised by a shorter drying time and higher durability. EB inks, hardened at reduced oxygen content, bring benefits in a form of high-quality print, which leads to further reduction of production costs. This technology eliminates loose remains of inks and chemical agents and this allows for reducing the amount of the resulting ozone. At the same time, the printed surface is smoother and glossier. Here it is vital to control the adhesion force or the amount or the used energy [17].

6 Conclusion

This article presents the subject of flexographic printing technology and proposes a model of its assessment. Thanks to the system approach, it enumerates key elements that comprise the technology and describes the parameters that most significantly affect its assessment. This has led to the development of a five-stage assessment process that allows for a multi-dimensional assessment of flexographic printing technology. The presented measures play a key role in the process of technology assessment and were chosen on the basis of knowledge and experience of people that for many years have been engaged in the printing industry.

Acknowledgements This work is partially supported by Prof. S. Marciniak from Warsaw University of Technology.

References

1. Bates I, Zjakic I, Budimir I (2015) Assessment of the print quality parameters' impact on the high-quality flexographic print visual experience. Imaging Sci J 63(2):103–110. https://doi.org/10.1179/1743131X14Y.0000000094
2. Dunn T (2015) Flexographic printing. Flex Packag 2015:27–37. https://doi.org/10.1016/B978-0-323-26436-5.00003-5
3. DuPont (2017) Life cycle assessment verified by independent third party. http://www.dupont.co.uk/products-and-services/printing-package-printing/flexographic-platemaking-systems/brands/cyrel/articles/life-cycle-assessment.html. Accessed 28 Aug 2017
4. Ejsmont K, Lipiak J (2015) Implementation of the SMED method in a printing company. Econ Organ Enterp 4(783):104–116
5. Ejsmont K, Lipiak J (2016) The problem of calculating the OEE for flexo machines—case study. In: Knosala R (ed) Innovations in management and production engineering. Opole

6. Geoff AG (2000) Design and technology of packaging decoration for the consumer market. Academic Press, Sheffield
7. Gordon LR (2012) Food packaging: principles and practice, 3rd edn. CRC Press, Boca Raton
8. Izdebska J (2016) Printing on polymers. Theory Pract. William Andrew
9. Janßen EM, Schliephacke R, Breitenbach A, Breitkreutz J (2013) Drug-printing by flexographic printing technology—A new manufacturing process for orodispersible films. Int J Pharm 441(1–2):818–825. https://doi.org/10.1016/j.ijpharm.2012.12.023
10. Katz S (2013) Shrink sleeve converting. Rodman Media, Ramsey USA
11. Kipphan H (2001) Handbook of print media: technol production methods. Springer, Germany
12. Kit LY (2009) The wiley encyclopedia of packaging technology, 3rd edn. Wiley, New York
13. Lipiak J (2017) Methodology for assessing the factors affecting the quality and efficiency of flexographic printing process. Procedia Eng 182:403–411. https://doi.org/10.1016/j.proeng.2017.03.122
14. Lorenz A, Senne A, Rohde J et al (2015) Evaluation of flexographic printing technology for multi-busbar solar cells. Energy Procedia 67:126–137. https://doi.org/10.1016/j.egypro.2015.03.296
15. Majid I, Nayik GA, Dar SM, Nanda V (2016) Novel food packaging technologies: innovations and future prospective. J Saudi Soc Agric Sci. https://doi.org/10.1016/j.jssas.2016.11.003
16. Marciniak S (1989) The integrated method of efficiency evaluation of technical and organizational projects. Warsaw University of Technology Publishing House, Warsaw
17. Sanatgar RH, Campagne C, Nierstrasz V (2017) Investigation of the adhesion properties of direct 3D printing of polymers and nano composites on textiles: effect of FDM printing process parameters. Appl Surf Sci 403:551–563. https://doi.org/10.1016/j.apsusc.2017.01.112
18. Smith RD (2009) Challenges in winding flexible packaging film. In: TAPPI 12th European PLACE conference, Budapest
19. Żołek-Tryznowska Z, Izdebska J (2013) Flexographic printing ink modified with hyperbranched polymers: Boltorn™ P500 and Boltorn™ P1000. Dyes Pigm 96(2):602–608. https://doi.org/10.1016/j.dyepig.2012.10.003

Using Building Information Modelling to Facilitate Decision Making for a Mobile Crane Lifting Plan

Ernest L. S. Abbott, Le Peng and David K. H. Chua

1 Introduction

Mobile cranes, because of their flexibility, are one of the most important pieces of heavy lifting equipment used in Singapore's construction industry. They are also one of the main sources of workplace incidents and accidents. The Singapore Workplace Safety and Health Council (WSHC) 2015 report [15] shows that across all industries that there were 66 workplace fatalities in 2015, an increase of 6 over the rate in 2014. In both 2014 and 2015, 27 of the fatalities were in the construction section, which accounted for the highest fatal accident rate across all industries. There were three crane related fatalities in 2014 and 5 in 2015. Crane related fatalities account for 18.5% of all construction industry fatalities.

The Singapore government is committed to work towards a zero-accident rate in the work place. Safe crane operations are one area where a significant contribution can be made. A major cause of crane related accidents has been identified by the Singapore's Ministry of Manpower (MOM) [8] as a lack of supervision and proper planning. To ameliorate this, the WSHC's Code of Practice on Safe Lifting Operations in Workplace [14] mandates the use of a permit to work system by the contractors. The permit to work system records the key procedures for mobile crane lifting operations.

The crane-lifting plan records the crane model, crane boom length and the intended load to lift. The critical areas checks of the lifting plan are: (1) the crane's

E. L. S. Abbott (✉) · L. Peng · D. K. H. Chua
Civil & Environmental Engineering, National University of Singapore,
Block E1A #07-03, 1 Engineering Drive 2, Singapore 117576, Singapore
e-mail: ceeaels@nus.edu.sg

L. Peng
e-mail: peng_le@u.nus.edu

D. K. H. Chua
e-mail: ceedavid@nus.edu.sg

S. Şahin (ed.), *8th International Conference on Engineering, Project, and Product Management (EPPM 2017)*, Lecture Notes in Mechanical Engineering,
https://doi.org/10.1007/978-3-319-74123-9_9

lifting capacity, (2) obstruction in lifting path. The current industrial practice is to carry out the checks and calculation manually, supplemented with the lifting engineer's experience. This leaves the young and inexperienced lifting engineer a difficult task to complete. Manual checking is a tedious and possibly error prone procedure, as each item to be lifted needs to undergo a check. Clearly, the use of an automated procedure would eliminate or at least reduce the number of workplace accidents and fatalities.

The use of Building Information Modelling (BIM) has been mandatory for all new developments in Singapore with a gross floor area greater than 5000 m^2 since the 1st July 2015 [2]. Such a requirement, coupled with the use of 3D BIM models and modelling software, gives the right conditions to develop an automated crane lifting plan for all objects in a construction project. The 3D BIM model facilitates the identification of: (1) the crane position, the objects loading and unloading point in 3D space; (2) the identification of any spatial constraints in the lifting path.

This paper proposes a method of automated crane positioning for lifting using 3D BIM, with AutoDesk's software Revit coupled with crane lifting and loading capacity provided by crane manufactures to derive automatically a safe lifting location.

The structure of the rest of this paper is a literature review followed by an explanation of the various tools and data used. It then moves on to the use of the tools. A worked exampled using a real situation is given.

2 Literature Review

The modelling of mobile crane operations needs to be cognisant of site impediments, which prevent easy lifting. Several researchers have proposed methods that mitigate site impediments [17], propose the use of a Rapidly-exploring Random Tree algorithm (RRT) to locate a safe lifting path. Lin et al. [6] improved on this by the use of a sampling strategy and an expansion strategy to improve the RRT to obtain an optimized path for a crawler crane lifting. The approach of the algorithm is to discretize the working envelope into different nodes, which are checked for the safe working load of the crane and being collision free of existing structures.

The focus of current research on collision detection tends to be between the lifting object and the existing building structure. This ignores any possibility of a collision by the crane with the existing building structure. However, Lei et al. [5], who take collision between the crane and the building into account, propose a generic method for mobile crane lifting binary (yes-or-no) path for mobile crane path planning and checking. The method, using the crane's capacity, calculates the maximum and minimum lifting radii, which are adjusted to take the site's constraints into account. The weakness of this approach is the simplification of the building as cuboid. While this may be a reasonable simplification for some buildings, it is incapable of handling irregular shaped buildings or structures such as those found in the petro-chemical industry. The simplified approach can easily ignore feasible lifting paths.

Distinct from the modelling of a mobile crane operation is the action of mobile crane planning, which considers the use of an appropriate crane model as well as the position of the crane relative to the building and the pick-up point of the object. Researchers have applied algorithms for this purpose.

Al-Hussein et al. [1], using manual placing of the crane, use a database of crane models to selected suitable cranes for an operation.

Tantisevi and Akinci [12] generate all possible crane-lifting locations, eliminating 3D spatial conflicts. The use of all 3D spatial conflict for every building component at each possible crane location is obviously computational intensive for anything but the simplest of buildings. To simplify this approach, Safouhi et al. [11] reduce the 3D problem to a 2D problem. The simplification reduces computational process, but it can easily miss some suitable crane locations.

BIM, as a repository, has the potential to be used for regulatory compliance checks. Such an idea was proposed by Malsane et al. [7]. Some research has also focused on utilizing BIM for construction safety. Nguyen et al. [9] have shown that it is a firm basis for validating a design for safety at all stages of a building's lifecycle. Yeoh et al. [16], have proposed a framework for tower crane lifting plans. This paper extend this work into mobile crane research planning.

Several researcher [4, 13], use BIM or IFC data in their work. They focus on Tower Crane positioning within a building site, taking similar factors into account as this research. However, with Tower Cranes boom angle and length are not taken into account. The difference between their research and this makes their's less significant to this research.

Fang and Cho [3] work is labour intensive and ignores the legal requirement of a lifting plan. Olearczyk et al. [10], inspite of encompasses several heavy lifting study results does not provide a lifting plan per se.

3 Research Tools and Method

3.1 Research Tools

The software chosen for this research was Autodesk's Revit. Revit is a BIM compliant software used by architects among other disciplines. It is the architectural aspects of Revit used here. The architect's drawing is composed of the building objects (beam, column, slab, wall etc.) that required lifting into place, if they are prefabricated objects. Revit supports a Microsoft.Net user link to access building components in an architect's drawing. Programs written, using the Net framework can extend the functionality of Revit. In this research, the programs were written using C# language, one of the family members of Microsoft's Visual Studio.

The various manufactures' crane load charts were used to create a database of crane information. Figure 1 is an example of a chart for the Liebherr LTF 1060-4.1

10,2 – 40 m T | 360° | 10,2 t | EN

m	10,2 m *	10,2 m	13,6 m	17 m	20,5 m	23,9 m	27,3 m	30,7 m	34,2 m	37,6 m	40 m	m
2,5	60											2,5
3	55,1	42,3	42,2	42,2	39,1	31,1	25					3
3,5	50,3	42,3	42,3	41,4	38,5	31,3	24,8					3,5
4	44,9	42	41,6	39,1	37,3	31,4	24,6	19,8				4
4,5	40,2	38,8	38,8	36	33,8	31,2	24,1	19,7				4,5
5	36,7	35,3	35,3	33,4	31,5	30,3	23,4	19,5	15,4			5
6	30,5	29,4	29,7	29,8	28,6	27,3	21,4	18,5	15,1	12,1	10,4	6
7	25,7	24,7	25	25,1	25,5	24,4	19,4	17,2	14,5	11,8	10,2	7
8			21,5	21,4	21,7	20,9	17,8	16	13,8	11,4	10	8
9			18,4	18,7	18,7	18	16,3	14,8	13,1	10,9	9,7	9
10			15,8	16,2	16,1	15,7	14,7	13,7	12,3	10,4	9,4	10
11			13,5	13,9	13,9	13,6	13,4	12,6	11,6	9,9	8,9	11
12				12,2	12,1	11,9	12,2	11,4	10,9	9,4	8,4	12
14				9,6	9,6	9,9	9,6	9,5	9,2	8,5	7,7	14
16					7,9	8,1	7,8	7,9	7,6	7,7	7	16
18						6,7	6,6	6,6	6,6	6,4	6,4	18
20						5,7	5,8	5,7	5,6	5,4	5,4	20
22							5	4,9	4,8	4,6	4,6	22
24							4,4	4,2	4,1	3,9	3,9	24
26								3,7	3,6	3,4	3,4	26
28								3,3	3,1	2,9	2,9	28
30									2,7	2,5	2,5	30
32										2,2	2,2	32
34										1,9	1,9	34
36											1,6	36

* nach hinten · over rear · en arrière · sul posteriore · hacia atrás · при выдвинутой назад стреле   t_199_00001_00_000 / 00101_00_000

Fig. 1 Example of a crane load chart

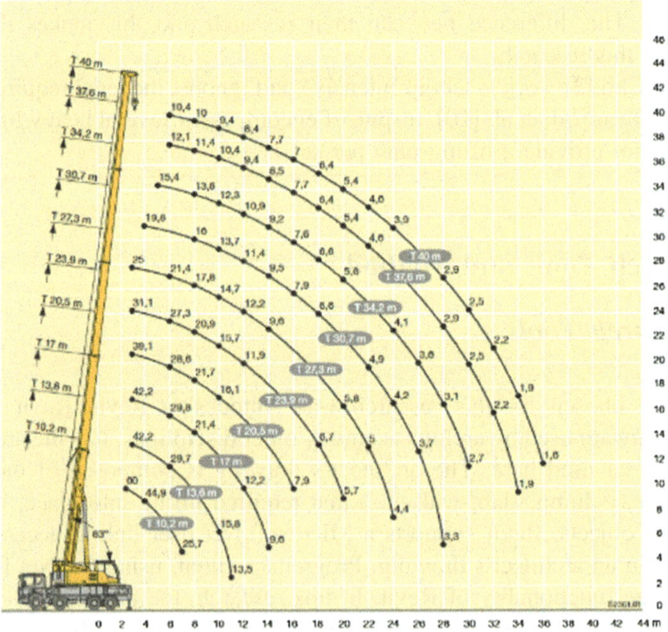

Fig. 2 Crane load chart in graphical format

The crane load chart shows the boom length in the column headings. It shows the boom reach, measured from the crane cab in the row headings. For example, a 17 m boom length with the hook distance of 5 m from the crane is able to lift a weight of 33.4 tonnes. However, at a hook distance of 14 m is only able to lift a weight of 9.6 tonnes.

The crane load chart in table format is also available in a graphical format, Fig. 2 illustrates this. Boom length, boom reach, and working angle of the boom are shown. The working angle is calculated from the distance of the hook from the crane and the boom length, using the table form of the chart.

The charts show the crane has physical lifting constraints. It is essential that these constraints are taken into account when devising a lifting plan. The loading charts only give the lifting constraints of the crane itself. Other constraints have to be considered when generating a lifting plan. Figure 3 illustrates a mobile crane's space constraints, i.e. the crane's circular clearance, since the crane can rotate a full 360°, and also the boom clearance, as the boom needs space when lifting a building object into position.

Figure 4 shows the working envelope of the crane. The concept of the working envelope is important in this research, as will be seen.

Using the manufacture's load chart, the crane's working envelope is divided into four regions, as illustrated in Figs. 5 and 6. The parameters in Fig. 6 are: l = boom length, R_I = Inner Radius, R_o = Outer Radius, W_{max} = Load Capacity (not illustrated).

As illustrated in Fig. 7, the representation used for the various building structure objects: beam, column and slab, is their bounding box. The bounding box for some objects may be marginally larger than the object itself. This is not a concern, since a marginally larger size does not detract from the method employed and is an accepted limitation on the side of caution.

Fig. 3 Crane's space constraints

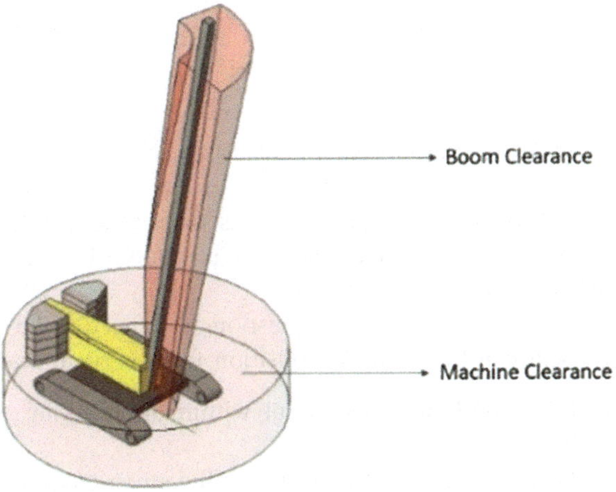

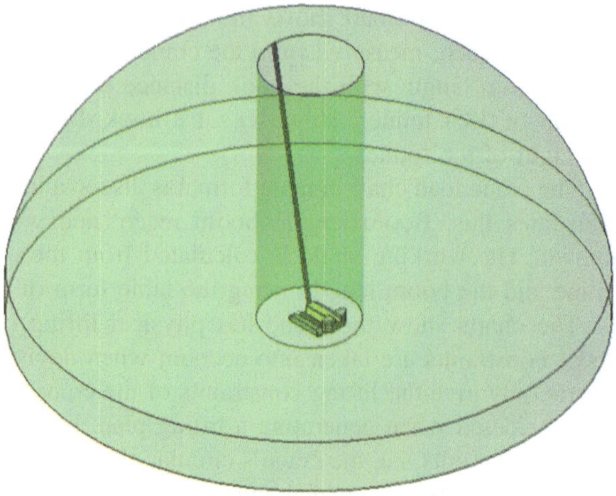

Fig. 4 Isometric view of a crane's working envelope

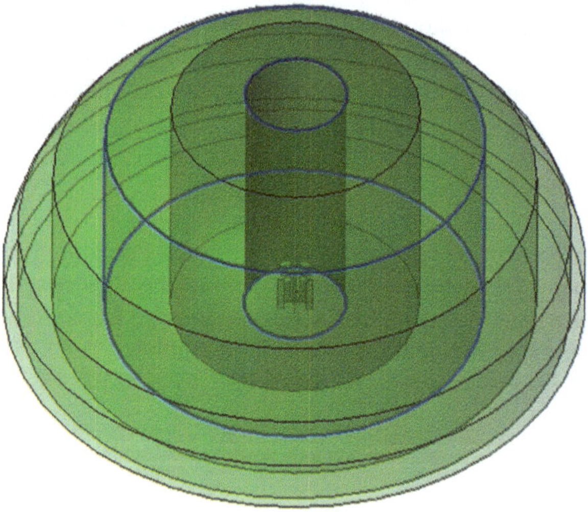

Fig. 5 Crane load capacity envelopes

The research approach deals with volumes and spaces, and does not discretize them. Starting with the position of the object or objects requiring lifting and the crane's position to satisfy this is determined. The position of the crane can be anywhere in the cranes working envelope that includes the object's loading and unloading point.

There are a number of questions that need to be satisfied to know if a building element can be lifted into position from a given crane position.

1. Is the building component within the working envelope of the mobile crane, both lifting and dropping?
2. Is the crane lifting capacity sufficient to move the building object?

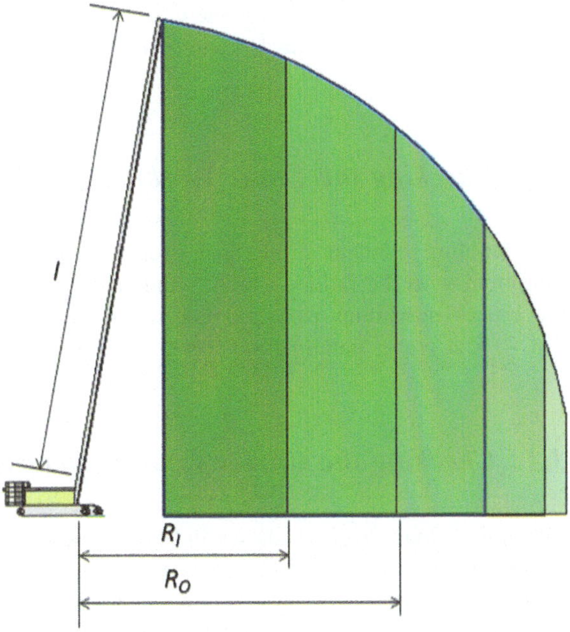

Fig. 6 Crane load envelope parameters

Fig. 7 Bounding box for a column

3. Are there any obstacles for the crane cab (machine clearance)?
4. Are there any obstacles for the crane boom?

3.2 Checking the Crane Working Envelope

The building element lifting and dropping points must be outside the crane's minimum working radius, which is indicated by the cylinder in Fig. 4 and within the cranes maximum working radius, which is indicated by the hemisphere in Fig. 4. The crane manufacturers' data is the source for these values.

3.3 Checking the Cranes Lifting Capacity

The crane's loading chart, shown in Fig. 1, gives the lifting capacity of the crane as various boom angles and distances from the crane's cab. Figure 6 serves as an example, using the variables as shown in Table 1. A boom length of 17 m, with the working radius between 10 and 11 m from the cab, has a safe lifting capacity of 16.9 tonnes.

3.4 Obstacle Checking

Obstacle checking for the crane's cab and its boom use the same procedure. The checking for any encumbrances for the crane's complete picking up and dropping down operation is accomplished by volume reduction. The working envelope of the crane searched for any obstacle and its volume is removed from the crane's working envelope. The volume is an extruded volume, i.e. the bounding box of any object within the crane's working envelope is extruded to the foot of the crane's working envelope. What remains is the feasible working envelope of the crane. Figure 8 illustrates a plan view of a crane's working envelope with obstacles indicated.

Table 1 Example of crane lifting envelope

Boom length (l)	Minimum radius (R_I)	Maximum radius (R_o)	Maximum lift tonnes (W_{max})
17 m	10 m	11 m	16.9 tonnes

Fig. 8 Crane working envelope with obstacles

3.5 Mathematics of a Complete Operation

Using with following notation:

B_{new}	*Location of where to add the new building component*
B_{exist}	*Volume occupied by existing building objects*
C_{crane}	*Clearance space of the crane*
E_{crane}	*Working Envelope of the crane*
W_{object}	*Weight of the new building component*
W_{crane}	*Load Capacity of the crane*

Working envelope requirement of the new building component has to satisfy: $B_{new} \subseteq E_{crane}$.

The load capacity requirement of the crane has to satisfy: $W_{object} < W_{crane}$.

Working clearance requirement for the operation has to satisfy: $\sum (B_{exists} \cap C_{crane}) = \emptyset$.

4 Illustrative Example

The Revit model, shown in Fig. 9 was used to demonstrate the effectiveness of the research. The user wishes to pick up a precast column, weighing 1.4 tonnes, from the supply point and install it on the sixth storey. The selected crane is a Liebherr LFT 1060-4.1 with a counter weight of 10.2 tonnes. The loading chart is that shown in Fig. 1.

Table 2 shows the input data and the results of the algorithms used in this research giving a safe and accurate crane load plan. The decision support algorithms have eliminated boom lengths are less than 30.4 m as the unloading position is outside it the total working envelope of the crane. Boom lengths of 34.2 and 37.6 m

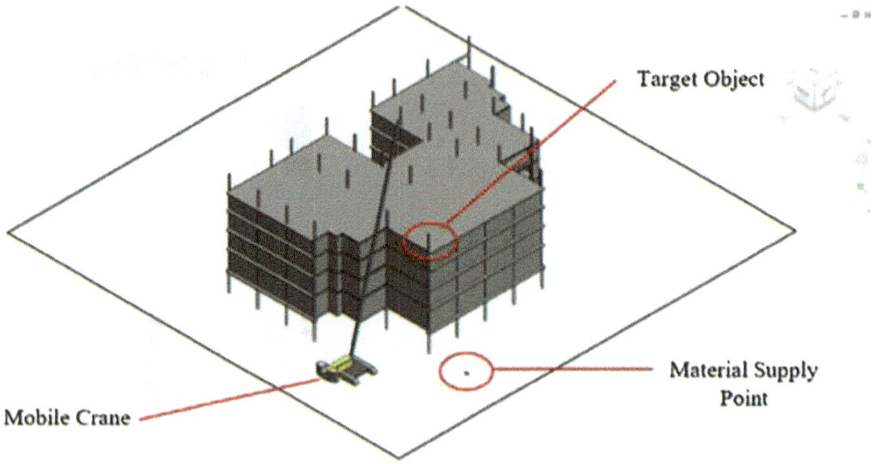

Fig. 9 Site layout of illustrative example

Table 2 Algorithm input data and results

Parameter	Value
Supply point: distance from crane	15.2 m
Unloading point: distance from crane	24.0 m
Minimum boom length	30.4 m
Clashing boom lengths	34.2, 37.6 m
Accepted boom length	40.0 m
Loading boom angle	67.6°
Unloading boom angle	55.2°
Swing boom angle	74.8°

have been eliminated as they result in a clash with existing structures. The boom length of 40 m results in both the pick-up and unloading points being within the working envelope and lifting capacity of the crane without any clashes.

The algorithm calculates the boom angle during the swing operation, with the result that a swing angle of 74.8° should be maintained to avoid any clash with existing building structures. This is acceptable since the maximum boom angle for this crane is 81.3°. In the case of this example, there is only one feasible lifting plan.

A second example, based on Fig. 10, is the positioning of a crane for construction. The building is a multi-storey car park. Using the same crane model, Liebherr LFT 1060-4.1, the algorithms detect 1124 building elements that are prefabricated and require a crane for installation. The crane, in the position shown in Fig. 10. Lifting Plan with Single Crane is able to install 996 of the 1124 element. In this case, using the crane for crane planning, a second crane of the same type would easily cover the whole building, as is shown in Fig. 11.

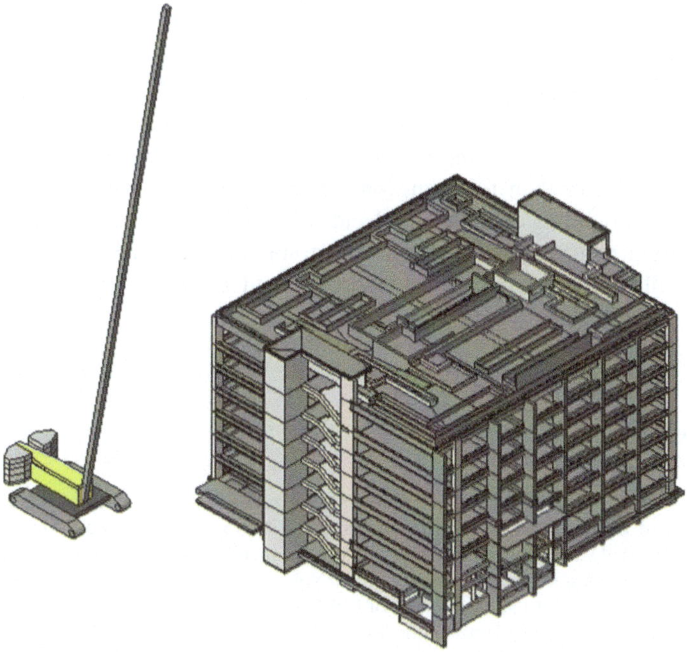

Fig. 10 Lifting plan with single crane

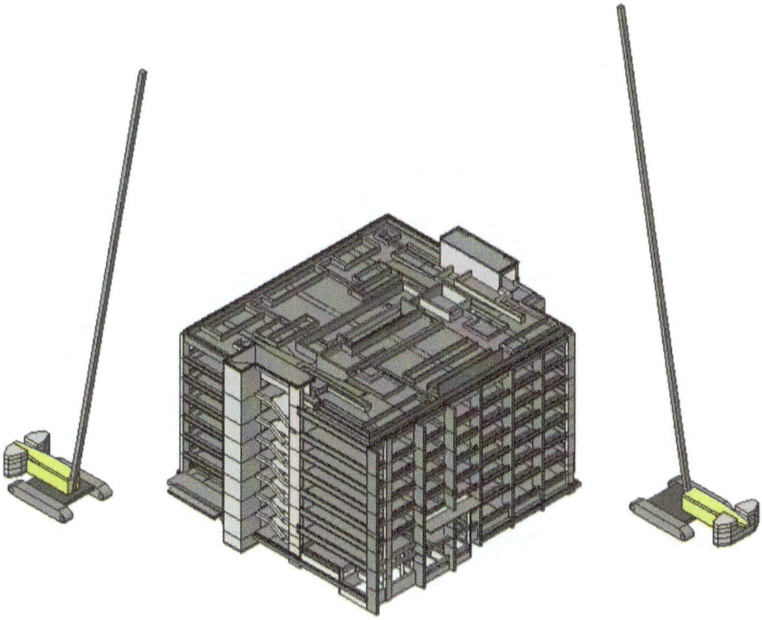

Fig. 11 Lifting plan using 2 cranes

5 Conclusion

The approach in this research is to leverage on BIM, using Revit, to derive a safe mobile crane lifting plan. The algorithms developed covers all the aspects of the crane's lifting operations, taking into account the crane's own working parameters. This research, rather than focussing on the crane and discretising working areas, begins with the building element and decides where it is possible to place a crane to meet all lifting requirements. This reduces computational time significantly.

The resultant crane lifting plan is speedy and accurate, providing the lifting engineers with a robust tool.

References

1. Al-Hussein M, Alkass S, Moselhi O (2005) Optimization Algorithm for Selection and on Site Location of Mobile Cranes. J Constr Eng Manage 131(5):579–590 (May 2005)
2. BCA (2015) Deadlines for projects requiring mandatory BIM (Building Information Modelling) E-submission for regulatory approval
3. Fang Y, Cho YK (2016) A Framework of Lift Virtual Prototyping (LVP) Approach for Crane Safety Planning. In: 2016 Proceedings of the 33rd ISARC, pp 291–297
4. Funtik T, Gašparík J (2016) Site Plan Development: Tower Crane Placement Using BIM Based on Data Obtained from IFC File. In: 2016 Proceedings of the 33rd ISARC, pp 36–41
5. Lei Z, Taghaddos H, Hermann U, Al-Hussein M (2013) A methodology for mobile crane lift path checking in heavy industrial projects. Autom Constr 31:41–53 (May 2013)
6. Lin Y, Wu D, Wang X, Gao S (2014) Lift path planning for a nonholonomic crawler crane. Autom Constr 44:12–24 (Aug 2014)
7. Malsane S, Matthews J, Lockley S, Love PE, Greenwood D (2015) Development of an object model for automated compliance checking. Autom Constr 49:51–58 (Jan 2015)
8. Ministry of Manpower (2000) Safe use of mobile cranes in construction site (OSD/ENG CIR/LE/4/01), pp 1–4
9. Nguyen TQ, Abbott ELS, Chua DKH, Goh YM (2014) Formalizing construction safety knowledge for intelligent bim-based review of design for safety. In: CIB W099 international conference on achieving sustainable construction health and safety, p 597–607
10. Olearczyk J, Lei Z, Ofrim B, Han SH, Al-hussein M (2015) Intelligent Crane Management Algorithm for Construction Operation. In: 2015 Proceeding of 32nd ISARC
11. Safouhi H, Mouattaid M, Hermann U, Hendi A (2011) An algorithm for the calculation of feasible mobile crane position areas. Autom Constr 20(4):360–367 (July 2011)
12. Tantisevi K, Akinci B (2007) Automated generation of workspace requirements of mobile crane operations to support conflict detection. Autom Constr 16(3):262–276 (May 2007)
13. Wang J, Liu J, Shou W, Wang X, Hou L (2014) Integrating building information modelling and firefly algorithm to optimize tower crane layout. In: 2014 Proceedings of the 31st ISARC, pp 321–328
14. Workplace Safety & Health Council (2014) Code of practice on safe lifting operations in workplaces
15. WSH Institute (2016) Workplace safety and health report 2015

16. Yeoh JKW, Wong JH, Peng L (2016) Integrating Crane Information Models in BIM for Checking the Compliance of Lifting Plan Requirements. In: 2016 Proceedings of the 33rd ISARC, pp 966–974
17. Zhang C, Hammad A (2012) Improving lifting motion planning and re-planning of cranes with consideration for safety and efficiency. Adv Eng Inform 26(2):396–410 (Apr 2012)

Critical Success Factors for Public Private Partnership in the Afghanistan Construction Industry

Ghulam Abbas Niazi and Noel Painting

1 Introduction

Partnership between the public and private sector is an approach in order to bridge the financial gap that the public sector is sometimes faced with. PPPs are collaborations in which the public and private sectors both bring their complementary skills to a project, with different levels of involvement and responsibility, for the sake of providing public services more efficiently [1]. The PPP is the arrangement that ensures that value-for-money is delivered in public infrastructure or services (Bing et al. 2004).

Developing countries like Afghanistan require significant infrastructure development, and similarly to most developing countries Afghanistan is getting benefit from public private partnerships (PPP). The profound advantages of PPP contracts (including boosting the local economy, cost reduction, promoting operational efficiencies, improving service quality by utilizing the private sectors' experience and knowledge) have gained the attention of industry practitioners, policy makers and researchers [2]. The public sector considers that using PPP contracts can enhance the social benefits, improve the capacity, reduce cost and time overruns, improve allocation of key risks to the private sector and ensuring value for money [3].

In Afghanistan, the government has recently enacted a public private partnership law, while formulation of policies and guidelines are underway. The main aim of the government is to take advantage of PPP procurement as the government

G. A. Niazi (✉)
Kabul Polytechnic University, Kabul, Afghanistan
e-mail: abbas.niazi@kpu.edu.af

N. Painting
University of Brighton, Brighton & Hove, UK
e-mail: n.j.painting@brighton.ac.uk

S. Şahin (ed.), *8th International Conference on Engineering, Project, and Product Management (EPPM 2017)*, Lecture Notes in Mechanical Engineering,
https://doi.org/10.1007/978-3-319-74123-9_10

struggles with limited capital financial resource. The aim of this research is to identify the critical success factors for implementation of public private partnership in the Afghanistan construction industry.

2 Literature Review

Critical Success Factors (CSFs) are defined as: "*Those few key areas of activity in which favorable results are absolutely necessary for a manager to reach his/her goals*" [4]. The CSFs are perceived as good outcomes for an organization that will assist the organizational survival and performance, or enhance the project performance [5]. Russell [6] states that understanding the CSFs could help organizations to improve the process and reduce the cost of project failure.

Various studies have been carried out to assess and identify the critical success factors for the implementation of the public private partnership in developed and developing countries. Critical success factors were evaluated by Hardcastle et al. [4] in the UK construction industry and factors were grouped under five principal headings: (Effective procurement, Project implement ability, Government guarantee, Favorable economic conditions and Available financial market).

Other research was conducted by Ismail and Ajija [7] to identity the critical success factors in PPP implementation in the Malaysian construction industry. The key success factors were determined as:—good governance, commitment and responsibility of public and private sectors, favorable legal framework, sound economic policy and available financial market.

Research was also conducted into the China construction industry and this identified 18 factors, and then categorized them into five main groups:—stable macroeconomic environment, shared responsibility between private and public sector, transparent and efficient procurement process, stable political and social environment, and judicious government control [8]. Stable political and social environment are the indicators for successful implement of PPP [9].

Helmy and Lindbergh [10] carried out a research to assess the development strategies of the government of Kuwait for the implementation of PPP in infrastructure projects, three critical factors were then identified, including effective procurement, project implement ability and government guarantees. Dulaimi et al. [11] did a study to ascertain the key success factor for the implementation of PPP in United Arab Emirates; these were determined as the political support of the government. A risk which might prevent success is lack of proper knowledge and skills of the consortium.

Additionally, Zhang [12] identified that the critical success factors in PPP contracts as: economic viability, appropriate risk allocation via reliable contractual arrangements, sound financial package, reliable concessionaire consortium, and favorable investment environment.

Table 1 CSF of PPP from published literature

Critical success factor	Sources
Strong private consortium	Jefferies et al. [5], Birnie [13]
Appropriate risk allocation and risk sharing	Qiao et al. [14], Grant [15]
Competitive procurement process	Jefferies et al. [5], Kopp (1997)
Commitment/responsibility of public/private sector	Stonehouse et al. [16], Kanter [17]
Realistic cost/benefit assessment	Qiao et al. [14]
Project technical feasibility	Qiao et al. [14], Tiong [18]
Transparency in the procurement process	Jefferies et al. [5]
Good governance	Qiao et al. [14], Frilet [19]
Appropriate legal framework	Bennett (1998), Stein [20]
Availability of financial market	Qiao et al. [14], Jefferies et al. [5]
Political support	Qiao et al. [14], Zhang [12]
Government involvement by providing guarantees	Kanter [17], Qiao et al. [14]
Sound economic policy	EIB [21]
Stable macro-economic environment	Qiao et al. [14], Dailami and Klein [22]
Well organized public agency	Jones et al. [23], Finnerty [24]
Shared authority between public and private sectors	Stonehouse et al. [16], Kanter [17]
Social support	Frilet [19]
Technology transfer	Qiao et al. [14]

On the basis of a thorough review of previous similar researches, initially eighteen critical success factors were identified that affect the public private partnership implementation in Afghanistan, the factors are presented in Table 1.

3 Research Method

After listing the possible critical success factors, a questionnaire was designed and sent out to 125 carefully selected stakeholders. The stakeholders were asked to rate the pre-determined factors having a significant contribution to the success of PPP projects in Afghanistan. To measure the data from the questionnaire, survey the ordinal scale was used. The Likert scale of five ordinal measures from 1 to 5 according to the level of contributing used where; 1 indicates not significant, 2 slightly significant, 3 moderately significant, 4 very significant, and 5 extremely significant. The Relative Importance Index (RII) was used to rank the factors; the RII is calculated as the following equation:

$$RII = \frac{\sum_{i=1}^{5} W_i \cdot X_i}{A \times N}$$

where:

RII Relative Importance Index
W Weighting given to each factor by the respondents and ranges from 1 to 5
X Frequency of ith response given for each cause
A Highest weight (i.e. 5 in this case)
N Total number of respondents.

4 Data Analysis

The structured questionnaire survey was carried out by distributing a total of 125 questionnaire sets. 76 of the 125 questionnaires distributed were returned. The summary of survey is illustrated in Table 2.

Table 3 presents the respondent's position indicating that the majority of respondents were project managers (53%) followed by construction manager (26%), followed by program managers (14%) and others (7%). Others included town planners, development managers and site officers. The data shows that 47% of the respondents have 5–15 years of construction industry, 25% less than 5 years, and 28% have more than 15 years of construction industry experience. Generally,

Table 2 Summary of survey carried out

Parameters	Values
Number of questionnaire distributed	125
Number of questionnaire received	76
Percentage of responses received (%)	60.8%

Table 3 Respondents' demographic

Parameters	Frequency	Percentage (%)	Cumulative (%)
Respondent's position			
Project manager	40	53	53
Construction manager	20	26	79
Program manager	11	14	93
Others	5	7	100
Working experience			
Less than 5 years	19	25	25
5–15 years	36	47	72
More than 15 years	21	28	100

the demographic of respondents demonstrates that respondents were suitable to respond to the survey.

From Table 4, it can be perceived that the six factors have the greatest importance in successful implementation of PPP in the Afghanistan construction industry:

1. an appropriate legal framework,
2. political support,
3. transparency in the procurement process,
4. good governance,
5. availability of financial market and lastly
6. appropriate risk allocation and risk sharing.

The respondents ranked that the appropriate legal framework is a crucial factor for successful implementation of PPP projects in Afghanistan. It can be noted that adequate legal and regulatory framework at reasonable cost should be available to ensure transparency in the entire bidding process. Dispute resolution is another main issue to be in place in order to maintain the stability of the PPP contracts.

Political support is ranked second by the participants, government support politically is essential to ensure development of PPP projects. Political support by the government can attract more national and international investors to participate in infrastructure development of the country. The government of Afghanistan should provide the political support to private investors to ensure the sound implementation of the PPP projects.

Table 4 Critical success factor

Critical success factor	RII value	
	RII	Rank
Appropriate legal framework	0.91	1
Political support	0.88	2
Transparency in the procurement process	0.86	3
Good governance	0.83	4
Availability of financial market	0.82	5
Appropriate risk allocation and risk sharing	0.79	6
Sound economic policy	0.75	7
Shared authority between public and private sectors	0.72	8
Stable macro-economic environment	0.71	9
Strong private consortium	0.69	10
Well organized public agency	0.67	11
Competitive procurement process	0.65	12
Government involvement by providing guarantees	0.63	13
Realistic cost/benefit assessment	0.62	14
Project technical feasibility	0.60	15
Commitment/responsibility of public/private sector	0.59	16
Technology transfer	0.56	17
Social support	0.51	18

The third most important success factor ranked by the respondents is transparency in procurement process. In Afghanistan, lack of transparency in the bidding process and bidding evaluation has had unfavorably impacts on procurement process, which most of the bidders are not satisfied with the process. In the country, most of the business transactions are carried out informally, so there is a need for a well-organized body to regulate and monitor the procurement process in a transparent manner. PPP is relatively a new concept in Afghanistan, still there should be a well-developed strategy to build the capacity of the staff that is involved in PPP implementation unit.

The participants ranked that good governance as another important success factor. Good governance and strong public institutions should be in place to ensure value for money of PPP projects. Good governance in PPP procurement should follow a process to promote fair competitions and reduce the time and the cost of the entire procurement process. The core aim of this process is to get the best proposal that can assure value for money. Availability of financial markets is also important success factor ranked by the respondents. Financing PPP projects is a main indicator for private investment in public infrastructure projects. Access to financial market with low interest cost and a wide range of financial products for private sector would be great incentives to implement PPP projects.

The other important success factor ranked by the respondents is appropriate risk allocation and risk sharing. One of the main principles in PPP contracts and arrangements is the allocation and sharing the risks to the party that is able to manage it. Generally, the public sector intends to transfer most of the risks to the private sectors that have adequate experience in managing the risks. However, the public sector should not allocate those risks that are beyond control of the private sector. Logically, the public sector should have the right measures in place to analyze the associate risks and share the risks in an effective manager between the public and private sectors, so as to ensure that each party can best manage it. On the other hand, the private sector should have a proper risk management strategy to identify and understand all the project risks and price them accordingly.

5 Conclusion

This paper has presented the findings of a study carried out in Afghanistan to identify the critical success factor for implementation of PPP projects. The questionnaire survey was employed to collect the data; the respondents were asked to rank the 18 factors that are important for successful implementation of PPP projects in the construction industry. The findings indicate that the most important success factor is favorable legal framework, followed by political support, transparency in the procurement process, good governance, availability of financial market and lastly appropriate risk allocation and risk sharing.

On the basis of the findings it can be concluded that the government of Afghanistan should develop the proper legal and regulatory framework to attract more national and international investors to participate in infrastructure development of the company. Additionally, more attention should be taken to transit from public planned economy to social market economy in order to boost the private sector involvement in the county economic development.

References

1. Efficiency Unit (2003) Serving the community by using the private sector—an introductory guide to public private partnerships (PPPs). Hong Kong Special Administrative Region Government, Hong Kong
2. Bing L et al (2005) The allocation of risk in PPP/PFI construction projects in the UK. Int J Proj Manage 23(1):25–35
3. Gruneberg S, Hughes W, Ancell D (2007) Risk under performance based contracting in the UK construction sector. Constr Manage Econ 25(7):691–699
4. Hardcastle C, Edwards PJ, Akintoye A, Li B (2005) Critical success factors for PPP/PFI projects in the UK construction industry. In: Proceedings of the conference on public private partnerships; opportunities and challenges. The Hong Kong Institution of Engineers Civil Division and The University of Hong Kong, Hong Kong, pp 75–83
5. Jefferies M, Gameson R, Rowlinson S (2002) Critical success factors of the BOOT procurement system: reflections from the Stadium Australia case study. Eng Constr Architect Manage 9(4):352
6. Russell RK (2008) Critical success factors for the fuzzy front end of innovation in the medical device industry. Eng Manage J 20(3):36–43
7. Ismail S, Ajija SR (2013) Critical success factors of Public Private Partnership (PPP) implementation in Malaysia. Asia Pac J Bus Adm 5(1):6–19
8. Chan APC et al (2010) Critical success factors for PPP in infrastructure developments: Chinese perspective. J Constr Eng Manage 136(5):484–494
9. Wong A (2007) Lessons learned from implementing infrastructure PPPs—a view from Singapore. Proceedings of seminar jointly organized by the Department of Civil Engineering of the University of Hong Kong and Civil Division of the Hong Kong Institution of Engineers, The University of Hong Kong and The Hong Kong Institution of Engineers, Hong Kong
10. Helmy MA, Lindbergh J (2011) Investigating the critical success factors for PPP projects in Kuwait. KTH Architecture and the Build Environment Department of Real Estate and Construction Management, Thesis No: 106, 46, urn: nbn:se: kth: diva- 77471. http://www.google.com/url?Investigating_the_Critical_Success_Factors_for_PPP_projects_in_Kuwait.pdf
11. Dulaimi MF et al (2010) The execution of public—private partnership projects in the UAE. Constr Manage Econ 28(4):393–402
12. Zhang XQ (2005) Critical success factors for public private partnerships in infrastructure development. J Constr Eng Manage 131(1):3–14
13. Birnie J (1999) Private finance initiative (PFI) UK construction industry response. J Constr Procurement 5(1):5–14
14. Qiao L, Wang SQ, Tiong RLK, Chan TS (2001) Framework for critical success factors of BOT projects in China. J Project Finan 7(1):53–61
15. Grant T (1996) Keys to successful public-private partnerships. Can Bus Rev 23(3):27–28
16. Stonehouse JH, Hudson AR, O'Keefe MJ (1996) Private-public partnerships: the Toronto hospital experience. Can Bus Rev 23(2):17–20
17. Kanter RM (1999) From spare change to real change. Harvard Bus Rev 77(2):122–132

18. Tiong RLK (1996) CSFs in competitive tendering and negotiation model for BOT projects. J Constr Eng Manage 122(3):205–211
19. Frilet M (1997) Some universal issues in BOT projects for public infrastructure. The Int Constr Law Rev 14(4):499–512
20. Stein SW (1995) Construction financing and BOT projects. Int Bus Law 23(4):173–180 (International Bar Association)
21. European Investment Bank (2000) The European Investment Bank and public private partnerships. The newsletter of the International Project Finance Association, vol 1, pp 3–4
22. Dailami M, Klein M (1997) Government support to private infrastructure projects in emerging markets. In: Irwin T (ed) World Bank Latin American and Caribbean studies viewpoints: dealing with public risk in private infrastructure. Washington, pp 21–42
23. Jones I, Zamani H, Reehal R (1996) Financing models for new transport infrastructure. OPEC, Luxembourg
24. Finnerty JD (1996) Project financing: asset-based financial engineering. Wiley, New York

The Reference Methodology of Prospective Analysis of Technology in Production Engineering

Katarzyna Halicka

1 Introduction

In-depth analysis of technology is difficult due to the cost, complexity of the problem and, above all, the pace of technological change on the global market.

The analysis of technology requires having adequate knowledge resources that are dispersed and at the same time cover many aspects of technological development. In the analysis technology, there is a need for tools and skills allowing for a substantive technical assessment [1] of the properties of its current state. Knowledge concerning the current trends in technology development is also necessary. The concepts allowing to predict the future state of technology include technology assessment, technological forecasting [2], technological foresight [3]. Each of these approaches has some important advantages, but also limits. The approach integrating the technology analysis and development tools discussed above is forward-looking technology analysis [4].

Prospective analysis of technology (PAT) allows, according to Saritas and Burmaoglu, in the long term, to take action to enable the development of new future technologies [5]. According to the author, PAT is a kind of a natural path of technological foresight evolution. However, these are not identical terms and technological foresight should not be identified with prospective technology analysis. According to the opinion of Boden et al. [6] as well as Eerola and Milesa [7], PAT facilitates decision-making and coordinating future actions, especially in the fields of science, technology, and innovation, as well as politics. Moreover, Haegeman and others [8] and Cagnin et al. [9] and others have argued that

K. Halicka (✉)
Faculty of Management, Bialystok University of Technology,
Wiejska Street 45A, 15-351 Bialystok, Poland
e-mail: k.halicka@pb.edu.pl

S. Şahin (ed.), *8th International Conference on Engineering, Project, and Product Management (EPPM 2017)*, Lecture Notes in Mechanical Engineering,
https://doi.org/10.1007/978-3-319-74123-9_11

prospective technology analysis is a single common term for a set of different tools that can be used to study and understand the future of technology from a variety of methodological perspectives.

The above definitions are general and do not reflect the nature of the PAT approach. Consequently, the author proposed her own definition of prospective technology analysis as a process whose main purpose is to anticipate the future of technology through detailed analysis (scanning) and the assessment of its current state, as well as the identification of strategic factors of its development in the future [10].

2 Proposition of the Research Process of Prospective Analysis of Technology

Critical review of literature, analysis of foreign and domestic PAT initiatives, and own research experience indicated the necessity of constructing a unified PAT research process, as is the case in forecasting, foresight, and technology evaluation. This process should have the nature of justified activities, covering the whole of the research procedure, being in line with the definition of PAT, to anticipate the future of technology. As a starting point for preparing the process of prospective technology analysis, the following processes were adopted: (1) prognostic process comprising two phases: Diagnosis of past and present, and defining the future; (2) technological foresight process consisting of research stages such as: preliminary, scanning, recruiting, main, planning, action and revision; (3) technology assessment process taking into account the identification, selection and assessment of the potential impact of technology on the environment.

Considering the definition of PAT, and the abovementioned assumptions, five successive phases of the PAT process can be identified (Fig. 1): (I) Concept, (II) Scanning, (III) Understanding, (IV) Anticipation, (V) Conclusions.

The first phase—Concept—focuses first on defining the premises, scope and objectives of the analyses. During this phase, the research problem is formulated and the data concerning the analysed technologies is collected. The formulation of the research problem means the definition of: technology being analyzed; purpose of technology analysis; time horizon; audience analysis; owned financial and human resources; anticipated effects; form of presentation of the results. In turn, within the framework of the collection of data regarding technology, basic data is collected concerning technologies, their properties, determinants, application possibilities, costs based on bibliographic databases, patent databases, reports, and websites. This phase also involves the collection of data on the institutions and centres using the analyzed technologies. The result of the work carried out in this phase, is the preparation of a set of data on the analyzed technology.

Then, in the second phase—scanning—the data collected in Phase I is sorted. Next, based on this data, the genesis of the technology, its previous properties and applications are analysed. The result of the research carried out within the second phase, is the collection of information on the past of the analysed technology.

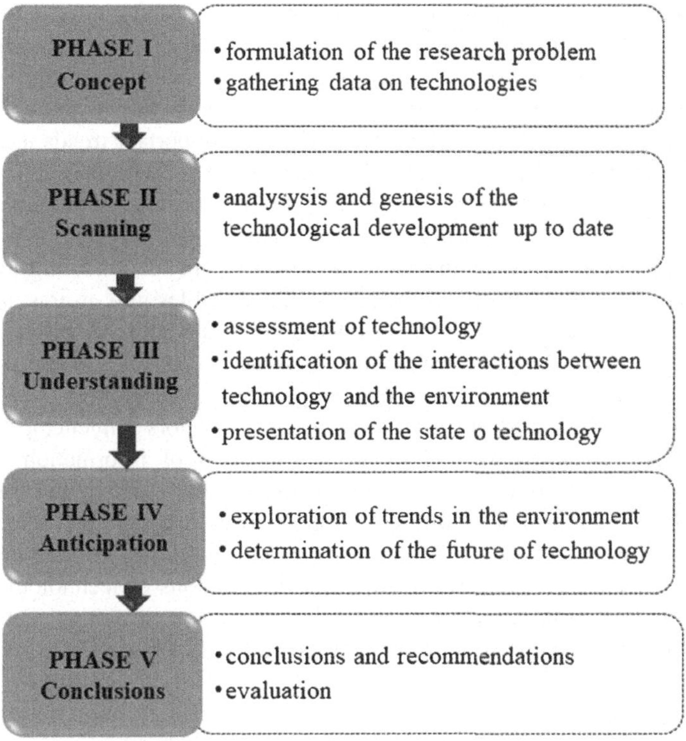

Fig. 1 Structure of the prospective analysis of technology (PAT)

The current level of technological development is also determined. Benefits of using the technology so far are also indicated—not available in the case of the application of alternative solutions.

As part of the third phase—understanding of technology—the information collected in phase II is initially supplemented. Among other things, the economic efficiency of the technology is determined [11], as well as its technical and implementation feasibility [12]. The level of technological maturity is assessed. Technology can also be evaluated taking account the criteria relating to its competitiveness, usability, as well as environmental, and various social-ethical criteria. This phase also involves the analysis of the current state of the technology in terms of its impact on the environment as well as of the environment on technology. The interactions between the technology and the environment are identified. The result of these activities may be a catalog of factors influencing the analyzed technology or a collection of information about the impact of technology on the environment [13]. As part of this phase, the comprehensive, current state of technology is also presented. Most often it comes down to the development of technology cards, the preparation of technology knowledge maps (patents maps, publications) or relational maps, or the creation of a knowledge base on technologies in the form of descriptions, tabular figures, visualization of certain characteristics of technologies.

Phase Four—anticipation—focuses on exploring trends and presenting the anticipated development of technology. Exploring the trends in the environment primarily involves the identification and analysis of important trends and changes occurring over time, which can affect the future of the technology being analyzed.

This phase initially identifies the factors influencing the development of the analyzed technology. These factors are then analyzed and selected. It is possible, inter alia, to reduce the available information on the factors affecting technology development. It is also possible to merge similar variables into groups and also to isolate the similarities between variables. In addition, the influence of factors on the development of the analyzed technology is assessed, as well as the trends and changes that may affect the development of the analyzed technology over time. The results of the work carried out may be a catalog of factors influencing the development of the analyzed technology or a collection of information on trends influencing the development of the technology. On the basis of the information obtained, the direct or indirect future and the changes and trends in selected areas are depicted. Future audiences, technology developers (companies that want to use or create a technology) are also identified. The directions of technological developments are also being developed, as well as activities promoting technological development and technology development scenarios. The results of work in phase four are routes, scenarios, and strategies.

The last phase (fifth) involves the summarization and application of the results of the conducted analyzes. The conclusions phase, among others, assesses the chances of achieving the objectives set and the possibilities for achieving them while choosing specific instruments, taking into account the resources allocated. The phase is aimed at optimizing resources, improving relevance in meeting the needs of stakeholders. The purpose of the study are the objectives, anticipated results, consistency with the development objectives of the technology, complementarity with other projects, as well as procedures for the implementation, monitoring, evaluation and financial management. This phase also includes verifying whether the expected results adopted in Phase I have been achieved. The results of the work in this phase are the conclusions and recommendations of the conducted analysis.

3 Identification of PAT Application Areas

In order to identify the areas of application of PAT, a detailed review of the literature has been carried out. Initially, the database of the foreign periodical Scopus was searched with the use of the term "future-oriented technology" contained in keywords, titles, and summaries. Over the past thirty years (from 1987 to 2015), there have been 94 publications, identified in the Scopus database related to prospective technology analysis. By analyzing the available publications, it has been noted that the interest in this topic has been systematically increasing.

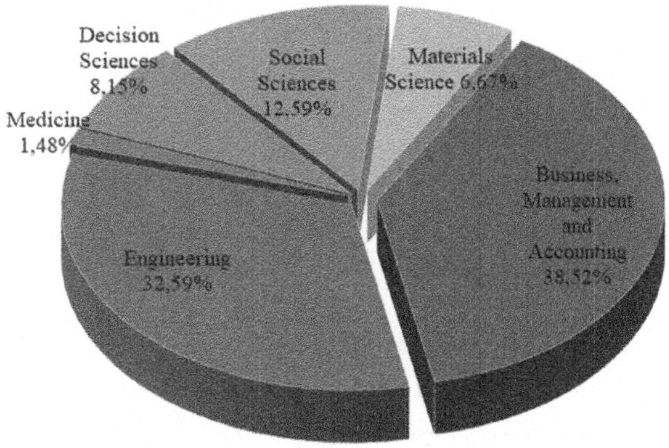

Fig. 2 Typology of PAT publications concerning the subject matter

The topics discussed in the PAT publications mainly concern business, management, and accounting (about 38% of publications), engineering (over 32% of publications), decision theory (about 8% of publications). The detailed typology of the publication in view of the subject matter in the field of prospective technology analysis is presented in Fig. 2.

Considering the potential of prospective technology analysis, it can also be used by manufacturing companies. Analysis may facilitate the completion of subsequent stages of the production process (Fig. 3). It can be used by manufacturing companies for the production preparation in the structural, technological and organizational area. PAT tools can also be useful at the stage of strategic planning, product design, or process design and manufacturing system. The methods of the prospective analysis of technology facilitate, among other things: (1) predicting the development of products, production methods and production capacity; (2) designing products and production methods; (3) constructing prototypes; (4) technical and economic analysis.

4 Example of Using Prospective Technology Analysis

The author applied prospective technology analysis to identify and demonstrate the development of pavement technologies in Poland with the greatest development potential. Using this methodology, among other things, it was possible to develop routes of the development of priority road paving technologies in Poland.

The process of prospective analysis of road paving technology in Poland consisted of five successive phases. The research methodologies specific to prospective technology analysis were used for the implementation of the various phases.

The methods were chosen based on the available resources, the subject matter of the research, and the experience and knowledge of the research team in the application of these methods. In phase 1—concept—the anticipated needs and requirements of road pavement technology in Poland and designated priority

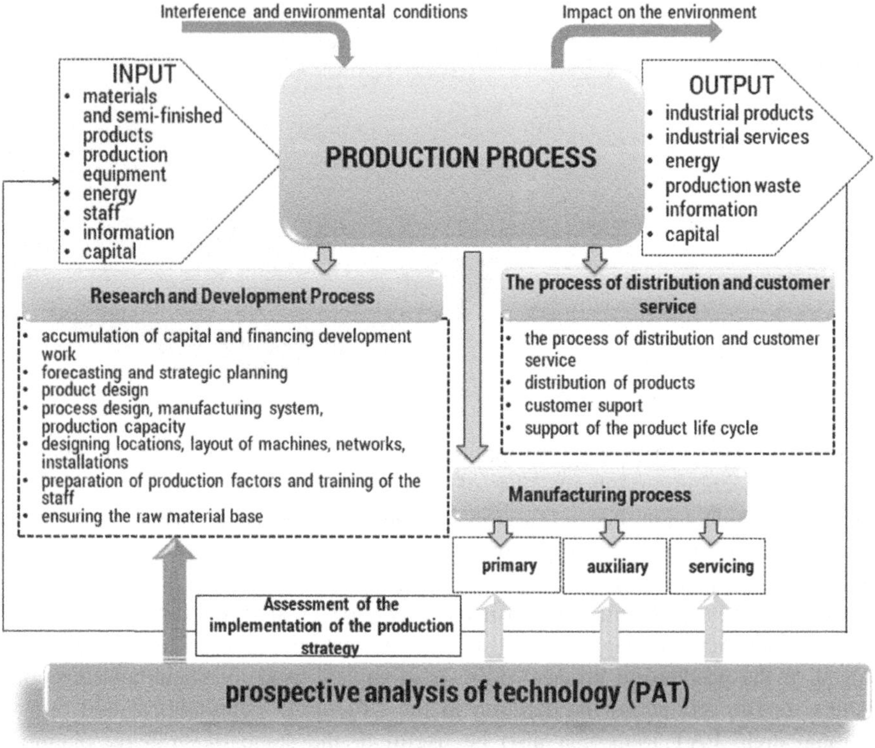

Fig. 3 The use of PAT in the production system

technologies have been presented. In this phase, webometric analysis, literature review, desk research, as well as Delphi and expert panels were used [14].

The result of the work was the identification of five priority road paving technologies in Poland (Tech1: mineral-asphalt mix with gum-asphalt binder; Tech2: porous mineral-asphalt mixes; Tech3: "perpetual" pavement; Tech4: traditional cement concrete; Tech5: Binders with increased Elastomer content) together with background data on these technologies [15].

In phase II—scanning—on the basis of previously acquired data, the analysis was conducted concerning the past of the technology, its origins, its current properties and applications. The result of the work in this phase was the development of a collection of information on the current development of the analyzed technologies, as well as the determination of the life cycle phase of technologies.

In this phase, methods such as S curve analysis and retrospective analysis were used. In Phase III—understanding—technology is judged in terms of its technological maturity and competitiveness. Social and ethical and ecological criteria were also taken into account during the analysis and evaluation. In the third phase of prospective technology analysis, the methods belonging to the accumulation class were taken into consideration—TRL readiness model and literature review, as well as the exploratory method—surveys, and from the typing class—statistic analyses. The result of work within the framework of this phase was the development of a card of priority road pavement technologies in Poland. The cards contained the collected most important information about the technology being analyzed, such as: technology characteristics, purpose of use, scope of use, approximate cost, degree of implementation, technological readiness levels, examples of current technology use, technology benefits, the level of social acceptance for the development of a given technology, expert assessment). The fourth phase is mainly focused on presenting the future development of the priority road paving technologies in Poland. Taking into account the assumptions adopted in the first PAT phase, the development of priority technologies has been presented in three time perspectives: 2015–2020, 2021–2030 and after 2030. The final result of the work carried out in the fourth phase was the preparation and presentation of the route of development for each of the priority technologies, in a clear and readable way. This phase of prospective technology analysis included exploratory methods such as surveys and brainstorming, and from projection class, methods such as technology development routes.

In the last phase—conclusions—a number of actions conducive to the expected development of priority technologies of road surfaces in Poland have been selected. Within this phase, it was also verified whether the anticipated revisions of phase I have been achieved. The final result of the work carried out within this phase was the presentation of a catalog of guidelines (recommendations) facilitating the development of road pavement technologies in Poland and presentation of conclusions from the analyses carried out. Expert panels were used during this phase of the prospective analysis of this technology. The results of the research work have been presented in detail in the publication [10].

5 Conclusion

Thanks to the prospective analysis of technology, it is possible to identify, examine in detail, assess, present the current state, as well as predict the future of priority technologies—with the greatest potential of development. With the use of PAT, it is also possible to build up their portfolios. The prospective analysis of technology also enables to distinguish the factors conducive to the development of these technologies, examine the impact on the environment, and also allows predicting the future of their development. An important advantage of this concept is the consideration of technology in various time perspectives: up to 5 years, from 5 to 10 years and above 10 years. Thus, PAT, can be the basis for decision-making in the selection of technology with the greatest potential for development.

The developed reference methodology will allow researchers to confront their research targets with the justified, theoretically and practically verified pattern of behaviour. It can also help to avoid many mistakes and contribute to the proper designing and carrying out of the process of the prospective analysis of technology. The methodology has a universal value expressed in the possibility of adapting the model to the various types of technology.

The undertaken subject is of an innovative nature, not yet described in the literature. The developed methodology also has application advantages. The possibility of its use has been demonstrated on the example of the road surface technology in Poland as well as the battery technologies [16].

Acknowledgements The research was conducted within S/WZ/1/2014 project and was financed from Ministry of Science and Higher Education funds.

References

1. Nazarko Ł (2015) Technology assessment in construction sector as a strategy towards sustainability. Procedia Eng 122:290–295
2. Ayres RU (1969) Technological forecasting and long-range planning. McGraw-Hill, New York
3. Georghiou L, Harper JC, Keenan M, Miles I, Popper R (2008) The handbook of technology foresight: concepts and practice. PRIME series on research and innovation policy. Edward Elgar Publishing, Cheltenham
4. Nazarko J, Kononiuk A (2013) The critical analysis of scenario construction in the Polish foresight initiatives. Technol Econ Dev Econ 19(3):510–532. https://doi.org/10.3846/20294913.2013.809030
5. Saritas O, Burmaoglu S (2015) The evolution of the use of Foresight methods: a scientometric analysis of global FTA research output. Scientometrics 105(1):497–508. https://doi.org/10.1007/s11192-015-1671-x
6. Boden M, Johnston R, Scapolo F (2012) The role of FTA in responding to grand challenges: a new approach for STI policy. Sci Public Policy 39:135–139. https://doi.org/10.1093/scipol/scs026
7. Eerola A, Miles I (2011) Methods and tools contributing to FTA: a knowledge-based perspective. Futures 43:265–278. https://doi.org/10.1016/j.futures.2010.11.005
8. Haegeman H, Marinelli E, Scapolo F, Ricci A, Sokolov A (2013) Quantitative and qualitative approaches in Future-oriented Technology Analysis (FTA): from combination to integration? Technol Forecast Soc Change 80(2):386–397. https://doi.org/10.1016/j.techfore.2012.10.002
9. Cagnin C, Keenan M, Johnston R, Scapolo F, Barré R (2008) Future-oriented technology analysis. Strategic intelligence for an innovative economy. Springer, Berlin
10. Halicka K (2016) Innovative classification of methods of the future-oriented technology analysis. Technol Econ Dev Econ 22(4):574–597. https://doi.org/10.3846/20294913.2016.1197164
11. Ejdys J, Ustinovicius L, Stankevičienė J (2015) Innovative application of contemporary management methods in a knowledge-based economy—interdisciplinarity in science. J Bus Econ Manage 16(1):261–274. https://doi.org/10.3846/16111699.2014.986192
12. Nazarko Ł (2016) responsible research and innovation—a new paradigm of technology management. In: 9th international scientific conference business and management 2016, Vilnius, Lithuania, 12–13 May 2016. https://doi.org/10.3846/bm.2016.71
13. Ejdys J, Matuszak-Flejszman A, Szymanski M, Ustinovicius L, Shevchenko G, Lulewicz-Sas A (2016) Crucial factors for improving the ISO14001 environmental management system. J Bus Econ Manage 17(1):52–73. https://doi.org/10.3846/16111699.2015.1065905

14. Nazarko J, Radziszewski R, Dębkowska K, Ejdys J, Gudanowska A, Halicka K, Kilon J, Kononiuk A, Kowalski KJ, Król JB, Nazarko Ł, Sarnowski Michał, Vilutienė T (2015) Foresight study of road pavement technologies. Procedia Eng 122:129–136. https://doi.org/10.1016/j.proeng.2015.10.016
15. Radziszewski P, Nazarko J, Vilutiene T, Dębkowska K, Ejdys J, Gudanowska A, Halicka K, Kilon J, Kononiuk A, Kowalski KJ, Król JB, Nazarko Ł, Sarnowski M (2016) Future trends in road pavement technologies development in the context of environmental protection. The Baltic J Road Bridge Eng 11(2):160–168. https://doi.org/10.3846/bjrbe.2016.19
16. Halicka K, Lombardi PA, Styczyński Z (2015) Future-oriented analysis of battery technologies. In: Proceedings of the IEEE international conference on industrial technology, pp 1019–1024. https://doi.org/10.1109/icit.2015.7125231

Menu Engineering in Jordanian Health-Care Centers: A Modified Balanced Scorecard Approach

Madher E. Hamdallah and Anan F. Srouji

1 Introduction

Globalization and the advances in the recent years, have motivated industries to make their work unique, it would also be logical to think that the adopting companies of both ME and BSC deliver some benefits from this implementation.

Developing an executive scorecard is not self-sustaining and requires constant oversight and maintenance in relation to ME. This requires management to have a concise list of success factors that lead to the right perspective and the right measures of an organization [1, 2]. Every measure used of the BSC has to be part of a cause-and-effect association which ends in improving long-term sustainable financial and non-financial performance [3].

2 Study Problem

Using the modified BSC approach, some deliberate verdict corporations should be made in relation to performance; either financial or non-financial; as the hefty proportion of competition also pushes companies to move from the traditional model to new ones. It is also required from health-care centers to work on their ME in relation to global usage and its easiness in facilitating the companies' transactions, in order to not lose market, share and expand earnings. The study problem

M. E. Hamdallah (✉) · A. F. Srouji
Accounting Department, Al-Zaytoonah University, P. O. Box 130, Amman 11733, Jordan
e-mail: M.Hamdella@zuj.edu.jo

A. F. Srouji
e-mail: anansrouji@zuj.edu.jo

S. Şahin (ed.), *8th International Conference on Engineering, Project, and Product Management (EPPM 2017)*, Lecture Notes in Mechanical Engineering,
https://doi.org/10.1007/978-3-319-74123-9_12

was indicated to study if there is a relationship between ME and a modified BSC approach in health-care centers in Jordan to ease such requirements, and whether there is a difference between employees' knowledge awareness on the importance of using ME and the modified BSC.

3 Theoretical Framework and Hypotheses Formulation

Hypothetical improvements and genuine practices with the solicitation of the BSC have revealed a need to acclimatize the standard tool in relation to the sector and company. It is the application of cost required to satisfy the patients' needs and enabling managers to have direct access to human resource and other workplace services for performance, reporting, knowledge management and learning as well as administration applications [4]. Foodservice quality and services may also have an impaction patients' contentment with their general hospital familiarity [5]. Food provisioning can be a key part of a hospital's fundamental healing mission [6]. Measures must be analyzed for mutual links in order to avoid reciprocated barring of the menu cost, in context of the entire perception of BSC.

3.1 Health-Care Centers in Jordan

Health-care centers are medical care corporations which store medical history and permit easy and fast retrieval of information on patients', equipped by different departments. Each health-care organization is different and it is important to consider the unique features of an organization when introducing standards. One size does not fit all.

"Within the Ministry of Health in Jordan, the main authorities are the Head of Mental Health Specialty and the Director of the National Center for Mental Health; the largest governmental psychiatric hospital includes the National Center for Addiction and outpatient clinics. In addition to mental health services provided under the Ministry of Health, universities and the private sector, there are also military services (Royal Medical Services), serving approximately 40% of the population" [7, p. 5].

"Exact numbers of human resources for mental health are unknown for both the public and private sectors. There are no coordinating bodies overseeing any mental health awareness campaigns and there is a lack of collaboration between the mental health and other relevant sectors in Jordan" [7, p. 6]. A registered dietitian will need to write menus for participants with special nutrition needs, once the menus are revised; they need to be evaluated for cost.

3.2 ME

ME is a procedure where management can assess contemporary and imminent menu rating, design and satisfied judgments. By using ME, items on the menu can individually be categorized, a more perilous inquiry of the plates may be accomplished, ME will facilitate and improve decision making for which items will be reserved or detached, others may need advertisement or explain the increase in costs [8].

The objective is to increase the influence of ME by introducing it to restaurant managers in the hope to be implemented to expand productivity and manageability [8]. ME theory obliges managing alignment of the profit the menu contributes [9]. A traditional evaluating strategy that is only based on costs and cost mark-ups may be hazardously imprecise and unintentionally obliging the food facility worker's aptitude to exploit revenues and profits. Patients and residents understanding increases in the service sector, when they have chances to select meals as in restaurants [10].

Customer-oriented service is being loomed by hospitals nowadays. Hospitals hire culinary-trained chefs to arrange pleasing and highly sophisticated food to augment patients' fulfillment [11]. In hospitals, staffs are liable on helping patients with their meal selections, compelling menu orders, and conveying food. Based on Fig. 1 the following terminologies could be clarified in relation to the menu items Profit Margin (PM) and selling rates. STARS: sell well and they bring the highest gross PM. PLOWHORSES: sell well, but don't do as well in PM. PUZZLES: don't sell as well as you'd like, but when they do, the PM is above average. DOGS: are items that don't produce for you at all, but are sometimes necessary to have on the menu.

Indulgence and observing costs and margins is required for any company owner and the inability to get it right indicates a lack of cost-effectiveness. In fact, the main reason of small businesses failure is mainly referred to low prices or high costs

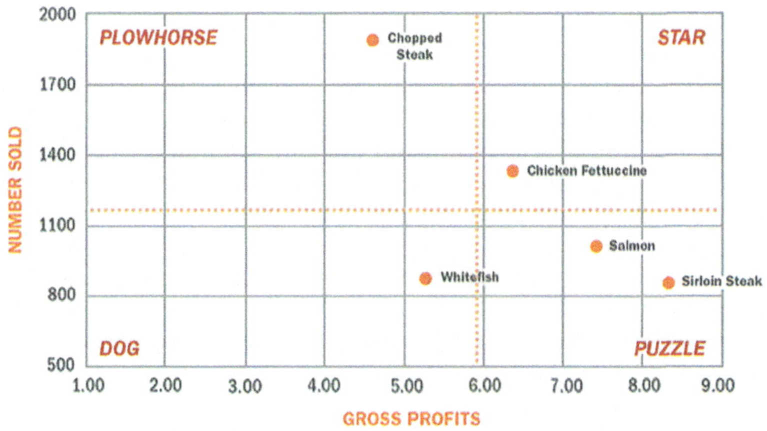

Fig. 1 Menu engineering [27]

which lack cost-effectiveness [8]. The PM designates the percentage profit companies make and specifies the influence a company has to earn profits. Profitability may be delineated as the investments ability to earn return [12].

3.3 BSC Approach

BSC leads to strategic management improvement and better strategy employment and assessment. The BSC is a new tool of measuring both financial and non-financial performance systems. As pointed out by Kaplan and Norton [13] the scorecard is accompanied gradually with strategy implementation. It performs as a management framework that helps in classifying and achieving companies' goals and strategies [14].

In general, with respect to competitiveness of organizational environment and survival in environment, application of BSC as a tool for evaluation of organizational strategies is a requirement in today organizations [15]. Term "score" was used in title of this approach because of the fact that this attitude is employed in order to allocate score to a group of performance evaluation parameters. While using the term "balanced" is due to point that this approach produces steadiness among financial and non- financial constraints, domestic and foreign benchmarks, internal and external recipients, performance leader and followers' parameters, and derives and barriers of strategy [16, 17].

In service sectors, as health-care where non-financial indicators are commonly used for effective managing, there are adverse constraints between the financial image and the medical outlook of health-care authorities, as measurement structures are still not able to efficiently assimilate between both visions [18]. As BSC has amplified in the health-care sector through the years, it also hit the ground running with an effective measurement system, for flexibility, openness, efficiency, and inclusiveness [19].

Duke University Hospital elucidated the four perceptions in health-care organizations as follow [20].

(a) Customer perspective: Bring qualified inventive care based on patients' preferences. (b) Financial perspective: Bring sturdy and reliable financial performance which satisfies all shareholders. (c) Learning (Clinical Quality) perspective: To help in improving worker's skills and enactment. (d) Business (Operational Effectiveness) perspective: To be accepted as a health-care service leader in the area.

Figure 2 illustrated the proxies of both the financial and non-financial performance of the centers.

Since the study added new BSC perspectives, the research came with a new template model to help health-care centers improve their performance, and the first main hypothesis was proposed that:

H1: There is no significant nexus between the modified BSC model and ME in health-care centers.

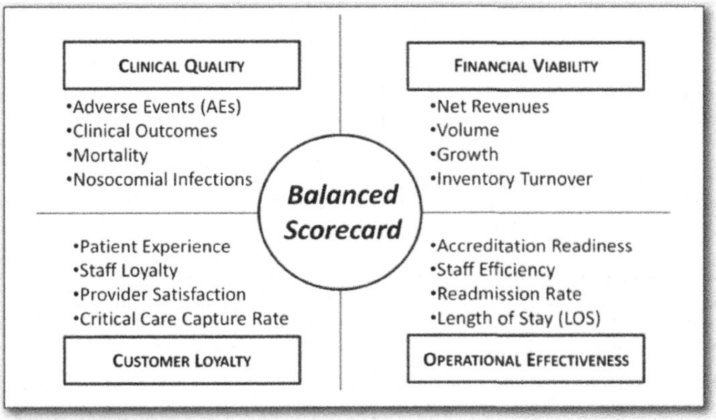

Fig. 2 Traditional BSC templates in health-care centers

First and in relation to the financial viability the research model came to add cost, and remove the volume. For volume here is the difference between both revenues and costs in the cost-volume profit analysis [21], so there is no need and impractical to repeat revenues in the two viabilities as in the traditional model (Net Revenues and Volume). This exposed the first sub-hypothesis:

H1a.: There is no significant nexus between financial perspectives and ME in the centers.

However, in the non-financial perspectives the research came to separate employee efficiency and morality from the operational effectiveness [22]. Operational effectiveness in new research is called Internal Business, while Quality is learning and innovation [23]. For internal business patient's lengths of stay and technical provisions were added to clinical internal business (while staff loyalty was shifted to the employee efficiency and morality, which included the staffs' length of stay and ethics). This exposed the second sub-hypothesis:

H1b.: There is no significant nexus between non-financial perspectives and ME in the centers.

Based on the idea whether there is a difference between respondent's awareness of the modified BSC perspectives in relation to ME in health-care centers, the second main hypothesis was positing that:

H2: There is no significant nexus between the respondents' different demographical factors and their awareness of the modified BSC perspectives and ME in health-care centers (gender, age, education, monthly salary and experience). Based on the previous literature the new BSC template model would be formulated as illustrated in Fig. 3.

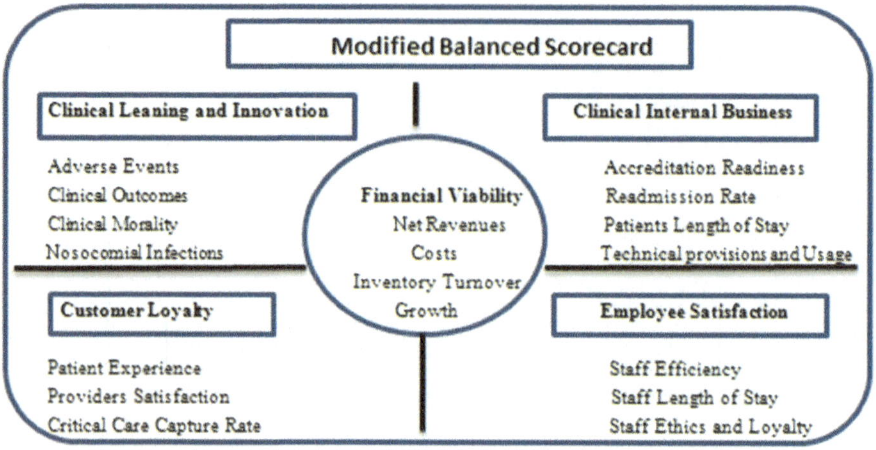

Fig. 3 Modified balanced scorecard templates in health-care centers

4 Research Method

4.1 Methodology

The study used primary data to collect the information needed, by using a structured questionnaire. Extensive stakeholders can benefit from the modified model as management, financial in addition to nutrition departments. Stratified random sampling was used as a type of probability sampling technique.

Information was collected through a questionnaire distributed in the different health-care centers; in Amman the capital of Jordan. Two hundred questionnaires were distributed among 45 centers, and 182 were recollected and valid for analysis by using SPSS 22 software.

4.2 Regression Equation

Multiple linear regression is used, by expressing y as the dependent variable, in relation of several independent variables x1, x2, …, xn [24]; as applied in study.

$$Yit = \beta 0 + \sum \beta KXit + \varepsilon it \quad (1)$$

where, Yit indicates the dependent variables (ME); of Health Center i for time period t. β0 is the intercept, βK epitomizes the coefficients of the Xit variables. Xit embodies the explanatory independent variables which are the modified BSC perspectives (financial and non-financial) of Health Center i for time period t. εit is the error term, which interpreters all indefinite factors not encompassed in t.

5 Results and Findings

5.1 Descriptive Statistic Results

After testing the reliability coefficient by alpha, the degree of reliability was (0.86), and deliberated to be suitable for the aims of study [25]; the study tested the correlation between the efficiency of BSC on ME. As in Table 1, a correlation of 0.873 and 0.795 was for financial and non-financial perspectives, respectively. Results showed a strong nexus between the variables, with a high correlation level of 0.000 significance.

5.2 Hypotheses Testing Results

Calculated (t) for hypotheses (H1a) and (H1b), has significance level of (0.00) which is greater than tabulated (t) value ($\alpha = 0.01$), by using multiple linear regression, as there were two independent variables (financial and non-financial perspectives). So both hypotheses were statistically not accepted indicating a relationship among the variables, as illustrated in Table 2.

One way Anova including Post Hoc tests was used to analyze the second hypothesis (Sheffe, and Dunett's C) to investigate the homogeneity [26], as illustrated in Table 3. The table indicated a positive significant nexus for age, as F was (6.875) at a significant level of (0.011), and outcomes indicated that those elder than 46 years were more realizable of the effect of the BSC than the others. Results also revealed that experience had a positive significant relationship, where respondents with experience more than 11 years (4.3135) were more realizable than others.

Table 1 Pearson correlation

Type of variables	Financial perspectives	Non-financial perspectives
ME (Pearson)	0.873	0.795
ME (Correlation)	0.000[a]	0.000[a]

Note [a]Significant at a level of 0.01

Table 2 Testing hypotheses

Hypothesis	H1a	H1b
Significance	0.007[a]	0.043[b]

Note [a]Significant at a level of 0.01
[b]Significant at a level of 0.05

Table 3 One way Anova (Second main hypothesis)

Demographic variable	Type of variable	Mean	df	F	Sig
Age	Younger than 25 years	4.0733	5	6.875	0.011[a]
	26–30	4.2833			
	31–35	4.2833			
	36–40	3.9753			
	41–45	4.0259			
	Elder than 46	4.5300			
Gender	Male	3.8154	1	9.163	0.064
	Female	4.1280			
Marital status	Single	3.9875	1	3.057	0.751
	Married	4.2536			
Education	Diploma	4.3340	2	2.479	0.122
	Bachelors	3.9839			
	Masters	4.0231			
Experience	Less than 5 years	4.0433	2	8.095	0.031[a]
	6–10	4.2433			
	More than 11 years	4.3135			

Note [a]Significant at a level of 0.05

6 Conclusions of Study

Multiple regression results indicated that the modified BSC template including both financial and non-financial perspectives affect the ME in health-care centers in Jordan, with a positive significant relationship. This is a good indication that all the perspectives included in the model affected the modified BSC template.

Age and experience of the respondents had a positive significant effect on their awareness of the relationships between the modified perspectives and ME, as they grew elder or experience period increased. Such a result verifies how the level of experience and age could truly affect the results in a positive manner regardless of the respondents' gender, marital status and education level.

7 Limitations and Recommendations

The study syndicates two different analytic systems (ME and BSC) into a new method; that exposes the performance image for menus' in Jordanian health-care centers. As it was found necessary to identify the following when dealing with health-care centers in order to increase efficiency and reduce limitations.

Financial and non-financial perspectives may differ from one point of view to another, so results may also differ widely. Depending on the proxies studied under each perspective. The research sample included only health-care centers in Amman (the capital) that covers more than 50% of the health centers population. Upcoming studies ought to relate the proposed model by including additional health-care centers located in different geographical areas to make sure it is accepted widely. Taking into consideration the type of centers may also vary the results. As hospitals for example, are the leading structured centers in Jordan, should be recommended as a future sample.

References

1. Kaplan R (2010) Conceptual foundations of the balanced scorecard. Harvard business school accounting & management unit working paper no. 10-074. Retrieved 24 Jan 2017 from: http://ssrn.com/abstract=1562586
2. Purnamasari V (2014) Environmental perspective: a new perspective in balance scorecard. S East Asia J Contemp Bus Econ Law 5(1)
3. Srouji A (2017) Corporate governance and bank performance of Islamic and conventional banks in Jordan: a balanced scorecard approach. Unpublished Dissertation, Universiti Malaysia Perlis
4. Warui C, Mukulu E, Karanja K (2015) Internal structure of the organization as a determinant of adoption of HRIS in the operations of Teachers Service Commission (TSC) in Kenya. Int J Acad Res Bus Soc Sci 5(2):61–77. https://doi.org/10.6007/ijarbss/v5-i2/1463
5. McLymont V, Sharon C, Stell F (2003) Improving patient meal satisfaction with room service meal delivery. J Nurs Care Qual 18(1):27–37. https://doi.org/10.1097/00001786-200301000-00005
6. Pederson M (2013) Menu of change- healthy food in health-care a 2013 program report with highlights, awards and survey results. www. https://noharm-uscanada.org. Accessed 14 Jan 2017
7. Abu Sleih N, Marini A, Morris J (2011) Who Aims-Jordan. http://www.who.int/mentalhealth/evidence/WHO-AIMS/en/index.html. Accessed 24 Dec 2016
8. Linassi R, Alberton A, Marinho S (2016) Menu engineering and activity-based costing: an improved method of menu planning. Int J Contemp Hospitality Manage 28(7):1417–1440. https://doi.org/10.1108/IJCHM-09-2014-0438
9. Taylor JJ, Brown DM (2007) Menu analysis: a review of techniques and approaches. FIU Hospitality Rev 25(2):74–82
10. Vasilion E (2004) Choice: a growing trend in food service. http://www.dmaonline.org/Publications/articles/2004_10_014Choice.pdf. Accessed 23 Jan 2017
11. Norton C (2008) Why room service? Is it for your hospital's foodservice operation? Manage Food Nutr Syst Diet Pract Group 27(Fall):1–11
12. Tulsian M (2014) Profitability analysis (A comparative study of SAIL & TATA steel). J Econ Finan 3(2:19–22. https://doi.org/10.9790/5933-03211922
13. Kaplan RS, Norton DP (1996) Balanced scorecard: translating strategy into action. Harvard Business School Press, Brighton
14. Murby L, Gould S (2005) Effective performance management with the balanced scorecard technical report. The Chartered Institute of Management Accountants, London
15. Khorshidi A, Mastaneh Z, Javidkar M (2013) The Balanced Scorecard (BSC), as a tool for evaluation of organizational strategies. J Educ Manage Stud 3(4):428–435
16. Khorshid S, Zabihi R (2010) A quantitative evaluation model for eigenvalue of trade name of products by means of fuzzy network analysis techniques- data envelop analysis based on BSC. Int J Ind Eng Prod Manage 1(21):11–30

17. Zarei MH, Jamparazmi M, Yazdani H, Biriaei H (2010) Review the relationship among corporative strategic tendency and organizational performance by means of BSC approach. J Commercial Adm 2(3):97–112
18. Bisbea J, Barrubés J (2012) The balanced scorecard as a management tool for assessing and monitoring strategy implementation in health-care organizations. Rev Esp Cardiol 65 (10):919–927. https://doi.org/10.1016/j.rec.2012.05.011
19. Senyđğđt Y (2009) The balanced scorecard in the health-care industry: a case study. international symposium on sustainable development, Sarajevo, 9–10 June 2009. www.prints.ibu.edu.ba/147/1/ISSD2009-MANAGEMENT_p139-p143.pdf. Accessed 11 Jan 2017
20. Jones M, Filip S (2000) Implementation and outcomes of a balanced scorecard model in women's services in an academic health-care institution. Qual Manage Health-care 8(4):40–51. https://doi.org/10.1097/00019514-200008040-00005
21. Siraj K, Pillai P (2012) Comparative study on performance of Islamic banks and conventional banks in GCC region. J Appl Finan Banking 2(3):123–161
22. Åströmz Z (2013) Survey on customer related studies in Islamic banking. J Islamic Mark 4 (3):294–305. https://doi.org/10.1108/JIMA-07-2012-0040
23. Abdullah A, Sidek R, Adnan A (2012) Perception of Non-muslims customers towards Islamic banks in Malaysia. Int J Bus Soc Sci 3(11):151–163
24. Eeckhout J, Pinheiro R, Schmidheiny K (2014) Spatial sorting. J Polit Econ 122(3):554–620. https://doi.org/10.1086/676141
25. Sekaran U, Bougie R (2015) Research methods for business: a skill building approaches, 2nd edn. Willey, Chichester
26. Hamdallah M (2012) Corporate governance and credibility gap: empirical evidence from Jordan. Int Bus Res 5(11):178–186. https://doi.org/10.5539/ibr.v5n11p178
27. Kasavana ML, Smith DI (1982) Menu engineering: a practical guide. Hospitality Publishers, Lancing, MI
28. Balanced Scorecard Template in Health-care Centers. https://www.google.jo/webhp?sourceid=chrome-instant&ion=1&espv=2&ie=UTF-8#q=balanced+scorecard+template. Accessed 4 Jan 2017
29. MenuEngineering. http://www.f4t.com.au/Documents.asp?ID=319&Title=Menu+Engineering. Accessed 8 Feb 2017

Data Analysis and Design of Construction Productivity Efficiency Multipliers

John-Paris Pantouvakis

1 Introduction

In every construction project the expected output per time unit (hour or day), usually termed productivity, determines the cost and the duration of construction activities [1] as such is of paramount importance. Productivity, on the other hand, is influenced by both managerial decisions and by the dynamically changing operational conditions [2]. As such, productivity estimation remains a difficult task to be performed during the pre-construction stage.

To this purpose various methodologies have been proposed, such as industry guidelines [3], construction equipment manufacturers' handbooks [4, 5], scientific publications [2, 6, 7] and other sources (e.g. [8, 9]). A comparative review of the various methods proposed for productivity estimation has been presented in [10, 11]. It should be noted that most of the methodologies used are based on deterministic mathematical models in which the effect of the operational conditions on productivity is expressed by the use of efficiency multipliers. In other words, the "ideal" productivity is multiplied by a number of factors to take into account the specific situation in hand (e.g. congested construction site, hard soil etc.). It has been argued [12] that although the deterministic nature of these methodologies does not portray accurately reality, the methods are still highly regarded because of their ability to capture the physical features of the problem in a way that is understandable to practitioners.

An interesting point, which formed the stimulus of the work presented in this paper, is the fact that different methodologies follow a somewhat different approach and propose different efficiency multipliers to be used in the various construction operations. As such, the results of the different methodologies are not directly comparable [11]. But what are the basic data requirements and their dependencies?

J.-P. Pantouvakis (✉)
National Technical University of Athens, Athens, Greece
e-mail: jpp@central.ntua.gr

S. Şahin (ed.), *8th International Conference on Engineering, Project, and Product Management (EPPM 2017)*, Lecture Notes in Mechanical Engineering,
https://doi.org/10.1007/978-3-319-74123-9_13

The question will be further analyzed in this paper for excavation operations based on two of the most well established productivity estimation methodologies, namely those proposed by Caterpillar and Komatsu [4, 5]. We will first present and compare the Komatsu and Caterpillar productivity estimation methodologies. We will then discuss briefly relational database concepts and we will then attempt to apply data analysis and design to the two methodologies with the purpose of designing a unified database for both methods. Finally, we will present the findings of this research. Although the scope of the paper is limited to excavation operations, the same principles may be applied to other construction operations.

2 Estimating Productivity

As mentioned in the introductory section, we will focus on two of the more widely accepted methodologies, namely the productivity estimation of excavation operations based on the 31st edition of the Komatsu Specifications & Application Handbook [5] and the 46th edition of the Caterpillar performance handbook [4]. More specifically, the general formula for excavation productivity determination in both methodologies is:

$$Q = q * \left(\frac{3600}{C_m}\right) * E \tag{1}$$

where: Q is the hourly production (m^3/hr), q is the production per cycle (m^3), C_m is the cycle time (sec) and E is the job efficiency, i.e. a factor multiplying the ideal productivity to arrive at a more realistic estimate.

More specifically, E according to Komatsu depicts such influences to the job in hand as topography, operator's skill and disposition of machines. According to Caterpillar it mainly depends on actual work time per hour (e.g. 50 min/h). E values range from 0.58 for poor to 0.83 for good operating conditions (Komatsu) and 0.67 to 0.91 (Caterpillar).

In both methodologies, q (the production per cycle) is calculated by:

$$\mathrm{q} = \mathrm{q}_1 * \mathrm{K} \tag{2}$$

where q_1 is the bucket capacity (m^3) and K is the bucket fill factor. Whereas q_1 is a physical characteristic of the excavator, K is a factor related to the general excavating conditions (easy, average, rather difficult and difficult) (Komatsu) and the type of material excavated (Caterpillar) as shown in Table 1.

The calculation of C_m is somewhat more complicated. More specifically, according to Komatsu C_m is estimated based on the standard cycle time which depends on the excavator model and the swing angle multiplied by a conversion factor which depends on the digging depth and the dumping conditions as shown in Eq. (3):

Table 1 Bucket fill factor

(a) Bucket fill factor Komatsu		
	Excavating conditions	K
Easy	Excavating natural ground of clayey soil, clay or soft soil	1.1–1.2
Average	Excavating natural ground of soil such as sandy and dry soil	1.0–1.1
Rather difficult	Excavating natural ground of sandy soil with gravel	0.8–0.9
Difficult	Loading blasted rock	0.7–0.8

(b) Bucket fill factor Caterpillar	
Material	Fill factor range (%)
Moist loam or sandy clay	A—100–110
Sand and gravel	B—96–110
Hard, tough clay	C—80–90
Rock—Well blasted	60–75
Rock—Poorly blasted	40–50

Table 2 Standard cycle time C (Komatsu)

Model	Swing angle	
	45°–90°	90°–180°
PC78	10–13	13–16
PW148	11–14	14–17
PC130, PC138US	11–14	14–17
PC160	13–16	16–19

$$C_m = C * t_C \tag{3}$$

where C is the standard cycle time (taken from Table 2) and t_C is the cycle conversion factor (from Table 3).

The calculation of C_m according to Caterpillar is given in Fig. 1 as a range depending on the model and the job conditions. The job conditions, in turn, depend on the soil being excavated (easy, medium, hard), the digging conditions (digging

Table 3 Cycle time conversion factor t_C (Komatsu)

Digging condition $\left(\frac{\text{Digging depth}}{\text{Specified max. digging depth}}\right)$ (%)	Dumping condition			
	Easy (Dump onto spoil pile)	Normal (Large dump target)	Rather difficult (Small dump target)	Difficult (Small dump target requiring maximum dumping reach)
Below 40	0.7	0.9	1.1	1.4
40–75	0.8	1	1.3	1.6
Over 75	0.9	1.1	1.5	1.8

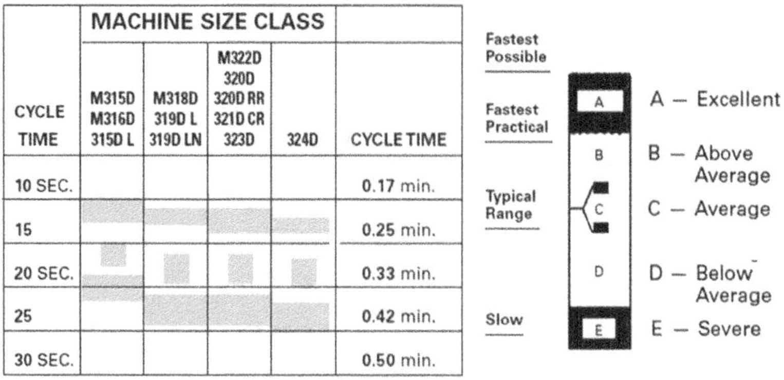

Fig. 1 Cycle time estimating chart (Caterpillar)

Table 4 Cycle time versus job conditions (Caterpillar)

A	Easy digging (unpacked earth, sand gravel, ditch cleaning, etc.). Digging to less than 40% of machine's maximum depth capability. Swing angle less than 30°. Dump onto spoil pile or truck in excavation. No obstructions. Good operator
B	Medium digging (packed earth, tough dry clay, soil with less than 25% rock content). Depth to 50% of machine's maximum capability. Swing angle to 60°. Large dump target. Few obstructions
C	Medium to hard digging (hard packed soil with up to 50% rock content). Depth to 70% of machine's maximum capability. Swing angle to 90°. Loading trucks with truck spotted close to excavator
D	Hard digging (shot rock or tough soil with up to 75% rock content). Depth to 90% of machine's maximum capability. Swing angle to 120°. Shored trench. Small dump target
E	Toughest digging (sandstone, caliche, shale, certain limestones, hard frost). Over 90% of machine's maximum depth capability. Swing over 120°. Loading bucket in man box. Dump into small target requiring maximum excavator reach. People and obstructions in the work area

depth/max digging depth expressed as a percentage), the swing angle, the dumping conditions and the possible obstructions of the operation. Table 4 summarizes the job conditions and their effect to cycle time.

3 Comparison of the Two Methodologies

The two methodologies (Komatsu and Caterpillar) use the same general formulae (1) for the calculation of the expected productivity (Q). They also both use the machine's bucket capacity (q_1). In addition, eight items of data are needed to complete the calculation as summarized in Table 5.

Table 5 Komatsu versus Caterpillar productivity estimation comparison

	Komatsu	Caterpillar
(i) Excavating conditions	$K_{Komatsu}$ [Table 1a]	C_m [Table 4 + Fig. 1]
(ii) Type of soil		$K_{Caterpillar}$ [Table 1b] C_m [Table 4 + Fig. 1]
(iii) Digging depth	$t_c \leftarrow (C_m)$ [Table 3]	C_m [Table 1 + Fig. 4]
(iv) Dumping conditions		
(v) Swing angle	For angles from 45°–180° $C \leftarrow (C_m)$ [Table 2]	For angles from 0°–180° C_m [Table 4 + Fig. 1]
(vi) Obstructions	Taken into account in (viii)	
(vii) Operator skill	Taken into account in (viii)	C_m [Table 4 + Fig. 1] Only for type A conditions
(viii) Job efficiency	$E_{Komatsu} \leftarrow (Q)$	$E_{Caterpillar} \leftarrow (Q)$

In Komatsu, the excavating conditions (easy, average, rather difficult and difficult) are related to the type of soil being excavated (soft clay, earth, earth with gravel etc.) and are used to determine the bucket fill factor. On the contrary in Caterpillar the soil type alone determines the bucket fill factor. As such, the values of the fill factor are different between the two methodologies (e.g. for easy excavating of clay Komatsu assumes a fill factor of 1.1–1.2 whereas Caterpillar 1.0–1.10).

For the calculation of the cycle time C_m, Komatsu is based on the digging and dumping conditions to calculate a standard cycle time (C) and then multiplies it with a conversion factor (t_c) depending on the swing angle.

Caterpillar depending on the excavating conditions, type of soil, digging and dumping conditions, swing angle, possible obstructions and operator skill classifies the job conditions in five categories named A, B, C, D and E. Once the job condition category is determined, the cycle time can be estimated using Fig. 1.

Finally, the job efficiency factor is considered absolutely differently in the two methodologies despite the same name and symbol used. In Komatsu it depends on a number of parameters as explained in Sect. 2 and summarized in Table 5. In Caterpillar it merely denotes the percentage of actual working time in an hour.

It follows that in both methodologies assumptions should be made based on judgment and experience with no specific analytical reasoning behind them. At the end of the day productivity estimations are merely approximations of the actual quantities achieved at the job site and this limitation should be understood.

A number of conclusions can be drawn from the analysis so far:

1. Both methodologies are deterministic, based on judgment and experience and estimate approximations of the actual productivity.
2. There are areas of operation, such as swing angles of less than 45° in Komatsu, or easy digging conditions and swing angles of more than 30° in Caterpillar where no direct results can be obtained. We will consider them as "dark areas" (i.e. areas with no data available) of the methodologies.

3. The job efficiency factor (E) although sharing the same name and symbol in both methodologies does not portray the same thing and as such it should be denoted differently as $E_{Komatsu}$ and $E_{Caterpillar}$.
4. Three physical characteristics of the excavator, namely model, bucket size and maximum digging depth and nine job characteristics, namely excavating conditions, type of soil to be excavated, digging depth, dumping conditions, swing angle, obstructions, operator skill, general site conditions and machine disposition (machine actual working time in an hour and general condition of the machine) are the required data for the estimation of construction productivity in both methodologies.

4 Relational Databases and Normalization

The purpose of this paragraph is to present the basic database concepts as used in this paper. Reference to the vast existing literature (e.g. [13]) is made for additional information. It should be noted that relational databases are based on four fundamental concepts, namely universe, domain, attribute and relation. Functional dependencies form an integral part as well. Figure 2 summarizes the concepts.

The columns in Fig. 2, namely "ITEM CODE", "DESCRIPTION" etc. are termed *attributes*. The set of all the possible values for an attribute is called the *domain* of the attribute. The union of all attributes is called the *universe*. All the attributes (columns) on the top row of the table in Fig. 2 define a template describing the kind of data stored in subsequent rows of the table. As such, all these attributes together are collectively referred to as the *relation scheme* or simply the *relation*. Each of the subsequent rows is called a *tuple* of the relation scheme. A minimal set of not null attributes of a relation whose values identify a unique tuple of a relation is called a *key* and is denoted by "#". The join of all relations is

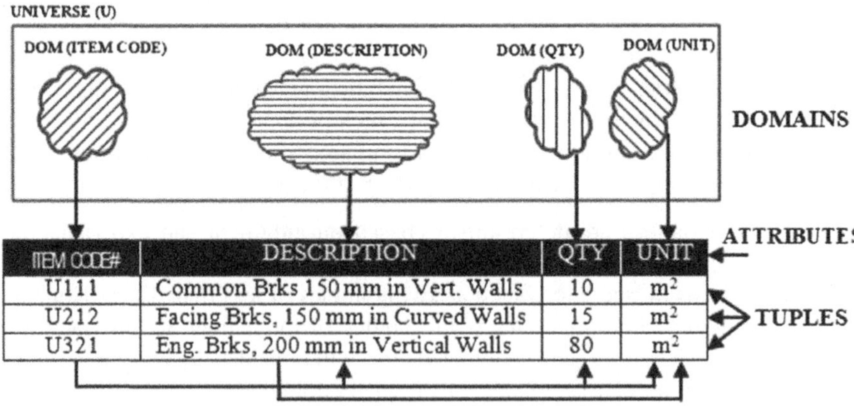

Fig. 2 Fundamentals of relational databases

called a *database scheme*. Furthermore, data dependencies are constraints imposed on data in a database. The primary kind of dependency is that between attributes X and Y in such a manner that each value of X uniquely identifies Y. For example, the attribute "ITEM CODE" in Fig. 2 uniquely identifies the attribute "DESCRIPTION", "QTY" and "UNIT". This type of dependency is termed *functional dependency* and is denoted X → Y. It should be noted that each of the attributes X and Y may be *simple* (i.e. consisting of a single attribute) or *composite* (i.e. consisting of more than one attributes).

Database normalization is a methodical process applying a set of rules in order to arrive at a robust table arrangement. With normalization the aim is to store data items only once within the system. With such an arrangement, it is ensured that any change made to a specific item is mechanically used by every part of the database. The process is based on the semantics of the data, i.e. what the data really means. There are several desirable forms of a database, denoted by different levels of normalization. So, we may have 1NF—First normal form, 2NF—Second normal form, 3NF—Third normal form etc. For the purposes of this paper we consider the 3NF as adequate. The necessary steps to achieve third normal form are (1) removal of repeating groups and multi-valued attributes, (2) removal of part-key dependencies and (3) removal of non-key dependencies. So, in a desired database scheme there are no repeating groups and all applicable functional dependencies rely on the key, the whole key and nothing but the key.

5 Database Analysis and Design

Having analyzed the productivity estimation methodologies as proposed by Komatsu and Caterpillar and being equipped with relational database theory, we will now analyze the data required with a view to determine if a single unified database can be designed.

Our Universe (U) consists of the Tables 1, 2, 3, 4 and Fig. 1. We need to analyze the dependencies between data items to arrive at a 3NF database scheme.

To this purpose, Table 5 is invaluable as it shows the necessary inputs in the first column and also assists us in clarifying certain concepts. More specifically:

1. We need data on three different entities; EQUIOMENT, JOB CHARACTERISTICS and SITE CONDITIONS.
2. EQUIPMENT refers to the particular model of equipment used (e.g. Komatsu PC78 or Caterpillar M316D). Based on manufacturer's data we can readily identify the "BUCKET SIZE" and the "MAXIMUM DIGGING DEPTH". Considering "EXCAVATOR MODEL" as the key, it can be seen that "BUCKET SIZE" and the "MAXIMUM DIGGING DEPTH" can be considered as functionally dependent only on the key and as such the relation is already in 3NF.
3. In case of different configurations (i.e. different booms, sticks or buckets) being used for a particular model, we need to identify each of the configurations with

an additional identifier or *attribute*, for example Komatsu PC78—bucket 1 or Komatsu PC78—boom 1 in order to maintain 3NF.

4. The entity JOB CHARACTERISTICS need a unique identifier (say "JOB CODE") to serve as the key. The specific job characteristics, namely "EXCAVATING CONDITIONS", "SOIL TYPE", "DIGGING DEPTH", "DUMPING CONDITIONS", "SWING ANGLE" and "SITE OBSTRUCTIONS" are all independent from one another and dependent only on the key (JOB NUMBER). As such the JOB CHARACTERISTICS relation is also in 3NF. There are no dependencies between this relation and the EQUIPMENT relation.
5. In case of a Komatsu machine, the relation "JOB CHARACTERISTICS" will be using attributes "EXCAVATING CONDITIONS" and "SOIL TYPE" to determine the fill factor as shown in Table 1a. In case of a Caterpillar machine, all non-key attributes will be used to determine the job condition class (i.e. A, B, C, …) as shown in Table 4 and then the EXCAVATOR MODEL and job condition class will be used to determine the cycle time based on Fig. 1. Finally, "SOIL TYPE" will identify the fill factor (Table 1b).
6. The SITE CONDITIONS relation will consist of a newly introduced key attribute (say "SITE CONDITIONS CODE") and the attributes "GENERAL SITE CONDITIONS" and "MACHINE DISPOSITION". The last attribute will determine the $E_{Caterpillar}$ whereas attributes "GENERAL SITE CONDITIONS", "MACHINE DISPOSITION" and the attributes "SITE OBSTRUCTIONS" and "OPERATOR SKILL" in the JOB CHARACTERISTICS relation will determine $E_{Komatsu}$. This relation is not associated to any of the previous ones and all applicable dependencies are with the key attribute only. SITE CONDITIONS is in 3NF.

So overall we have designed a database scheme consisting of three relations each of which is in 3NF and as there are not dependencies between the relations the whole scheme is also in 3NF. This scheme may now be summarized in Fig. 3:

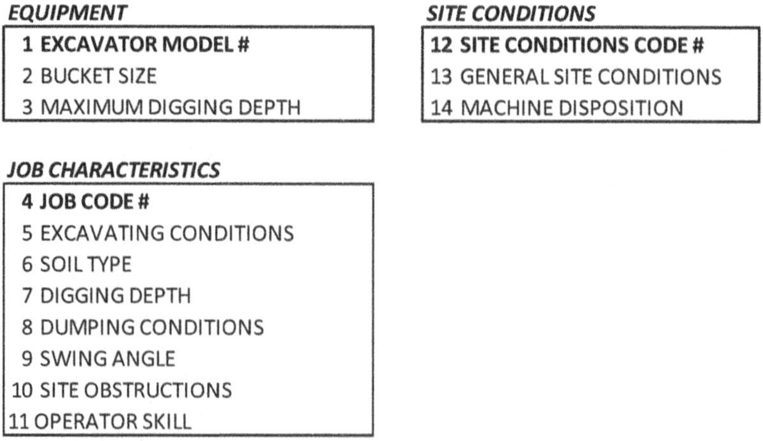

Fig. 3 Productivity database scheme in 3NF

Of course this scheme is at a logical level and minor adjustments may be needed to cater for the job efficiency factors according to Komatsu and Caterpillar.

Also, certain assumptions should be made to accommodate multi-valued attributes as, for example, the range of swing angles from 45° to 90° in Table 2 to atomic values.

The most important implementation step, however, remains the conversion of data contained in Tables 1, 2, 3, 4 and Fig. 1 to a homogeneous database following the relational principles. For this step we need to understand the real significance of the values displayed in the Universe of the database scheme and make appropriate assumptions. For example, it is sensible to accept that swing angles range in multiples of 15° with values in certain ranges missing, as for example, for swing angles of less than 45° for Komatsu excavators. These areas with missing values have been characterized as 'dark' in this paper. For the dark areas, productivity estimation is not possible based on the published data. Engineers, however, may arrive at estimations based on values selected based on their expert judgment.

As such the common database designed and proposed herein forms a useful tool for those dealing with construction productivity as it helps understand deeper the underlying semantics of the available data.

6 Conclusion

Data analysis of construction productivity methodologies based on relational theory assists in the development of a unified database of efficiency multipliers across different manufacturers, identifies "dark" areas which are not covered by a particular methodology (e.g. swing factors of less than 45° for Komatsu), highlights and explains differences in values given (e.g. the values of "E" between Komatsu and Caterpillar) and helps the better understanding of existing methodologies. Although only excavation has been analyzed in this paper, the same principles can be applied in other construction operations also in order to develop a unified approach to all construction productivity problems.

The real importance of this research however, is the understanding that despite the different data of different manufacturers, the similarity of the underlying principles may lead to a common treatment of the productivity estimation problem based on the same set of inputs which depend on the job characteristics and the specific site conditions. So this research complements previous comparison studies found in literature as it analyzes deeper the relationships of the underlying data based on the relational theory.

As such, further research will develop manufacturer independent software for commercially available models without the need of making simplifications as those found in many books on the subject. The software will deal initially with excavation operations but later will encompass other construction activities. As data is made available from other manufacturers or relevant websites, additional makes of equipment may be added to enhance the usability of the system.

References

1. Park H (2006) Conceptual framework of construction productivity estimation. KSCE J Civil Eng 10(5):311–317
2. Kotte G (1997) Ermittlung der Nutzförderleistung von Hydraulikbaggern. Tiefbau Ingenieurbau Strassenbau 9:18–28
3. BML (1983) Handbuch BML: Daten für die Berechnung von Baumaschinen-Leistungen. Zeittechnik Verlag, Neu-Isenburg
4. Komatsu (2013) Specifications and application handbook, 31st edn. Komatsu, Japan
5. Caterpillar (2016) Caterpillar performance handbook, 46th edn. Caterpillar Inc, Illinois USA
6. Peurifoy RL, Schexnayder CJ, Shapira A (2006) Construction planning, equipment, and methods, 7th edn. McGraw-Hill Higher Education, New York
7. Nunally SW (2007) Construction methods and management. Prentice Hall, Upper Saddle River
8. Voutzi M (2017) Comparative study of construction equipment productivity. M.Sc. Dissertation, Hellenic Open University
9. RSMeans (2017) Building construction cost data: 75th annual edition. R. S. Means Company Inc, Kingston
10. Panas A, Pantouvakis JP (2010) Evaluating research methodology in construction productivity studies. Built Hum Environ Rev 3(1):63–85
11. Panas A, Pantouvakis JP (2010) Comparative analysis of operational coefficients' impact on earthmoving operations. Eng Constr Architect Manage 17(5):461–475
12. Schexnayder CJ (1997) Analysis of earth-moving systems using discrete-event simulation: discussion. J Constr Eng Manag 123(2):199
13. Date CJ (2012) Database design and relational theory. O'Reilly Media, Sebastopol, USA

Identification and Tracking of Process Inconsistencies in Manufacturing Enterprises

Arkadiusz Jurczuk

1 Introduction

The current environment is characterized by dynamic changes and growing business complexity. Under these circumstances, enterprises more often consider an implementation of a process-based methodology to adapt to the requirements of the market and the customers [1, 21]. Generally, a process approach means "end-to-end" concentration on the business processes that enables an organization to recognize and satisfy the needs of their internal/external customers. It may have a positive influence on improving an enterprise's performance thanks to a better understanding of the present situation and the efficiency of the entire organization [1, 19, 26]. One of the most important factors for improving an enterprise's capacity is realizing and acknowledging that any process inconsistencies actually exist [6, 32]. The results of their identification and prioritization justify the aim and scope of process improvement initiatives [11, 23]. Moreover, elimination of process inconsistencies can improve an enterprise's ability to react to changing requirements [31].

The cognitive purpose of this paper is to identify the main sources of process inconsistencies in manufacturing enterprises, as well as to analyse attitudes towards their elimination. The paper presents an attempt to use the classification tree analysis for clustering and tracking process inconsistencies. Therefore, the research questions (RQ) addressed here are as follows:

- RQ1. What kinds of process inconsistencies prevail in the chosen manufacturing enterprises?

A. Jurczuk (✉)
Faculty of Engineering Management, Bialystok University of Technology, Białystok, Poland
e-mail: a.jurczuk@pb.edu.pl

S. Şahin (ed.), *8th International Conference on Engineering, Project, and Product Management (EPPM 2017)*, Lecture Notes in Mechanical Engineering,
https://doi.org/10.1007/978-3-319-74123-9_14

- RQ2. What attitudes towards process inconsistencyelimination predominate in the chosen manufacturing enterprises?

The research findings are based on a literature review and an empirical study. To achieve the set objectives, interviews were carried out among managers and managing directors of three leading manufacturing enterprises from North-Eastern Poland. Moreover, a classification tree was used to analyse and track the organizations' attitudes towards process inconsistency elimination. This approach allowed the author to identify omitted areas of process improvement in the analysed enterprises. The main contribution of this paper is to enhance the understanding of the situation in manufacturing enterprises as regards the occurrence and elimination of process inconsistencies.

2 Conception and Sources of Process Inconsistencies

The concept of a process inconsistency has been developed by scholars in the field of software engineering [7, 23, 28]. A process inconsistency could be seen as a discrepancy between a process description and the actual way in which it is executed or organised. Furthermore, an inconsistency may arise from the differences between the perceptions of a process by its participants [25, 28]. However, thanks to the universal background of this idea, it might be adopted and considered from the perspective of an enterprise's capability. Therefore, in the wider context, a process inconsistency can be defined as a lack of the ability to adapt to present or future requirements, which is the result of the influence of specific factors [7, 11, 16]. An inconsistency and its sources could be considered from an inside-out perspective, which is grounded in process-based methodologies [7, 16], or from a situational perspective, which is reflected in the contingency theory [5].

From the inside-out perspective, a process inconsistency may be regarded as a negative effect of internal factors on an organization's capability. The main sources of process inconsistencies are related to the execution and organization of the process, as well as its perception by the stakeholders [16, 28]. A general symptom of the occurrence of a process inconsistency is failure to fulfil the requirements of internal and external customers. Moreover, a process inconsistency might appear as a result of using inappropriate methods for measuring the degree of goals achievement, as well as the improper use of company resources [7, 16, 27, 29]. The occurrence of an inconsistency triggers changes which aim at improving an enterprise's capability. The improvement initiatives are mainly connected with performance issues, leadership, and organizational changes. Moreover, changes result from inconsistencies caused by the end of IT life-cycle or its incompatibility with new functionalities and business requirements [13, 24]. According to [10], inconsistencies in process execution could be caused by inefficient domain training of process participants and inadequate resources, or faulty equipment used for process execution.

From a situational perspective an occurrence of a process inconsistency could be perceived as an effect of organizational misfit [5, 30]. In this context, organizational effectiveness is based on the fit between organizational characteristics and contingencies. An organization's adaptation to contingencies is triggered by market dynamics, competitive threats, requirements of information transfer and global standards, government regulations, regulatory and technological advancements [17, 18]. It requires defining and implementing a unique way of managing business processes [31]. This, in turn, requires identification of potential areas for change and improvement—the source of the process inconsistency.

A process inconsistency is perceived as undesirable, and should be avoided if it is possible or reasonable. The elimination of an inconsistency makes a process more prescriptive and compliant with its description, model or process stakeholders' expectations [10, 25, 28]. Analysis of the current state that includes identifying inconsistencies is a key aspect of a business process improvement [23, 24]. The role of this aspect is strongly highlighted by many scholars in the context of organizational development and accreditation or certification requirements [10, 13, 21, 26]. Among the critical success factors of process improvement initiatives and problem solving attempts are the attitudes of managers and employees. It is underlined that process inconsistencies are not successfully identified and removed because of a lack of common process understanding, lack of an efficient inter-functional communication and top management support [14, 30]. Furthermore, it is a matter of employee and corporate knowledge related to process orientation maturity level [4, 27].

3 Data Collection and Respondents' Profile

In order to identify process inconsistencies, interviews were carried out among managers and managing directors of three leading manufacturing enterprises from North-Eastern Poland. The general selection of the enterprises followed non-probability sampling technique and was based on purposive sampling [3]. The choice of the sampling method was dictated by the availability of a company and its willingness to participate in the survey. The primary criterion of company selection was the affiliation to industrial processing groups according to the classification of business activities in Poland (C group). Enterprises represent firms from the private sector (domestic capital or primarily domestic capital) and, respectively, the sectors of agricultural machinery and equipment, manufacturing of machine parts, and manufacturing of electromagnetic drilling/milling machines. The number of employees in these companies ranged from 100 to 999. They had been active in the market for more than 25 years.

A triangulation of observers was used [8] and conducted with three managers (representing different functional areas) from each company. Each respondent had to have a minimum of three years' experience working for the company. The survey was conducted from October to December 2016.

3.1 Research Model

Qualitative and quantitative approaches were used to identify and track the elimination of process inconsistencies. Table 1 shows the research model.

3.2 Phase I

The research objectives were reached through an empirical survey and literature studies. The empirical survey used the technique of a semi-structured interview. It consisted of several key questions that helped to identify process inconsistencies in an enterprise. This method allowed the respondents to describe each problem and the approaches to solving it in more detail, including:

- character of process inconsistency,
- source of information about process inconsistency,
- result of undertaken activities to solve given problem.

The character of a process inconsistency was defined in the survey by means of the following four statuses related to the necessity to remove it:

- it needs removal because it interrupts, disturbs work,
- it needs removal, but removal would be costly,
- it needs removal, but not urgently,
- it does not require removal at present.

Moreover, respondents were asked to identify the source of information about each process inconsistency in their companies. It was assumed that the occurrence of a process inconsistency could be reported by employees, customers, or suppliers. In order to track the status of a process inconsistency, the interviews focused on achieving information from process participants about the success or failure in their elimination.

Table 1 The research model (adopted from [15])

Research model	Phase I	Phase II
Purpose	Identification of process inconsistencies Identification criteria for inconsistencies clustering	Classification of process inconsistencies Identification of a company's attitude towards elimination of an inconsistency
Survey and applied data	9 business cases, domain papers	Identified process inconsistencies
Method	Interviews Literature studies	Classification tree analysis

This enabled the author to describe each company's attitude towards the elimination of inconsistencies. Furthermore, the results of the analysis made it possible to track process inconsistencies in Polish manufacturing enterprises.

According to the assumption of the research model, process inconsistencies identified during the interviews were clustered in the following categories: Organization (O); Management (M); Employees (E); Customers and co-operators (CO); Products and services (PS); Processes (P) and Information technology (T).

These categories represent the triggers and drivers of Business Process Orientation implementation and development [13, 17]. It was assumed that the existence of a process inconsistency might trigger a change in an organization and its processes towards a process improvement approach, prompting it to become more process-oriented. Such a change may result in reducing the organization's misfit for present or future contingencies and requirements.

3.3 *Phase II*

The data mining technique of classification and regression trees (C&RT) was applied in the second phase of the research. This approach was successfully used for detecting turning points in business process orientation and maturity [9, 22]. The C&RT analysis is based on an algorithm known as recursive partitioning. Recursive partitioning is a step-by-step process by which a decision tree is constructed by either splitting or not splitting each node on the tree into two child nodes. A classification tree is the result of asking an ordered sequence of questions, and the type of question asked at each step in the sequence depends upon the answers to the previous questions of the sequence. The sequence terminates in a prediction of the class. The C&RT method produces a visual output in the form of a hierarchical, multilevel structure which resembles the branches of a tree [2, 12].

The proposed approach to C&RT implementation concentrates on analysis of the paths represented by the tree's branches. A path is formed by defined splitting conditions (determined by categorical predictors). A terminal node and a parent node (tree root) mark the boundaries of a path. It was assumed that analysis of a sequence of splits would allow for tracking attitudes to problem solving and elimination of process inconsistencies. The set of questions defining splitting rules included independent variables (categorical predictor variables). They were defined in *Phase I* of the research process. The independent variables represent:

- character of process inconsistency (need, or lack thereof, to remove),
- source of information about process inconsistency (employees, customers, suppliers),
- result of undertaken activities to solve problem (solved, returning).

Moreover, a dependent variable is represented by a category of process inconsistency (see *Phase I*). With the C&RT algorithm, all possible splits for each predictor variable at each node are examined to find the split producing the largest improvement in goodness of fit. In the presented approach, the Gini measure and the FACT stopping rule for the C&RT analysis were used. **STATISTICA** software (version 13.1, Dell Inc.) was applied for solving the research problem.

4 Research Results and Discussion

The empirical results helped to identify a set of existing process inconsistencies in the manufacturing enterprises. The conducted interviews with managers and managing directors made it possible to identify 27 process inconsistencies. As the research method required, the process inconsistencies were clustered using the defined categories. The majority of the identified process inconsistencies belonged to three categories: Processes (33%), Organization (26%) and Management (19%). Backlogs and production delays were the main sources of the inconsistencies occurring in the enterprises. Moreover, the respondents underlined their problems with communication and gaps in processes. The organizational inconsistencies (O) were caused by unclear business goals and low ability to respond to the requirements of new markets and customers. Furthermore, managers pointed out that other important sources of problems included failure to keep up with changes in the external regulations and non-compliance with quality standards requirements. Process inconsistencies grouped in the "Management" category were related to a lack of reliable or conflicting management information, a lack of control over processes, and poor organization of work resulting in low performance.

The results of the C&RT analysis enabled the author to classify process inconsistencies according to the attitude to problem solving. The origin of each inconsistency, its character and the result of its elimination were used as predictor variables. The categorized process inconsistencies were used in the model as a dependent variable. The classification tree was created using the CR&T algorithm delivered by STATISTICA software. The model contains 9 splits and 10 terminal nodes, as shown in Fig. 1.

Taking into account the guidelines of the C&RT analysis [12], only three paths (sequences of splits) were analysed. They included the terminal nodes (3, 10, 14) that contained a significant number of elements. Thus, it can be seen that there are two paths (P1, P2) through this tree for process inconsistencies reported by customer and employees and one path (P3) for process inconsistencies which generally needed removal but in the respondents' opinion it was not urgent. The most important predictor variables in the analysis included the character of a process inconsistency (100%), its reoccurrence (86%), and the source of information (inconsistencies reported by customers, employees, and suppliers—84%).

Analysis of path P1 (node 10 → node 1) indicates that process inconsistencies belonging to the three categories P, M, PS were equally reported by customers and employees. They disturbed, or even interrupted, work in the enterprises. In the context of process improvement, it is important to notice that despite signals both

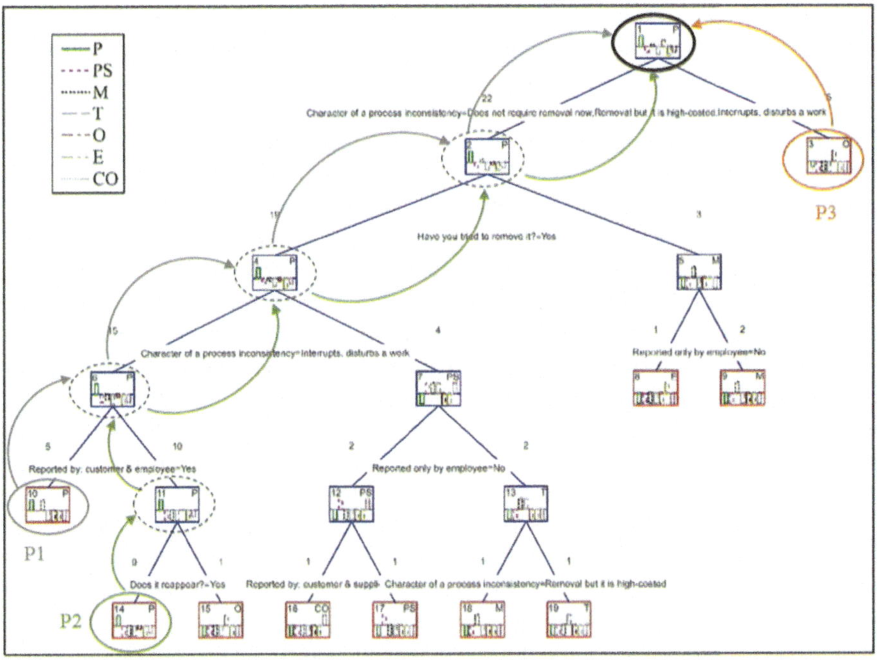

Fig. 1 Classification tree of process inconsistencies

from inside and outside the enterprise, these inconsistencies kept returning. The terminal node No. 14 includes reoccurring process inconsistencies signalized by the employees (path P2). The majority of the unsolved problems represented by this path belong to the “Process” category. As it was mentioned above, they are related to backlogs and production delays, low quality of communication, and unclear roles and responsibilities in a process. The P3 path (node 3 → node 1) shows process inconsistencies from the organizational area (O). Inconsistencies of this kind strongly determine the ability to respond to market and customer requirements. But in the analysed cases, they were perceived by managers and managing directors as important ones, although their elimination was not regarded as urgent.

5 Conclusions

The paper proposes the usage of the classification tree analysis for clustering and tracking process inconsistencies. The features of the CR&T method, and the predefined set of the splitting conditions, enable and support recognizing attitudes towards the elimination of process inconsistencies.

As regards the results of the survey, it can be concluded that processes are the most prevalent source of inconsistencies in the chosen manufacturing enterprises. Inconsistencies arise from knowledge gaps and insufficient understanding of end-to-end processes by its participants. Moreover, despite the fact that managers perceive them as important ones, no improvement initiatives are undertaken. In other words, the enterprises do not respond to the requests of their customers and employees. It may suggest that the analysed companies represent low levels of process maturity [17, 30]. The existence of process-based inconsistencies can strongly constrain the implementation of the process orientation. McCormack et al. [22] claim that process documentation and process structure are the leading factors at the initial stages of an organization's maturity. Glavan et al. [9] present similar research findings. From a situational perspective, managers' decisions to omit this type of inconsistencies may lead to an organization's misfit [30]. An enterprise will not be able to meet customer needs and/or seize new opportunities as they arise [13]. This factor is of the utmost importance for a holistic implementation of the process orientation by business companies [10, 18].

Further, the CR&T analysis allowed the author to identify the path representing an "abandonment attitude" to problem solving. The managers' attitude to these aspects could be considered from the perspective of problem solving paradoxes. It is worth considering whether some inconsistencies are not used as shields, i.e. clinging to certain problems might protect managers from facing even more serious ones. This paradox means that it is sometimes easier to complain about a company's dysfunctionality than it is to face the existing inconsistencies [20].

The paper argues that a systematic approach to probing an inconsistency could be a valuable element of an "as-is" analysis in enterprises that implement a process orientation. It helps develop a reactive attitude to process improvement [16]. The contribution of the paper lies in presenting a new approach to analysis of process inconsistencies, with a focus on the omitted areas of enterprises' problems. Therefore, the presented approach may support the configuration of a roadmap of business process improvement. Through identification of process inconsistencies, organizations can reduce resources wasted when trying to solve inappropriately defined projects. Moreover, inconsistency management is worth considering as a key activity throughout process improvement and the development of a contemporary organization.

The main limitation of this research is the size and structure of the sample. There is no possibility to generalize the research findings to a larger population. However, the achieved results may provide a base for further research, implying the directions of future investigations in the field. Additional studies could allow researchers to validate the presented approach and the achieved results.

Acknowledgements The research was conducted within S/WZ/1/2017 project and was financed from Ministry of Science and Higher Education funds.

References

1. Bhargav D (2017) Business process management—a construction case study. Constr Innov 17(1):50–67. https://doi.org/10.1108/CI-10-2015-0055
2. Breiman L, Friedman J, Stone CJ, Olshen RA (1984) Classification and regression trees. CRC
3. Cooper DR, Schindler PS (2014) Business research methods, 12th edn. The McGraw-Hill/ Irwin, New York
4. Davenport TH (2010) Process management for knowledge work. In: vom Brocke J, Rosemann M (eds) Handbook on business process management 1 international handbooks on information systems. Springer-Verlag, Berlin, Heidelberg
5. Donaldson L (2006) The contingency theory of organizational design: challenges and opportunities. In: Burton RM, Håkonsson DD, Eriksen B, Snow CC (eds) Organization design the evolving state-of-the-art, Springer, US. https://doi.org/10.1007/0-387-34173-02
6. Emery CR (2009) A cause-effect-cause model for sustaining cross-functional integration. Bus Proc Manage J 15(1):93–108. https://doi.org/10.1108/14637150910931488
7. Gulledge TR, Sommer RA (1999) Process coupling in business process engineering. Knowl Process Manage 6(3):158–165
8. Glaser B, Strauss A (1999) The discovery of grounded theory: strategies for qualitative research. Aldine Transaction, New Brunswick
9. Glavan LM, Vukšić VB, Vlahović N (2015) Decision tree learning for detecting turning points in business process orientation: a case of Croatian companies. Croatian Oper Res Rev 6(1):207–224. https://doi.org/10.17535/crorr.2015.0017
10. Hammer M (2010) What is Business Process Management? In: vom Brocke J, Rosemann M (eds) Handbook on business process management 1 international handbooks on information systems. Springer-Verlag, Berlin Heidelberg
11. Ho SKM (1993) Problem solving in manufacturing. Manage Decis 31(7)
12. Izenman AJ (2008) Modern multivariate statistical techniques. Regression, classification and manifold learning. Springer, New York. https://doi.org/10.1007/978-0-387-78189-1
13. Jeston J, Nelis J (2008) Business process management: practical guidelines to successful implementations, 2nd edn. Elsevier, Burlington
14. Jørgensen F, Boer H, Gertsen F (2003) Jump-starting continuous improvement through self-assessment. Int J Oper Prod Manag 2(10):1260–1278
15. Jung M, Lee Y, Lee H (2015) Classifying and prioritizing the success and failure factors of technology commercialization of public R&D in South Korea: using classification tree analysis. J Technol Transf 40(5):877–898. https://doi.org/10.1007/s10961-014-9376-5
16. Jurczuk A (2016) Reactive approach for cause analysis of business processes inconsistency. Organ Rev (Przegląd Organizacji) 3:42–48 (in Polish)
17. Jurczuk A (2016) Towards process maturity—triggers of change. In: Proceedings of 9th international scientific conference business and management. 20 Nov 2016
18. Karagiannis D (2013) Business process management: a holistic management approach. In: Mayr HC, Kop Ch, Liddle S, Ginige A (eds) Information systems: methods, models and applications, lecture notes in business information processing 137:1–12
19. Kohlbacher M, Reijers HA (2013) The effects of process-oriented organizational design on firm performance. Bus Process Manage J 19(2):245–262
20. Lowy A (2011) Nine paradoxes of problem solving. Strategy Leadersh 39(3):25–31
21. Matthews RL, Tan KH, Marzec PE (2015) Organisational ambidexterity within process improvement: an exploratory study of four project-oriented firms. J Manuf Technol Manage 26(4):458–476. https://doi.org/10.1108/JMTM-12-2013-018
22. McCormack K, Willems J, van den Bergh J, Deschoolmeester D, Willaert P, Indihar Štemberger M, Škrinjar R, Trkman P, Ladeira MB, Valadares de Oliveira MP, Bosilj Vuksic V, Vlahovic N (2009) A global investigation of key turning points in business process maturity. Bus Proc Manage J 15(5):792–815

23. Nuseibeh B, Easterbrook SM, Russo A (2001) Making inconsistency respectable in software development. J Syst Softw 58(2):171–180. https://doi.org/10.1016/S0164-1212(01)00036-X
24. Phogat S, Gupta AK (2017) Identification of problems in maintenance operations and comparison with manufacturing operations: a review. J Qual Maintenance Eng 23(2):1–16. https://doi.org/10.1108/JQME-06-2016-0027
25. Saaty TL (1994) How to make a decision: the analytic hierarchy process. Interfaces 24(6):19–43
26. Sanders Jones JL, Linderman K (2014) Process management, innovation and efficiency performance: the moderating effect of competitive intensity. Bus Proc Manage J 20(2): 335–358. https://doi.org/10.1108/BPMJ-03-2013-0026
27. Škrinjar R, Trkman P (2013) Increasing process orientation with business process management: critical practices'. Int J Inf Manage 33(1):48–60
28. Sommerville I, Sawyer P, Viller S (1999) Managing process inconsistency using view points. IEEE Trans Software Eng 25(6):784–799
29. Talib F, Rahman Z (2015) Identification and prioritization of barriers to total quality management implementation in service industry. TQM J 27(5):591–615
30. Trkman P (2010) The critical success factors of business process management. Int J Inf Manage 30(2):125–134
31. vom Brocke J, Zelt S, Schmiedel T (2016) On the role of context in business process management. Int J Inf Manage 36(3):486–495
32. Walker ED, Cox JF (2006) Addressing ill-structured problems using Goldratt's thinking processes: a white collar example. Manag Decis 44(1):137–154

Risk Management: The Relationship Between Perceived Risk Factors of Crowd Disaster and Perceived Safety in Large Buildings

Mohammed Alkhadim, Kassim Gidado and Noel Painting

1 Introduction

Safety in built environment is classified into objective safety and subjective safety (perceived safety). In an organizational context, objective safety is measured as the actual number of or the risk of incidents or injuries occurring. Subjective safety is intangible and refers to the feeling or perception of being safe or unsafe within a specified period. Numerous studies have been undertaken on objective safety in the built environment, but there has been a lack of research on subjective safety (perceived safety) particularly in large buildings where large numbers of users attend an event.

Dickie [5], confirmed that poor risk management in large buildings or spaces during an event has led to many crowd disasters across the world. Booty [4], stated that each large building used by large numbers of people (crowd) normally has diverse types and levels of risk requiring an effective management approach. Experts have defined risk management as a proactive approach to eliminate threats to an organization through anticipating, identifying, assessing and mitigating the possible risks. The British Institute of Facilities Management [3], have classified Risk Management (RM) as one of the 24 key components of Facilities Management (FM). FM covers all aspects of planning, managing space, designing, environmental control, health and safety and support services. It significantly contributes to the delivery of strategic and operational objectives on a day-to-day basis.

M. Alkhadim (✉) · K. Gidado · N. Painting
School of Environment and Technology, University of Brighton,
BN2 4GJ Brighton, UK
e-mail: M.Al-Khadim@brighton.ac.uk

K. Gidado
e-mail: K.I.Gidado@brighton.ac.uk

N. Painting
e-mail: N.J.Painting@brighton.ac.uk

S. Şahin (ed.), *8th International Conference on Engineering, Project, and Product Management (EPPM 2017)*, Lecture Notes in Mechanical Engineering,
https://doi.org/10.1007/978-3-319-74123-9_15

When events are held in large buildings, research highlighted that facilities managers must be involved before, during and after the event to reduce risk and enhance safety. Experts have established that organizations that own large space assets for public use often make the risk reduction strategy a priority to enable them gain advantage over their competitors.

It is therefore concluded that FM of large buildings and spaces used by large crowds must involve effective risk management as a key component. In current practice, emphasis is placed on addressing objective safety.

Fruin [6], studied this issue and established some of the key factors that influence objective crowd safety however there is a lack of understanding of the issue relating to subjective safety. This study has adopted the factors used by Fruin and used them to investigate whether they affect subjective safety in large buildings and spaces. These factors are referred to by the acronym FIST: Force, Information, Space, and Time. The paper argues that there is a strong relationship between FIST and perceived safety in large buildings by studying the extent to which perceived force, perceived poor information, perceived insufficient space, and perceived poor real time management influence perceived safety.

For this research the Holy Mosque in Makkah, Saudi Arabia was chosen having the largest crowd of any event within a large building.

2 The Holy Mosque as a Large Building

The Hajj is a religious event that includes a large number of pilgrims with diverse cultures, ages, genders, nationalities and languages. It is one of the five pillars of Islam and an obligation for Muslims who are capable financially and physically to perform Hajj at least once in their lifetime. Annually around two million pilgrims visit Makkah (also called Mecca) to perform the Hajj, for between 4 and 6 days. This is considered one of the largest gatherings in the world, and the number of people who wish to perform Hajj is increasing yearly. The rituals of the pilgrimage run between 8th and 13th Dhul Hijjah (Islamic Calendah) and are mainly concentrated in four holy places: The Holy Mosque, the Mina, Muzdalifah and Arafat. The first holy place the pilgrims visit when they arrive is the Holy Mosque to perform Tawaf and Saee. Tawaf is a movement of the pilgrims around the Kaaba, which is situated in the centre of the Holy Mosque. In Tawaf, pilgrims move around the Kaaba seven times in an anticlockwise direction. While in the Saee, pilgrims walk seven rounds between two points in the Holy Mosque called Safa and Marwah. After visiting the other 3 holy places, they return to the Holy Mosque for another Tawaf and Saee. The Holy Mosque is a large building which can accommodate around 1.2 million worshipers at the same time. The current area of the Holy Mosque is about 356,800 m^2 and still expanding.

3 The Conceptual Model

According to Fruin [6], the FIST elements were derived from personal experiences, analyses of major crowd incidents and traffic flow principles. Indeed, the FIST model was developed based on the real conditions and objective safety. The proposed conceptual model replaces tangible items used by Fruin with perceived situations and their effect on perceived safety. The conceptual model includes one dependent variable (perceived safety) and four independents variables (perceived force, perceived poor information, perceived insufficient space, perceived poor real time management). For this study perceived safety is defined as the feeling (or perception) of an **unsafe situation** at an event over a specified period. These variables will be empirically and statistically measured.

3.1 Perceived Safety (PS)

Perceived safety refers to the feeling (or perception) of an unsafe situation existing during an event. If people feel unsafe for some reason, they panic and often attempt to escape from the real or perceived danger by acting abnormally and/or chaotically by pushing and shoving. Studies in urban design have shown that perceived safety can be affected by the characteristics of the environment, the physical condition, and the configuration of spaces [9]. These perceptions can vary with age, sex, culture, and familiarity with the environment, for example: women and older people have a more diverse sense of safety compared with others. Similarly, someone new to an environment may find it safe because they may not be familiar with specific cues within the context. In crowding studies, it has been highlighted that the perceived safety is closely tied to perceived crowding. Perceived crowding is defined as "the psychological counterpart to population density" Kim et al. [8]. Perceived safety is negatively affected by the perceptions of crowding—and research has shown that when perception of crowding increase, people's sense of safety declines.

3.2 Perceived Force (PF)

Perceived force is the feeling or perception of force by an individual within a crowd which may originate by either seeing, hearing or feeling. There are several consequences that may result from the perceived force which have been termed as indictors (items) within this study. Research has established that force within the crowd is usually created from the interior of the crowd and has two main forms: the self-driven force; and the leaning force that comes from the weights of the bodies. It can reach a high level such that it cannot be easily controlled or resisted due to high pressure of the crowd [6, 10]. It is argued by researchers, that the force among people is a significant factor that leads to casualties. Still [10], stated that most of the deadliest event disasters involved large crowds (those with the largest number of injuries and fatalities) originate from crowd compression and the subsequent loss of footing or inability to move. Fruin [6], pointed out that "horizontal forces sufficient to cause compressive asphyxia would be more dynamic as people push off against each other to obtain breathing space". He confirmed that news media have reported that compressive asphyxia is the main reason of deaths not trampling. Experts argued that when people in a crowd are being swept along with movement and compressed, it can lead to serious injuries and fatalities from suffocation. There is evidence to suggest that most of the people who die from suffocation die because of the enormous pressure on their chests [up to 4500 N (1000 lbs)]. Forces among people in a crowd are generally created when the density is higher than a certain level, and disaster can occur when the crowd density reaches a critical density.

From the above review, it is clearly demonstrated that 'Force' is a critical factor to crowd safety. This study will test the effect of perceived crowd force on perceived safety in the Holy Mosque during the Hajj. Hence, one hypothesis is formulated as follows: **H1**: Perceived force has a significant influence on perceived safety.

3.3 Perceived Poor Information (PPI)

Before attending an event an individual may consider a wide range of information with regards to the venue and the type of crowd. This information could include the nature of the group, experience with similar groups, familiarity with the venue, crowd behaviour, signage and means of communication between those managing the crowd and the crowd. Fruin [6], pointed out that the information includes the means of communication, sights and sounds influencing the perceptions of the group, public address, signs, ticketing, actions and training of personal. It is already established that poor communication before or during an event is characterized as one of the causes of crowd disaster. Obtaining real time information about the situation of the crowd in large assembly spaces including crowd actions, reactions, real or perceived is therefore essential. Experts have underlined that communication and real time information are key factors in preventing crowd disasters. It is a good practice to set up a communications centre and a centralized crowd management system. Information communicated to—or withheld from—the crowd can influence their perceived safety. Experts have established that communicating with the crowd is essential in maintaining order and managing behavior. Based on the above literature review, the following hypothesis is formulated: **H2**: Perceived poor information has a significant influence on perceived safety.

3.4 Perceived Insufficient Space (PIS)

Space in built facilities includes physical facilities, seating areas, corridors, stairs, escalators, standing areas and lifts. Architects and engineers will typically pay attention to local building codes but may have little regard for people's movement and perception. It has been shown that when the venue does not have enough space to accommodate the crowd and the capacity becomes high, say eight people per square meter, human psychology will usually undergo a change. It is also argued by experts that when people attempt to escape from a possible disaster they rush to an exit ignoring alternative exits made available. Fruin [6], has indicated that it is hard to describe the psychological and physiological pressure within a high-density crowd. Additionally, individuals may lose their control; several studies have shown that crowd density can influence the perceived safety and behaviour Alnabulsi and Drury [1]. The bodies within a crowded space are surrounded by heat and thermal insulation to the extent that some people may be weakened and faint. Ineffective or poor use of space is one of the key risk factors in crowd disaster [10]. Critical crowd density is around 1.5 ft^2 (i.e. 7 persons/m^2). This also relates to the crowd control failure which creates psychological and physical scenarios that causes human stampede. Therefore, the following hypothesis: **H3**: Perceived insufficient space has a significant influence on perceived safety.

3.5 Perceived Poor Real Time Management (PPRTM)

Fruin [6], argued that real time information and intervention are a key factor for preventing crowd disasters. Time plays an important role in an event, for instance the density of the crowds before the event is much less compared to the rapid egress and heavy crowd densities leaving an event. Research has shown that failure to detect crowd behavior at the right time can lead to injuries and fatalities. Crowd management literature has made it clear that it is a requirement to ensure the flow of the pedestrians does not exceed the capacity of the spaces through which they are flowing or the capacities of the space in which they are congregating. There is evidence to suggest that lack of consideration is sometimes given to how crowd flow and density can be successfully managed by controlling timings. Hence, the hypothesis is formulated based on the literature review as follows: **H4**: Perceived poor real time management has a significant influence on perceived safety.

4 Method

Primary data used for this study was obtained using group-administered questionnaire. This method is chosen instead of the self-administered approach because of the following: the respondents (pilgrims from all over world) have limited time at Makkah; the need to achieve a high rate of return; and to enable the respondents to seek clarification if any misunderstanding arises [11].

The questionnaire encompassed six sections: section one is background information; sections two to five are designed to measure the user (pilgrim) perception with reference to the independent variables (perceived force, perceived poor information, perceived insufficient space, perceived poor real time management), section six refers to the dependent variable (perceived safety).

The items included in the questionnaire were adapted from [1, 6]. Several items have been modified to achieve the aim of the study. All the items were measured on a 5-point Likert scale (1 = strongly disagree; 5 = strongly agree) and some items (1 = never occur; 5 = almost always occurs).

A pilot study was carried out in Makkah in 2016 before the Hajj began in order to evaluate the validity and reliability of the questions and instructions. It aimed to check the clarity of instructions and the items of the questionnaire, to determine the time needed to complete the questionnaire, to ensure the statements were clear and easy (not difficult or complex) to understand and to gain any other useful comments.

The population sampling for this study targeted all the pilgrims (local and coming from outside of the country specifically for the Hajj) during Hajj within the zone of Makkah. A total population size of 1,942,946 pilgrims was determined based on the report provided by Ministry of Hajj and General Authority for Statistics. 1940 participants were surveyed with an estimated confidence level of 95% and a 2% margin of error during the Hajj 2016 (1437 Arabic Calendar). The questionnaire statements were programmed into iPad devices and linked to the database centre at the Hajj and Umrah Research Institute. Twelve experienced and trained postgraduate research students collected the data at different locations in Makkah. The majority were collected at the pilgrims' camps or accommodation.

Structural Equation Modelling (SEM) was used as statistical test for this study. SEM was chosen as a statistical technique for several reasons: Firstly, this study attempts to establish the interrelationship between the FIST factors and perceived safety; latent variables are encountered that cannot be measured directly. Secondly, SEM is a powerful tool that is able to test the model fit to the data and at the same take into account any measurement error (unreliability) for each latent variable of the constructs being estimated.

5 Results

The data was analysed using the Analysis of Moment Structure (AMOS) for Structural Equation Modeling (SEM). Prior to undertaking the SEM analysis, checks were made to ensure that the collected data is clean and normally distributed. The Confirmatory Factor Analysis (CFA) was also was performed for all latent constructs prior to modelling the interrelationship in SEM. All the measurement items for the latent constructs should have an acceptable factor loading of 0.60 and above in an effort to improve the fitness level of the measurement model [2]. Initially, the total number of all items was 29 before conducting the CFA procedures. With the purpose of achieving the minimum fitness index, 10 items were deleted one at a time and some were linked. After conducting the CFA procedures, the model contains 19 items with an acceptable factor loading of at least 0.60 as detailed in Table 1.

Research has shown that to establish convergent validity, the model fit must be adequate, and the average variance extracted (AVE) must exceed 0.50. The results of AVE for all constructs as presented in Table 1 have achieved the standard minimum required level of 0.50.

Two reliability tests have been undertaken for this study: composite reliability (CR) and Cronbach's alpha. This paper has used both tests to guarantee the reliability of the data before conducting any further analysis. Other researchers have claimed that CR is more accurate than Cronbach's alpha because it does not assume that the loadings or error terms of the items are equal. Both tests CR and Cronbach's Alpha have met the standard minimum threshold of 0.60 and 0.70 respectively. This indicates the acceptability of internal consistency and confirms that all the items used in the model are technically free from the errors.

Overall, the result of the assessment of the measurement model shows solid evidence of validity and reliability. It clearly shows that the items on each construct of the study are considered reliable.

The structural model as shown in Fig. 1 presents the interrelationship among the variables. It consists of 4 unobserved exogenous constructs (perceived force, perceived poor information, perceived insufficient space, and perceived poor real time management) and one unobserved endogenous constructs (perceived safety). Based on the fit indexes, the model is a good fit and all measures of comparative fit index (CFI) = 0.979, standardized root mean square residual (SRMR) = 0.032, and root mean square error of approximation (RMSEA) = 0.043 have achieved the required level. Hu and Bentle [7] and Awang [2], recommend a CFI $\geq$ 0.90, SRMR $\leq$ 0.08, and RMSEA $\leq$ 0.06 for acceptable model fit. Consequently, the model is accepted for further analysis and testing the hypothesis.

Table 2 shows the path for the construct and its coefficient as well as the significance for that particular path. It presents the effect of each exogenous construct on the respective endogenous construct. The results reveal that all the independent variables have significant effect on perceived safety. The path coefficient of perceived force to perceived safety is 0.189. This value indicates that for every

Table 1 Factor loading for items and reliability test

Constructs	Items	Factor loading	CR > 0.6	AVE > 0.5	Cronbach > 0.7
PF			0.877	0.640	0.886
	Breathing difficulties	0.77			
	Crowd pushing	0.82			
	Movement difficulties	0.79			
	Crowd pressure	0.82			
PPI			0.920	0.698	0.922
	Health and safety information	0.80			
	Communication	0.81			
	Availability of all types of signs	0.88			
	Signs visibility	0.86			
	Warning signs	0.83			
PIS			0.814	0.598	0.824
	Activities areas densities	0.71			
	Availability and distribution of stairs, escalators and lifts	0.91			
	Entrances and exits densities	0.67			
PPRTM			0.870	0.691	0.866
	Crowd flows control	0.81			
	Real time information and intervention	0.85			
	Waiting time	0.84			
PS			0.808	0.514	0.804
	Perceived risk of fatalities	0.68			
	Perceived risk of damaged facilities	0.72			
	Perceived risk of falls, slips and trips	0.76			
	Perceived risk of trampling or stampede	0.71			

one-unit increase in the perceived force, its effect on perceived safety would increase by 0.189 units. And more importantly there is a significant effect ($p < 0.001$) of all constructs on perceived safety. The impact of perceived poor information on perceived safety is 0.088. In contrast, perceived insufficient space has a negative impact on perceived safety by −0.193. The perceived safety is also affected by perceived poor real time management by 0.305.

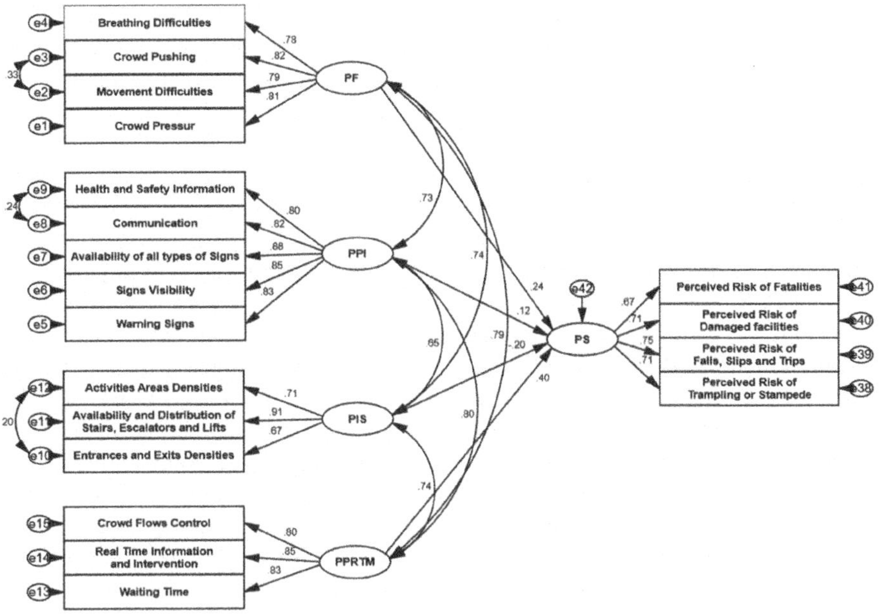

Fig. 1 Regression path coefficient for the structural model

Table 2 Regression weight for path coefficient and it's significant

Path construct relationship	Estimate	p-value	Hypothesis
H1: Perceived force → Perceived safety	0.189	***	Supported
H2: Perceived poor information → Perceived safety	0.088	0.010**	Supported
H3: Perceived insufficient Space → Perceived safety	−0.193	***	Supported
H4: Perceived poor real time management → Perceived safety	0.305	***	Supported

$^{*}p < 0.05$, $^{**}p < 0.01$, $^{***}p < 0.001$

6 Discussion

The findings clearly supported our hypotheses and the proposed conceptual model. It has been confirmed that the perception of the users (pilgrims) about safety in large buildings (Holy Mosque) during religious event (Hajj) are strongly affected by four main factors: perceived force, perceived poor information, perceived insufficient space and perceived poor real time management.

The analyses have revealed that the perceived force has a positive impact on perceived safety by 0.189. Increase in level of feeling (or perception) of breathing difficulties, crowd pushing, difficulty in movement, and crowd pressure make the pilgrims feel unsafe.

Perceived poor information also has a positive effect on perceived safety by 0.088, which means that the perception of pilgrims about safety can be effected by poor information. The results could be interpreted to mean that pilgrims felt unsafe due to a number of reasons including: insufficient health and safety information provided to them, poor communication between the pilgrims and police/security at the Holy Mosque, inadequate signs, unclear signs, and insufficient warning signs.

Thirdly, perceived poor real time management has a positive effect on perceived safety by 0.305. This is because of three main reasons: loss of crowd control, poor real time information and intervention, and waiting time to use the facilities was unacceptable.

In contrast, there is a negative relationship between perceived insufficient space and perceived safety by −0.193. This is an interesting finding; it means that the more the pilgrims perceive space to be insufficient the less they feel unsafe. This confirms the findings of Alnabulsi and Drury [1], which found that increasing level of crowd density reduces the pilgrim's feeling of being unsafe. They argued this to be the case because the pilgrims were high in social identification as Muslims. They are of the view that social identification should moderate the negative effect of crowd density on reported safety at the Holy Mosque. The finding in this paper may suggest that the negative effect of perceived insufficient space on perceived safety may only be applicable to conventional dense crowds made up of individuals in unity of purpose with a common social identity—social category membership with high expectations of social support from others in the crowd.

7 Conclusion

This study used Structural Equation Modeling technique to examine the relationships between perceived force, perceived poor information, perceived insufficient space, perceived poor real time and perceived safety factors in a large building (Holy Mosque) during the Hajj event. It was found that all perceived FIST factors have significant effect on perceived safety.

After conducting the CFA procedures, 19 items with an acceptable factor loading of at least 0.60 were identified (as detailed in Table 1). In order to reduce the risk or potentiality of a crowd disaster at the Holy mosque, those items under Perceived Force (PF), Perceived Poor Information (PPI), Perceived Insufficient Space (PIS) and Perceived Poor Real Time Management (PPRTM) need to be considered by facilities managers. The result of the assessment of the measurement model has shown solid evidence of validity and reliability. It also clearly confirms that the items on each construct of the study are reliable.

References

1. Alnabulsi H, Drury J (2014) Social identification moderates the effect of crowd density on safety at the Hajj. Proc Natl Acad Sci USA 111:9091–9096. https://doi.org/10.1073/pnas.1404953111
2. Awang Z (2015) SEM made simple: a gentle approach to learning structural equation modeling. MPWS Rich Publication
3. BIFM (2014) The facilities management professional standards
4. Booty F (2009) Facilities management handbook, 4th edn. Library of Congress
5. Dickie JF (1995) Major crowd catastrophes. Saf Sci 18:309–320
6. Fruin JJ (1993) The causes and prevention of crowd disasters. Eng Crowd Safety: 1–10
7. Hu L, Bentler PM (1999) Cutoff criteria for fit indexes in covariance structure analysis: conventional criteria versus new alternatives. Struct Equation Model: A Multidisciplinary J 6:1–55
8. Kim D, Lee K, Sirgy J (2016) Examining the differential impact of human crowding versus spatial crowding on visitor satisfaction at a festival. J Travel Tourism Mark 33:293–312. https://doi.org/10.1080/10548408.2015.1024914
9. Mehta V (2013) The street: a quintessential social public space. Routledge
10. Still GK (2016) Crowd safety and risk analysis. http://www.gkstill.com/ExpertWitness/CrowdDisasters.html
11. Zohrabi M (2013) Mixed method research: instruments, validity, reliability and reporting findings. Theory Pract Lang Stud 3:254–262

Software Project Management: Resources Prediction and Estimation Utilizing Unsupervised Machine Learning Algorithm

Mohammad Masoud, Wejdan Abu-Elhaija, Yousef Jaradat, Ismael Jannoud and Loai Dabbour

1 Introduction

A project is defined as a correlated tasks executed over time with different constrains and limitations [1]. These constrains vary from virtual resources, such as, time period, programming tasks and computation power to physical resources, such as, workers, budget and tools. This definition treated resource management as one of the main parts of any project, including software projects.

In recent years, software projects proliferated. These projects vary from small programming projects with thousands' lines of codes, such as, IPhone games and UNIX ver1, to massive projects of millions' lines of codes, such as Google, Windows and Facebook (www.informationisbeautiful.net/visualizations/million-lines-of-code/). This diversity requires managing software projects in the development phase. Software project management is a challenging task. In software projects, many resources are hard to measure for two reasons. First, many resources are virtual resources, such as, IDEs, programing languages, programmers' efficiency, development tools and efforts. The cost, efficiency and quality of these

M. Masoud (✉) · W. Abu-Elhaija · Y. Jaradat · I. Jannoud
Electrical Engineering Department, Al-Zaytoonah University of Jordan,
130, Amman 11733, Jordan
e-mail: m.zakaria@zuj.edu.jo

W. Abu-Elhaija
e-mail: w.elhaija@zuj.edu.jo

Y. Jaradat
e-mail: y.jaradat@zuj.edu.jo

I. Jannoud
e-mail: Ismael.jannoud@zuj.edu.jo

L. Dabbour
Architecture Engineering Department, Al-Zaytoonah University of Jordan,
130, Amman 11733, Jordan

S. Şahin (ed.), *8th International Conference on Engineering, Project, and Product Management (EPPM 2017)*, Lecture Notes in Mechanical Engineering,
https://doi.org/10.1007/978-3-319-74123-9_16

resources are hard to be estimated or measured. Second, software projects require development of new algorithms. These algorithms require deep research in different areas before deploying. The question is how to manage and estimate resources in software projects?

Effort estimation and prediction has emerged as a method to tackle this issue. Many effort estimation models have been proposed to predict required resources and time periods. These models are classified into two main classes; statistical methods and artificial intelligent models. In both models, features are extracted from enterprises' records to estimate the development capacities and manage software production process. Nevertheless, optimization is required to enhance prediction efficiency.

In this work, an effort estimation clustering method based on estimation maximization soft-clustering unsupervised machine learning algorithm is proposed. The new model is utilized to classify software projects into four different classes. Each class has different requirements in the development and deploying process. This model is used to predict if an enterprise has the ability to handle a project or not. Unlike other effort estimation models, this model helps in decision making rather than estimate the required efforts. In this way, enterprises may support their decision in the development procedure of the project. Moreover, the classifier predicts if the resources of an enterprise are capable of developing of the new proposed project or not.

The rest of the paper is organized as follows. Section 2 shows the related works that have been conducted in this area. Section three introduces the clustering model. Section 4 demonstrates the conducted experiment and obtained results. Finally, this paper is concluded in section five.

2 Related Works

Software effort estimation and prediction has attracted researches in the last decade. This process can be categorized in three main categories; statistical, supervised, unsupervised methods. Statistical models are the oldest methods in effort estimation. These models have low accuracy. However, they are simple and do not require complex programming skills to derive them. Constructive cost model (COCOMO) [2] is the oldest statistical software effort estimation model proposed in 1981. In 2000, a successor of COCOMO was published under the name COCOMO II [3]. This model has enhanced the accuracy of its predecessor COCOMO I, which is named COCOMO 81. Nevertheless, the accuracy of these statistical methods required enhancement. To improve effort estimation, the authors of Song et al. [4] proposed a statistical model based on linear regression (LR). The model utilized the same dataset of COCOMO 81 and COCOMO II, which is known as COCOMO dataset these days. More LR models have been proposed, such as, Sentas et al. [5], Mittas and Angelis [6], Chiu and Huan [7] and Aroba et al. [8]. However, it has been reported that all of these works have issues in the reported results [9]. It worth mentioning that, COCOMO 81 and COCOMO II datasets have been utilized in more than 99 projects and 1000's research papers (Galorath 2008).

In the second category, supervised machine learning algorithms have been utilized. In Du et al. [10], the authors proposed a supervised model based on neuro-fuzzy algorithm to enhance the accuracy of SEER-SEM project [8], which has been proposed to estimate efforts in software projects. In Gharehchopogh et al. [11], the author compared the accuracy of different artificial neural networks' models with COCOMO. Multi-Layer Perceptron (MLP), Radial Basis Function (RBF), Wavelet Neural Network (WNN), Functional Link Artificial Neural Network (FLANN) and Generalized Regression Neural Network (GRNN) models have been implemented. Different performance criteria have been utilized, such as, Magnitude of Relative Error (MRE), Mean of Absolute Errors (MAE), Correlation Coefficient (CC) and Root Mean Square Error (RMSE). Their results show that neural networks performance exceeds the estimation performance of statistical methods. In Kumari and Pushkar [12], the authors proposed an estimation model based on genetic algorithm to enhance the estimation process according to Analogy Based Estimation (ABE) performance. All of these works have utilized COCOMO dataset in the evaluation process of the new model.

In the final category, unsupervised machine learning has been used. However, this method is not popular in software effort estimation since the clustering process does not estimate a value for the required effort. It can cluster new features in a group of projects similar with the new project. This method shows the similarity of projects and does not predict a certain effort value. In Nagarajan and Narayanan [13], the author utilized a hybrid K-mean, Particle Swarm Optimization (PSO), neural network and ABE method has been proposed. This system is complex method and hard to implement. Moreover, it can lead to over-fitting. In Goyal and Srivastava [14], K-means algorithm has been utilized to cluster projects of COCOMO dataset into two clusters. The author has shown that clustering projects in K-means provides insights of effort estimation for projects and companies.

This work defers from all of the above conducted works in two main folds. First, unsupervised clustering estimation maximization soft clustering algorithm is used. This algorithm will cluster projects in four different clusters. However, the extracted clusters are soft clusters with projects that will be classified in two or three clusters. This process is impossible in K-mean algorithm. Second, this work aims to answer the question that can a company takes a project or not based on the project features. Finally, the proposed model is simple and easy to be implemented unlike most of the previous artificial intelligent models.

3 The Proposed Soft-Clustering Model

Soft-clustering is an unsupervised machine learning algorithm that consists of two main components; estimation and maximization. This estimation technique has been utilized in different areas [15]. In the estimation process, the harvested features are clustered in different m clusters according to their probabilities in each cluster. To calculate this probability for each project in each cluster, a probability model should be adopted. Our data in this work follows a normal distribution model. Gaussian Distribution (GD) will be used. However, GD deals with one random variable. Our harvested data consists of z number of projects. Each project has n number of random features that impact its probability to belong to a certain cluster. The question is how to extend GD for multi-random features? Fortunately, GD has a multivariate version that deals with any number of random variables. This method has been adopted in this model. Equation 1 shows the multivariate GD (MGD) process.

$$P(x_n|\mu,\varepsilon) = \frac{1}{(2\pi)^{\frac{1}{2}}|\varepsilon|^{\frac{n}{2}}} e^{\left(-\frac{1}{2}(x-\mu)^T\varepsilon^{-1}(x-\mu)\right)} \quad (1)$$

The (MGD) process requires two main variables; μ and ε as can be observed from Eq. 1 that. These variables should be estimated for each random variable 'feature' in each cluster. For example, if each project has 5 features and these project are going to clustered in two different clusters, 10 values of μ and 10 values of ε should be estimated. The question how to estimate these values? The answer is maximum likelihood process (MLP). MLP has been proposed to estimate unknown variables in probability distribution functions. Equation 2 shows the definition of MLP. Finally, Eqs. 3 and 4 shows the estimated values of μ and ε after solving MLP with GD.

$$\begin{aligned}\{\mu,\sigma^2\}_{MLH} &= \max\{l(\mu,\sigma^2;x_1,x_2,\ldots x_n)\}\\ &= \max\{f(x_1,x_2,\ldots x_n;\ \mu,\sigma^2)\}\\ &= \max\{P(x_1|\mu,\sigma^2) * P(x_2|\mu,\sigma^2)\ldots P(x_n|\mu,\sigma^2)\}\\ &= \max\left\{\prod_{i=1}^{n} P(x_i|\mu,\sigma^2)\right\}\end{aligned} \quad (2)$$

$$\mu = \frac{1}{n}\sum_{i=1}^{n} X_i \quad (3)$$

$$\sigma^2 = \frac{\sum (X_i-\mu)^2}{n} \quad (4)$$

One thing to be mentioned is that ε is a square matrix of σ^2. As mentioned n*m σ^2 values exist. These values will be arranged in m scaling matrices as in Eq. 4

$$\begin{bmatrix} \sigma_{12}^2 & \cdots & 0 \\ \vdots & \ddots & \vdots \\ 0 & \cdots & \sigma_{1n}^2 \end{bmatrix} \quad (5)$$

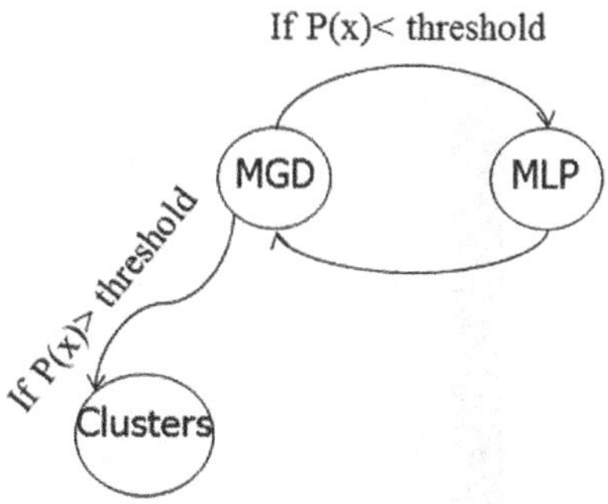

Fig. 1 FSM of soft-clustering

After maximization these values, new probabilities will be estimated for each project again using MGD. Subsequently, MLP will be utilized again to maximize new estimation for GD μ and ε values. This loop will stop when the probability of each project belongs to a specific cluster is maximized and exceeds a certain threshold value. Figure 1 shows finite state machine (FSM) of this process.

To start soft-clustering, random values will be added as the cluster for each project. These values are known as hidden variables. It worth mentioning that proposed model cluster projects into four classes; massive, big, normal and small. After implementing the model, any new project will be classified in one of these classes on the fly without requiring the estimation maximization to run again.

4 Experiment

To start the clustering process, COCOMO 81 dataset has been downloaded from (http://promise.site.uottawa.ca/SERepository/datasets/cocomo81.arff). The dataset consists of 63 projects. Each project has 16 harvested features and an output to show the effort value. The demonstration of these features can be found in (http://promise.site.uottawa.ca/SERepository/datasets/cocomo81.arff). In supervised machine learning, this effort output value is used to train the machine learning models. However, in this work, this value is compared with the obtained clusters. A matrix of 63 × 16 has been generated from these values in MATLAB. Features in each column have been normalized for a value in the range {0–1}. Subsequently, the distribution of data in each column has been plotted as histogram to study its distribution. If the histogram does not follow a normal distribution, log and square roots have been used to convert these data into a semi-normal distribution. For example, Fig. 2a shows the histogram of the 16th column ‘feature 16’. As can be observed, the histogram does not follow a normal distribution. However, after obtaining the log values of these data, Fig. 2b shows a semi-normal distribution histogram.

MATLAB has been utilized in this work to implement soft-clustering algorithm using estimation maximization technique. In addition, K-mean algorithm has been

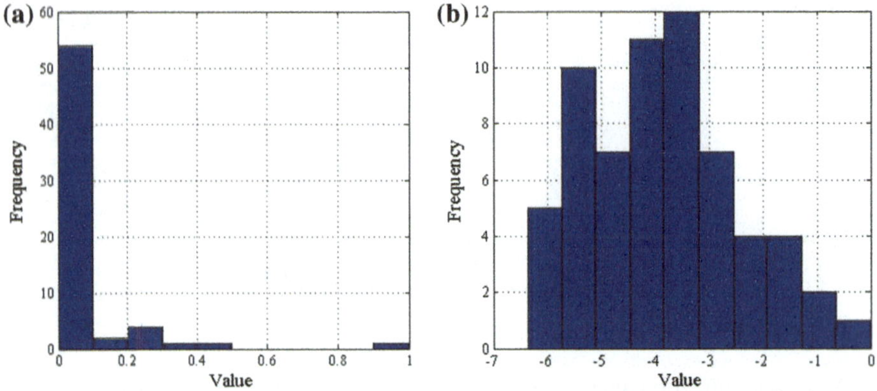

Fig. 2 Feature 16 histogram **a** does not follow normal distribution **b** semi-normal distribution

utilized to show the differences between soft and hard clustering. One thing to be mentioned that the time required for soft and hard clustering to converge is 10 and 7 iterations. This shows that both algorithms are approximately equal in speed.

5 Results

Figure 3 shows matrix plot of the obtained clusters against the input features. As shown in the figure, feature 1 has a massive impact on the first three clusters. If this feature is decaying, the probability of clustering the project in the first or the third clusters increases. This observation is also true for the forth and the fifth features. However, it is hard to observe such observation for cluster 4 or 2 as these two clusters depend on the sixteen features.

Figure 4 shows the estimated effort clustered in four different clustered. These data have been taken from COCOMO 81 project as the output real effort values of each project. These values have been deleted from the data in the clustering process. These values are utilized in the result section to show how the clustering process based on the harvested features behaved. Figure 4a shows K-mean algorithm clustering. As can be shown, higher effort values have been clustered in one cluster and how low values have been clustered in the second cluster.

Nevertheless, we can observe overlapping between the four clusters in the middle of the figure. The reason behind this is the size of dataset that has been used in this work. 63 projects are not enough to generate an accurate estimator in machine learning field. However, the same model can be used with other dataset by tuning the number of features only.

Finally, Fig. 4b shows soft-clustering process. The figure showed We five different colors. Four colors are for our clusters. However, the fifth color is for point or projects that may belong to two different clusters. This process enhances the

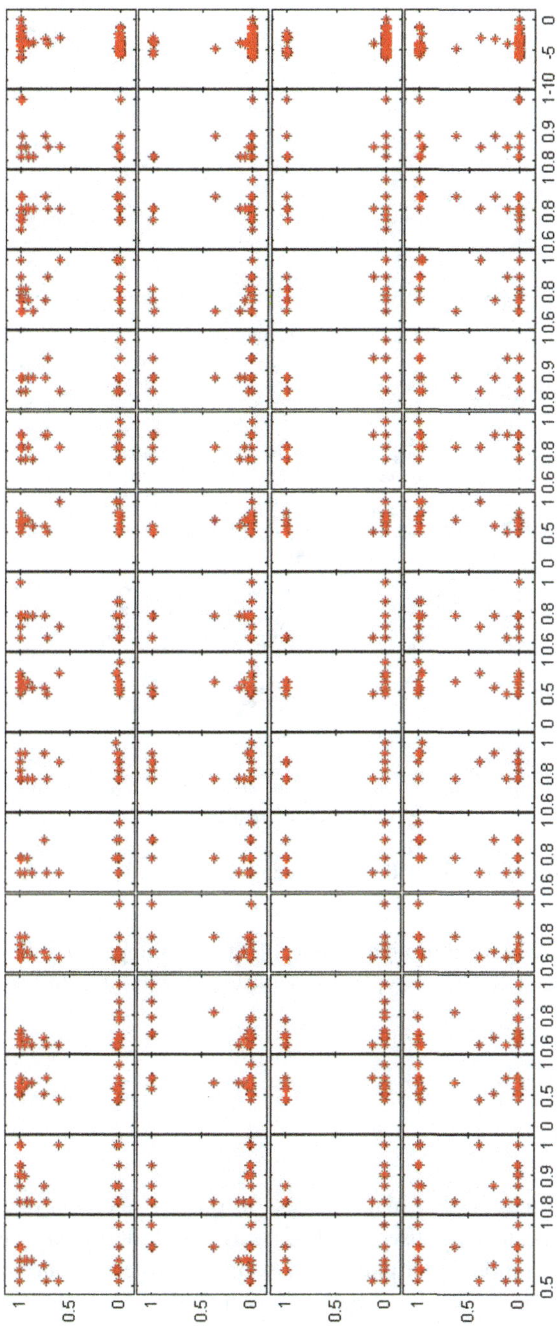

Fig. 3 Clusters versus features

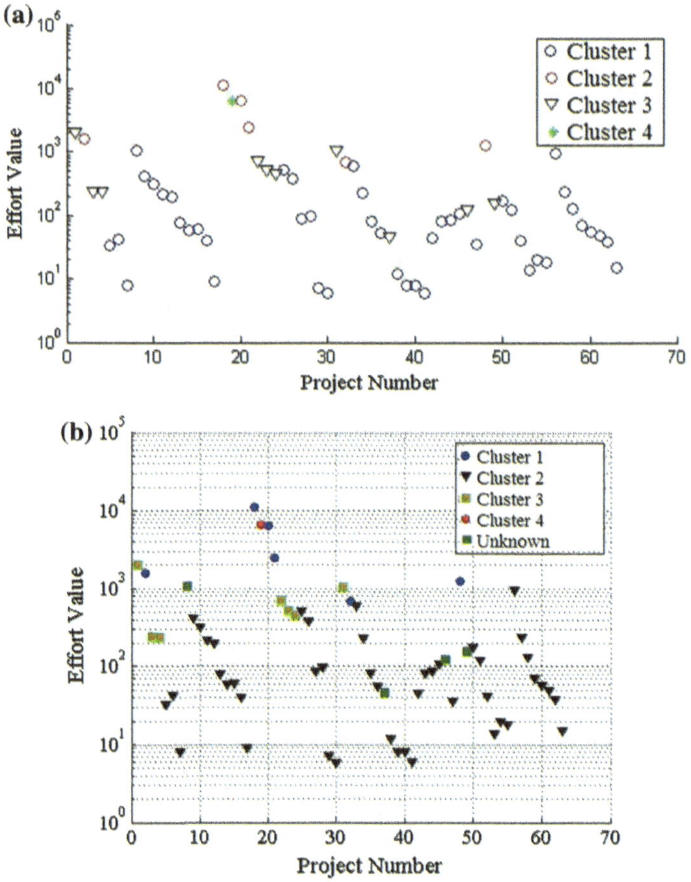

Fig. 4 Clustering process. **a** K-mean **b** soft-clustering

accuracy of K-mean algorithm since some of the overlapped points may occur in two different clusters. In this way, users deploy a threshold value to cluster these points.

6 Conclusion

Software Effort estimation is a main process in project selection decision making process. In this work, unsupervised machine learning clustering algorithm has been utilized to generate a model that places any software project in one of four classes; massive, medium, normal and small. Subsequently, an organization decides according to these classes if a project can be handled or not. COCOMO 81 dataset has been utilized to generate and test the proposed model. Our results show that the accuracy of the model requires improvement through utilizing another bigger dataset. Moreover, the results show how soft-clustering gain more insight on the points that may belong to more than one cluster. Attempts to utilize another dataset will be investigated in future works. Moreover, feature selection process will be utilized to study the efficiency of the features that have been used in other datasets.

References

1. Turner JR (2014) The handbook of project-based management, vol 92. McGraw-hill, New York, NY
2. Boehm BW (1981) Software engineering economics, vol 197. Prentice-hall, Englewood Cliffs (NJ)
3. Boehm BW, Madachy R, Steece B (2000) Software cost estimation with Cocomo II with Cdrom. Prentice Hall PTR
4. Song Q, Shepperd M, Mair C (2005) Using grey relational analysis to predict software effort with small data sets. In: Software metrics, 2005. 11th IEEE international symposium, IEEE, p 10
5. Sentas P, Angelis L, Stamelos I, Bleris G (2005) Software productivity and effort prediction with ordinal regression. IST 47:17–29
6. Mittas N, Angelis L (2008) Comparing cost prediction models by resampling techniques. JSS 81:616–623
7. Chiu N-H, Huan S-J (2007) The adjusted analogy-based software effort estimation based on similarity differences. JSS 80:628–640
8. Aroba J, Cuadrado-Gallego JJ, Sicilia M-Á, Ramos I, García-Barriocanal E (2008) Segmented software cost estimation models based on fuzzy clustering. JSS 81:144–1950
9. Kitchenham B, Mendes E (2009) Why comparative effort prediction studies may be invalid. In: Proceedings of the 5th international conference on predictor models in software engineering, ACM, p 4
10. Du WL, Ho D, Capretz LF (2015) Improving software effort estimation using neuro-fuzzy model with SEER-SEM. arXiv preprint arXiv:1507.06917
11. Gharehchopogh FS, Maleki I, Sadouni S (2014) Artificial neural networks based analysis of software cost estimation models. Algorithms 20:15
12. Kumari S, Pushkar S (2016) A framework for analogy-based software cost estimation using multi-objective genetic algorithm. In: Proceedings of the world congress on engineering and computer science, vol 1
13. Nagarajan S, Narayanan B (2016) K-means clustering algorithms to compute software effort estimation. J Comput Theor Nanosci 13(10):7093–7098
14. Goyal R, Srivastava DK (2016) Evaluation of effort and time reduction through cluster analysis techniques. Imperial J Interdisc Res 2(12)
15. Masoud M, Jaradat Y, Ahmad A (2017) Machine learning approach for categorizing internet autonomous systems' links. ICGHIT, Hanzhou, China

BIM Based Bridge Management System

Mahmoud Dawood

1 Introduction

Due to the age, environmental conditions, and overloading, many structures are deteriorating with time [1]. Maintaining good performance by following a reasonable repair strategy for different infrastructures projects is costly for any municipality and transportation agency. Under limited budget, the governments all over the world are under pressure to ensure long term sustainability for public infrastructure projects. Bridges are the most observable and costly infrastructure projects. Partial or complete failure for bridges has a direct negative impact on the performance of the highways network and it will dramatically increase the public wastage in time and cost. In this regard, concrete and steel bridge networks need to be timely monitored and managed in a way to ensure their continuous acceptable public performance.

Bridge Management Systems (BMS) has been introduced to manage bridges' networks and to keep it in healthy performance along its service life. With increasing number of bridges with limited budget and resources, it becomes mandatory to apply and to implement BMS for and country especially in Middle East region. Almost all literature has been introduced Bridge Management Systems (BMS) to address three major aspects to manage bridges: current condition assessments, future deterioration monitoring, and the decision strategies for repairing, maintenance, and rehabilitation. Although many studies discussed the three aspects, few literatures have been integrating Building Information Modeling (BIM) with Bridge Management Systems (BMS) to enhance the decision strategy.

In deterioration models, the a priori classification of bridges and bridge components commonly used in deterministic to incorporate explanatory factors may overlook the impact of unobserved factors that influence deterioration rates.

M. Dawood (✉)
Middle East College, Muscat, Oman
e-mail: mdawood@mec.edu.om

S. Şahin (ed.), *8th International Conference on Engineering, Project, and Product Management (EPPM 2017)*, Lecture Notes in Mechanical Engineering,
https://doi.org/10.1007/978-3-319-74123-9_17

Stated another way, the statistical model may ultimately predict the average deterioration for a group of bridges well but inaccurately predict the deterioration of the bridges individually. In this paper, BIM will be integrated with BMS to make sure that the bridge is in a serviceable condition in a visual way [2].

2 Components of Bridge Management System: Overview

Bridge Management System (BMS) is defined as a rational and systematic process to manage individual ad well as a network of bridges. The aim of the BMS is to maximize the benefits, to maintain the bridges in healthy condition while minimizing the life cycle cost [3]. Ryall [4], confirmed that BMS is the mean by which bridges as a network are cared for from conception to the end of its useful life. As a result, BMS supports decision makers to select optimum strategy and cost effective alternatives for managing bridges at all levels. The purpose of a BMS is to combine management, engineering, and economic inputs in order to determine the best actions to take on a network of bridges over time [5].

Hudson et al. [3] demonstrated that the activities of a BMS should: define bridge condition; allocate funds for maintenance and improvement actions; prioritize bridges for improvement actions; identify bridges for posting; find cost-effective alternatives for each bridge; account for actual bridge expenditures; track minor maintenance; inspect bridges; and maintain an appropriate database of information. Figure 1 shows the main components of the BMS. These components are: data storage, cost models, deterioration models, and optimization models [6].

As shown in Fig. 1, the heart of a BMS is a database that is derived from the regular inspection and maintenance activities. The integrity of a BMS is directly related to the quality and accuracy of the bridge inventory and physical condition data obtained through field inspections (Manual for condition Evaluation of bridges 1994). Information such as bridge name (ID), location, and construction are stored. It is considered the starting point for the system, where drawings, maintenance records and surveys are reviewed. The database and inventory allow bridge

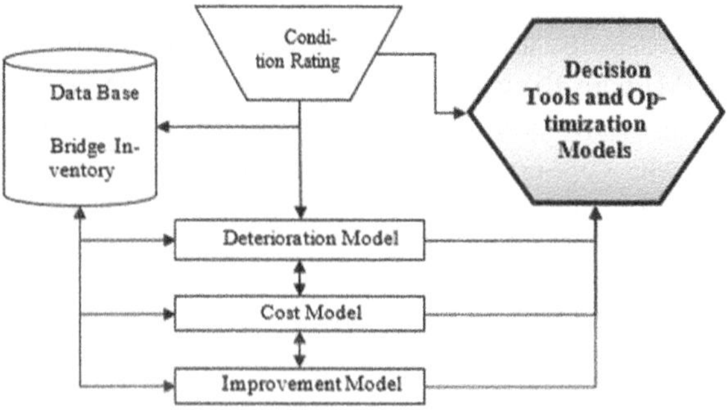

Fig. 1 Components of BMS [6]

managers to be fully informed of the bridge stock under their control so that they can make informed decisions about future maintenance and repair activities [6].

3 Condition Rating Tools

As shown in Fig. 2, the bridges' elements (whether substructure or superstructure) deteriorate with time due to many factors. Materials characteristics, environmental effects, loads, the traffic flow, etc. are recorded as the major factors that have a direct impact on the bridges condition [7–9]. Bridges should be timely monitored to evaluate its performance to ensure its safety and its functionality. Having adequate data such as traffic volume statistics, structural characteristics, and weather information, as well as reliable data gathered through inspection processes are essential requirements for the condition rating of bridges [10–13].

The condition of any components of a bridge may have an impact on the integrity of the structure, performance, usage or the safety (Laman and Guyer 2010; Sutton et al. 2013). Austroads (2004), mentioned that the primary goal of a bridge inspection is to find out the current condition of its components and for the whole bridge as well [8].

The inspection process involves the used materials and the physical condition of bridges components. Consequently, accurate condition assessment must include both the severity of the deterioration or disrepair and the extent to which it is widespread in the component being rated. Aktan et al. (1996) investigated that the condition rating process can be summarized in the following steps: measure damage

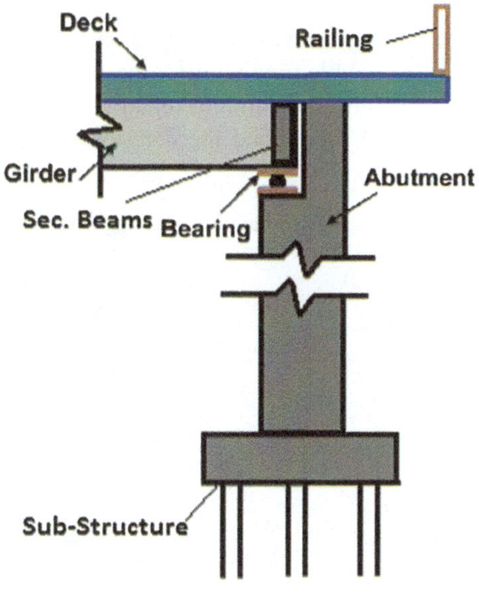

Fig. 2 Main components of concrete bridges

and deterioration, determine the effect of that damage/deterioration on the condition of the facility, set a scale of parameters that describe the condition of the facility as a whole, and compare the existing damage/deterioration with the previous records of condition assessment [6]. Zanini [14] identified three major different tools for the assessment of technical indicators for bridge management as follows: visual surveys; non-destructive techniques; and probing.

Visual techniques, in the Bridge Management System (BMS), are the most common method for assessing structural condition. Visual inspection is carried by expertise and to be recorded manually. To avoid transcription errors and reduces the workload, speech recognition is an implemented method to improve the effectiveness of such tool [15]. This tool is based on converting the recorded audio data directly into text data which omits the need for handwriting on paper. Also, robots are used to improve the process for the detection of deterioration [14].

Due to cost, time waste and difficulties in data interpretation, non-destructive techniques methods are usually less adopted. Goangseup et al. (2008) illustrated the use of infrared thermography, useful for the detection voids located few centimeters below the surface. The main issue when using these tools is how to translate non-destructive results in a proper condition rating. Zanini [14] presented the results of a study that examined the condition of a reinforced concrete bridge deck using multiple sensors. Finally, probing is one of the condition rating tool that focus on the definition of main materials properties, through the removal of samples of materials from the existing structures and the subsequent testing in laboratory or with the execution of situ tests.

4 Bridge Deterioration Models

Forecasting bridges' deterioration is a challenge tackled by many municipalities. The reasonable prediction of the future condition of bridges' elements is necessary for selecting the optimum repair strategy. As shown in Fig. 3, Abed-Al-Rahim and Johnston [16] defined the deterioration of bridges with the time as the process of

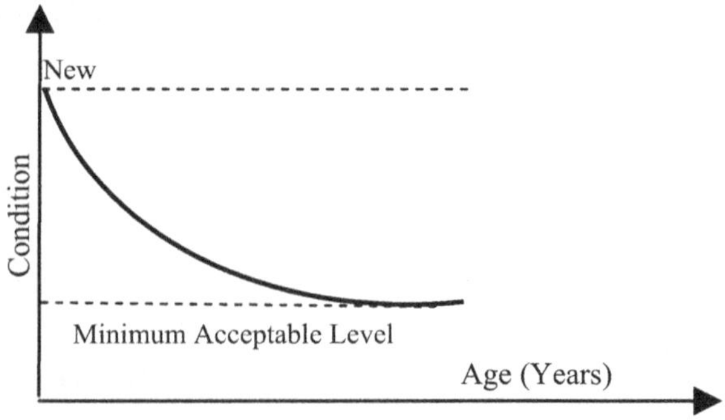

Fig. 3 Bridge deterioration with time (adopted from Elbehairy [6])

decline in the condition of the bridge resulting from normal operating conditions, excluding damage from such events as earthquakes, accidents, or fire. The deterioration process exhibits the complex phenomena of physical and chemical changes that occur in different bridge components [6]. One of the main challenges in bridge deterioration assessment is that each element has its own unique deterioration rate [17]. Accurately predicting the rate of deterioration for each bridge element is, therefore, crucial to the success of any BMS. One of the most common used method for condition forecasting is the Markov Chain method.

Markov process, a stochastic technique for infrastructure deterioration is based on the concept of probabilistic cumulative damage and currently it is commonly used in deterioration forecasting of bridges' components and bridges' network [18]. Markovian bridge deterioration models are based on the concept of defining states in terms of bridge condition ratings and obtaining the probabilities of a bridge condition changing from one state to another [19].

$$\begin{array}{c} \textbf{\textit{Condition}} \quad \textbf{\textit{Rate}} \\ \begin{array}{cc} & \begin{array}{ccccc} 5 & 4 & 3 & 2 & 1 \end{array} \end{array} \end{array}$$

$$\boldsymbol{TPM} = \begin{array}{c} 5 \\ 4 \\ 3 \\ 2 \\ 1 \end{array} \begin{bmatrix} P_{55} & 1 - P_{55} & 0 & 0 & 0 \\ 0 & P_{44} & 1 - P_{44} & 0 & 0 \\ 0 & 0 & P_{33} & 1 - P_{33} & 0 \\ 0 & 0 & 0 & P_{22} & 1 - P_{22} \\ 0 & 0 & 0 & 0 & 1 \end{bmatrix} \tag{1}$$

The Markov model, incorporated in the current model was developed by Jiang [19]. The proposed transition probability matrix (TPM) condition ratings (1–5) are adopted from [18]. The probability matrix was formulated taking into consideration different bridges components. Equation (1) represents the deterioration transition matrix. Each row in the matrix has two values that represent the probability remaining in its current condition state and the probability moving to the next lower condition state (Fig. 4). By applying Eq. (1), Fig. 5 presents the deterioration curve for the concrete deck in interstate bridge.

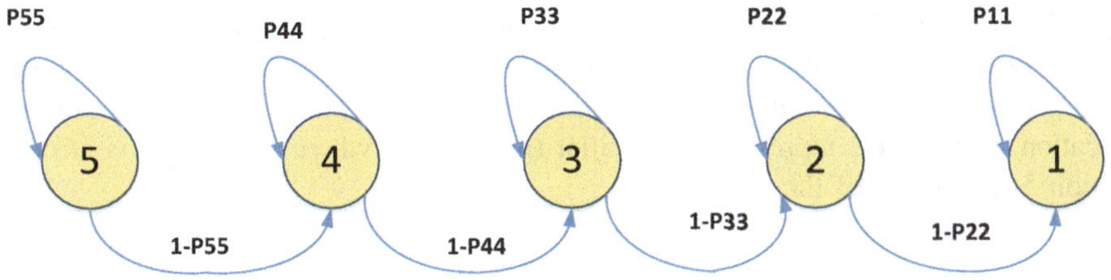

Fig. 4 Probability from condition state to the other

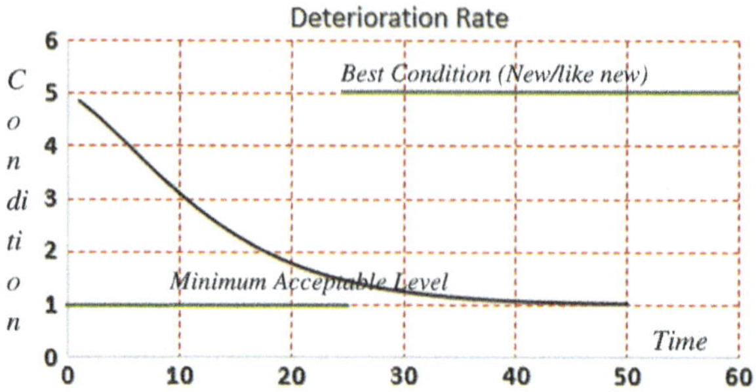

Fig. 5 Deterioration curve for concrete deck

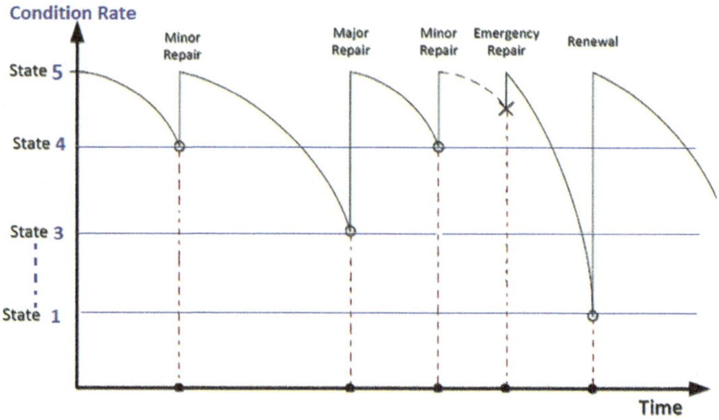

Fig. 6 Sample of deterioration pattern along with historical work done on a bridge (adopted from Le [20])

The typical deterioration process of any asset can be mapped as in in Fig. 6. It is observed that the time gap between major repair and minor maintenance is representing the full life time of the component reaching the state where minor repair is required from the 'as new' state. Table 1 represents the transition probability matrix (TPM) for the bridge's components and for the whole bridge as well. The location is one of the main factors that has a tangible impact on the TPM. Obviously, the traffic flow and the environmental conditions are depending on the geographic location. Equation 2 represents the initial (new state), where the bridge is at condition 5, condition of the bridge.

Table 1 Detailed transition probability matrix (adopted from Li [18])

Category	Transition probability matrix (TPM)		
	Entire city	Central city	Suburbs
Bridge deck system	$\begin{bmatrix} 0.856 & 0.144 & 0 & 0 & 0 \\ 0 & 0.67 & 0.33 & 0 & 0 \\ 0 & 0 & 0.771 & 0.229 & 0 \\ 0 & 0 & 0 & 0.865 & 0.135 \\ 0 & 0 & 0 & 0 & 1 \end{bmatrix}$	$\begin{bmatrix} 0.86 & 0.14 & 0 & 0 & 0 \\ 0 & 0.693 & 0.308 & 0 & 0 \\ 0 & 0 & 0.773 & 0.227 & 0 \\ 0 & 0 & 0 & 0.887 & 0.113 \\ 0 & 0 & 0 & 0 & 1 \end{bmatrix}$	$\begin{bmatrix} 0.847 & 0.153 & 0 & 0 & 0 \\ 0 & 0.598 & 0.402 & 0 & 0 \\ 0 & 0 & 0.797 & 0.203 & 0 \\ 0 & 0 & 0 & 0.887 & 0.113 \\ 0 & 0 & 0 & 0 & 1 \end{bmatrix}$
Super-structure	$\begin{bmatrix} 0.945 & 0.055 & 0 & 0 & 0 \\ 0 & 0.789 & 0.211 & 0 & 0 \\ 0 & 0 & 0.803 & 0.197 & 0 \\ 0 & 0 & 0 & 0.836 & 0.164 \\ 0 & 0 & 0 & 0 & 1 \end{bmatrix}$	$\begin{bmatrix} 0.945 & 0.055 & 0 & 0 & 0 \\ 0 & 0.764 & 0.236 & 0 & 0 \\ 0 & 0 & 0.81 & 0.19 & 0 \\ 0 & 0 & 0 & 0.824 & 0.176 \\ 0 & 0 & 0 & 0 & 1 \end{bmatrix}$	$\begin{bmatrix} 0.943 & 0.057 & 0 & 0 & 0 \\ 0 & 0.823 & 0.177 & 0 & 0 \\ 0 & 0 & 0.821 & 0.179 & 0 \\ 0 & 0 & 0 & 0.807 & 0.193 \\ 0 & 0 & 0 & 0 & 1 \end{bmatrix}$
Sub-structure	$\begin{bmatrix} 0.943 & 0.057 & 0 & 0 & 0 \\ 0 & 0.903 & 0.097 & 0 & 0 \\ 0 & 0 & 0.982 & 0.018 & 0 \\ 0 & 0 & 0 & 1 & 0 \\ 0 & 0 & 0 & 0 & 1 \end{bmatrix}$	$\begin{bmatrix} 0.932 & 0.068 & 0 & 0 & 0 \\ 0 & 0.914 & 0.086 & 0 & 0 \\ 0 & 0 & 1 & 0 & 0 \\ 0 & 0 & 0 & 1 & 0 \\ 0 & 0 & 0 & 0 & 1 \end{bmatrix}$	$\begin{bmatrix} 0.956 & 0.044 & 0 & 0 & 0 \\ 0 & 0.892 & 0.108 & 0 & 0 \\ 0 & 0 & 0.944 & 0.056 & 0 \\ 0 & 0 & 0 & 1 & 0 \\ 0 & 0 & 0 & 0 & 1 \end{bmatrix}$
Whole bridge	$\begin{bmatrix} 0.92 & 0.08 & 0 & 0 & 0 \\ 0 & 0.885 & 0.115 & 0 & 0 \\ 0 & 0 & 0.923 & 0.077 & 0 \\ 0 & 0 & 0 & 0.925 & 0.075 \\ 0 & 0 & 0 & 0 & 1 \end{bmatrix}$	$\begin{bmatrix} 0.9 & 0.1 & 0 & 0 & 0 \\ 0 & 0.898 & 0.102 & 0 & 0 \\ 0 & 0 & 0.952 & 0.048 & 0 \\ 0 & 0 & 0 & 0.973 & 0.027 \\ 0 & 0 & 0 & 0 & 1 \end{bmatrix}$	$\begin{bmatrix} 0.917 & 0.083 & 0 & 0 & 0 \\ 0 & 0.904 & 0.096 & 0 & 0 \\ 0 & 0 & 0.9 & 0.1 & 0 \\ 0 & 0 & 0 & 0.986 & 0.014 \\ 0 & 0 & 0 & 0 & 1 \end{bmatrix}$

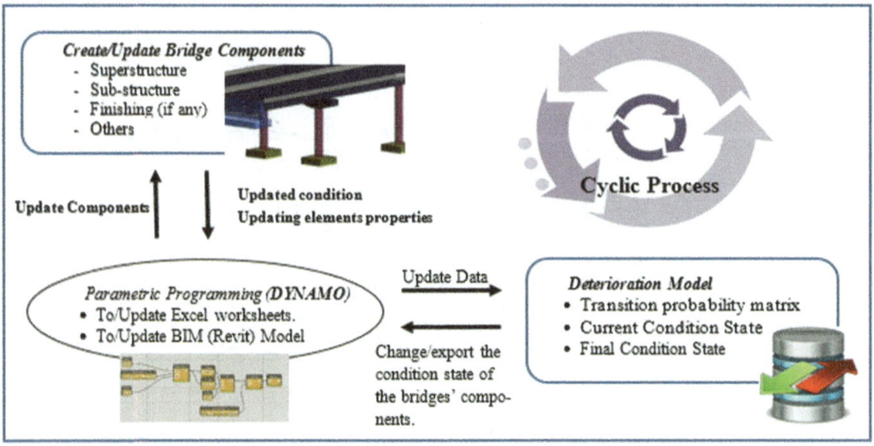

Fig. 7 The process work flow

5 BIM Integration Framework

This research introduces a Building Information Model (BIM)-based integrated model that can be used to visualize the condition state of the bridge as whole as well as its components. Figure 7 shows the process workflow. Proposed model might support the decision makers to visually show the conditions in color scheme. As shown in Fig. 8, different conditions will be viewed in different colors (Green, Light Green, Yellow, Amber, and Red). The framework is depending on the collaboration and integration between Microsoft Excel and Autodesk Revit. Decision maker has to start by creating the BIM model. After that, the bridge and its components will be exported to Excel environment. Excel is used to prepare the deterioration model by evaluation of the condition state. Finally, condition rate will be exported to Revit to visualize the bridge state in different color.

6 Summary and Conclusion

Generally, bridges in have a direct impact on the economic income of any country due to its high asset value. Based on the previous discussion, it is a challenge to maintain the bridges and its components at a high level of working standard due to the limited budget and costly resources. The proposed method will improve the BMS process to manage bridges in an effective way. The core of this model is the condition level forecast using Markov chain. Based on Markovian model can recognize the bridges in the dangerous condition among its components and the whole bridge in a network. This research aimed to introduce the framework to integrate BIM and Excel worksheet models. This integration helps the decision

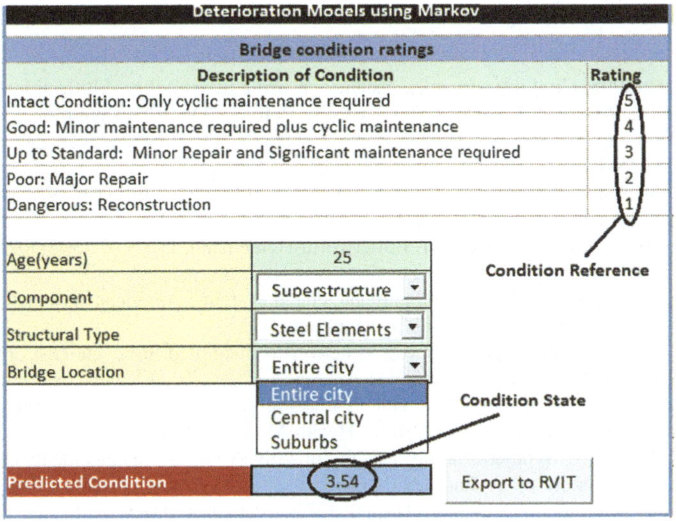

(a) Excel workspace

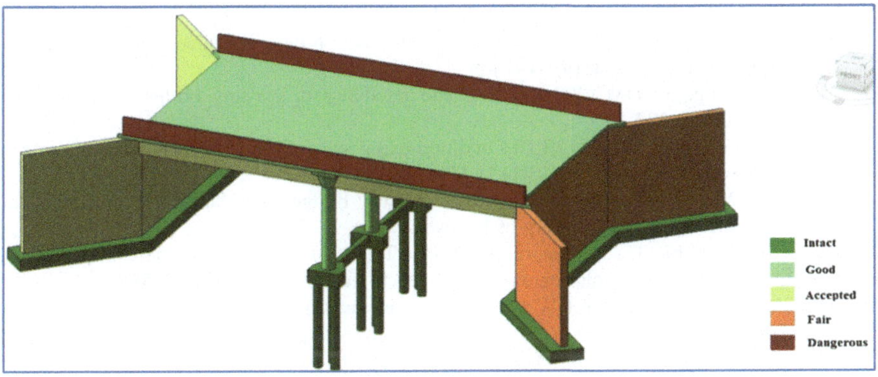

(b) Revit workspace

Fig. 8 Bridge elements condition rating in color pattern

makers to visualize the rating of the bridges components as well as the whole bridge. Generally, this research resulted in the development of a practical, easy-to-use visualized deterioration model. DYNAMO is used as a visual programming tool to automate the integration of data, thereby improving infrastructure condition assessments.

References

1. Bordogna J (1995) Civil infrastructure systems: ensuring their civility. J Infrastruct Syst 1:3–5
2. Cavalline TL, Whelan MJ, Tempest BQ, Goyal R, Ramsey JD (2015) Determination of bridge deterioration models and bridge user costs for the NCDOT bridge management system (No. FHWA/NC/2014-07)
3. Hudson RW, Carmichael RF III, Hudson SW, Diaz MA, Moser LO (1993) Microcomputer bridge management system. J Transp Eng 119(1):59–76
4. Ryall M (2001) Bridge management. Butterworth-Heinemann, ISBN (0-7506-5077), Oxford, UK
5. AASHTO (2001) Guidelines for bridge management systems. American Association of State Highway and Transportation Official, Washington, D.C
6. Elbehairy H (2007) Bridge management system with integrated life cycle cost optimization. Dissertation, University of Waterloo
7. Aflatooni M, Chan TH, Thambiratnam DP, Thilakarathna I (2013) Synthetic rating system for railway bridge management. J Civil Struct Health Monit 3(2):81–91
8. Aflatooni M (2015) Synthetic rating system for railway bridge management. Dissertation, Queensland University of Technology
9. Shih HW, Thambiratnam DP, Chan TH (2009) Vibration based structural damage detection in flexural members using multi-criteria approach. J Sound Vib 323(3):645–661
10. Adhikari RS, Moselhi O, Bagchi A (2012) Automated prediction of condition state rating in bridge inspection. Gerontechnology 11(2):81
11. Agrawal AK, Kawaguchi A, Chen Z (2010) Deterioration rates of typical bridge elements in New York. J Bridge Eng 15(4):419–429
12. Akgül F, Frangopol DM (2003) Rating and reliability of existing bridges in a network. J Bridge Eng 8(6):383–393
13. Nukul S, Bonaventure HWH (2010) Condition rating system for Thailand's concrete bridges. J Constr Developing Countries 15(1):1–27
14. Zanini M (2016) Quality specifications for roadway bridges, standardization at a european level. Scientific Report, University of Padova
15. Kenmotsu A, Ushio H, Tsugimura H, Oyama T (2006) A system for field inspection of infrastructures in snowy cold regions using speech recognition. IABMAS 2006
16. Abed-Al-Rahim IJ, Johnston DW (1995) Bridge element deterioration rates. Transp Res Rec 1490:9–18
17. Thompson P (2001) Condition predictive model results. Memorandum
18. Li L, Sun L, Ning G (2014) Deterioration prediction of urban bridges on network level using Markov-chain model. Math Prob Eng
19. Jiang Y (1990) The development of performance prediction and optimization models for bridge management systems. Dissertation, Purdue University
20. Le B (2014) Modelling railway bridge asset management. Dissertation, University of Nottingham
21. Zi G, Sim J, Oh H, Lee J (2008, June) Optimum NDT using infrared thermography for defected concrete. In: Bridge maintenance, safety management, health monitoring and informatics-IABMAS'08: proceedings of the fourth international IABMAS conference, Seoul, Korea, 13–17 July 2008, p 278 (CRC Press)

Using Hollow Concrete and Thermostone Blocks in Sound Isolation System

Mousa Bani Baker and Raed Abendeh

1 Introduction

Noise is one of the most spread pollutions that influence human beings [1]. Shootings at test shooting sites indoors can cause a risk of hearing loss for the workers at the site. The noise can also disturb other persons working in nearby rooms [2]. At test shooting sites, noise can be as high as 130–170 dB [3, 4].

Acoustically insulating an indoor shooting range may be a difficult and expensive, and there are often sound leakages from the shooting site to nearby office rooms like ventilation, openings and electrical cables [2].

The peak SPLs (Sound Peak Levels) at the shooter's ear ranged from 132 to 183 dB. The spectral content of the main part of the acoustic energy was less than 400 Hz for large-caliber weapons and 150–2500 Hz for small-caliber weapons (rifles). The safe distances from the noise source (less than 140 dB peak SPL) were 50–200 m for large-caliber weapons. Earmuffs are ineffective against impulses from large-caliber weapons [4].

National Institute for Occupational Safety and Health-USA(NIOSH) has recommended that all worker exposures to noise should below a level of 85 dBA for eight hours to minimize noise induced hearing damage. Occupational Safety and Health Administration-USA (OSHA) in their Occupational Safety and Health Standards, states that the permissible noise exposure for 8 working hours (Table 1) should be less than 90 dB [5].

M. B. Baker (✉) · R. Abendeh
Department of Civil and Infrastructure Engineering, Al-Zaytoonah University of Jordan, P.O. Box 130, Amman 11733, Jordan
e-mail: m.banibaker@zuj.edu.jo

S. Şahin (ed.), *8th International Conference on Engineering, Project, and Product Management (EPPM 2017)*, Lecture Notes in Mechanical Engineering,
https://doi.org/10.1007/978-3-319-74123-9_18

Table 1 Permissible noise exposures [5]

Duration per day (h)	Sound level (dB)
8	90
6	92
4	95
2	97
1 1/2	102
1	105
1/2	110
1/4 or less	115

2 Materials and Methods

Sound absorbing materials are used to control noise in which the sound wave fade effectively when their waves hits their surfaces. Some of the sound energy will be reflected from the surface and the rest will influence the solid material to vibrate. The sound travels in solid structure or the air. The solid materials allow low frequency sounds to go through, however, air allows high frequency sounds to travel through it. Building materials in general have different frequency responses to sound waves.

The weight, density or thickness of a wall are important factors to block sound. An air space within a wall can also help to enhance sound isolation. In fact, it creates two independent walls. Installing sound absorptive material to the wall enhances the ability to decrease the transmitted sound. Locally produced 150 mm thermostone blocks made by combining a small amount of cement and lime with sand (AAC), locally produced 100 mm HCB, 6 mm facing felt, 25 mm fiberglass, 30 mm polystyrene sheets, and mortar, were used to build two sandwich walls. Tables 2 and 3 show the properties of HCB and thermostone blocks used in this research.

There are national and internatinal guidelines for the noise isolation properties of wall structures [7]. Sound insulating property of a partition element is expressed in terms of sound transmission loss. The procedure to measure this property is to construct the partition between two reverberation rooms. The two walls were constructed separately as follows:

Table 2 Properties of the 100 mm HCB [6]

Fire rating (h)	1
Void (%)	37
Compressive strength (N/mm^2)	3.5
Density (kg/m^3)	2000
Average weight (kg/block)	11
Average thermal conductivity at 15 °C (W/mK)	0.9
Dimensions (mm)	400 × 200 × 100

Table 3 Properties of the 150 mm thermostone blocks

Fire rating (h)	6
Average compressive strength (N/mm^2)	4.2
Average density (kg/m^3)	510
Average mass (kg/block)	11.5
Dimensions (mm)	600 × 250 × 150
Average thermal conductivity at 15 °C (W/mK)	0.77

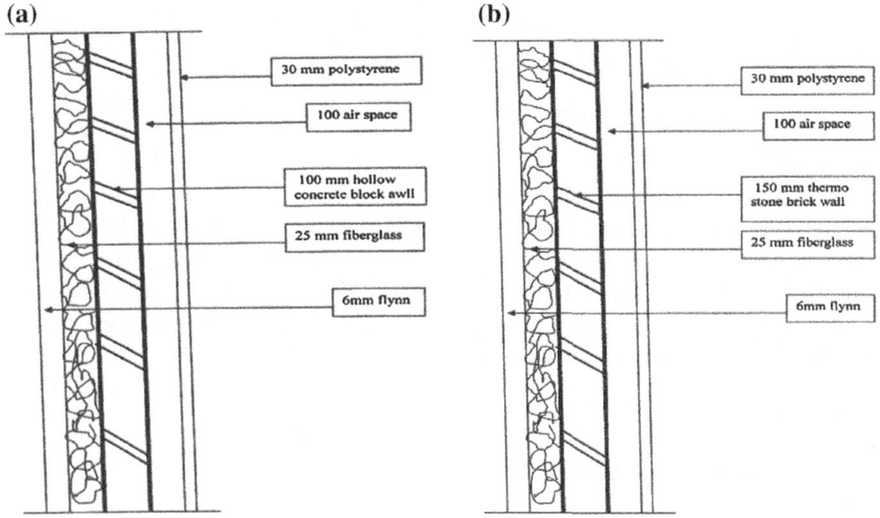

Fig. 1 Sections of both walls: **a** (Left), **b** (Right)

- *Wall A:* was constructed from 6 mm of facing felt, 25 mm fiberglass, 100 mm HCB, 100 mm air space and 30 mm polystyrene as illustrated in Fig. 1.
- *Wall B:* similar to the first wall except for 100 mm HCB were replaced by 150 mm thermostone blocks (Fig. 1).

Both walls were constructed consequently in an opening (2 m × 2 m) between two chambers in the laboratory among the source and the receiving wall. The two chambers (Fig. 2) were completely isolated, they contain reflectors that are not parallel to the wall to ensure that sound is distributed uniformly to the tested walls and the only significant sound transmission between the two rooms is through the test specimen. Fine mortar was employed to mount the blocks and to seal any empty spaces, the curing time of the sealant was for one week, and the room temperature and humidity were: 25 ± 2 °C, 45 ± 5%. The specimens were tested in accordance with the American Society for Testing and Materials designation ASTM E 90-2004, "Standard Test Method for Laboratory Measurement of Airborne Sound Transmission Loss of Building Partitions".

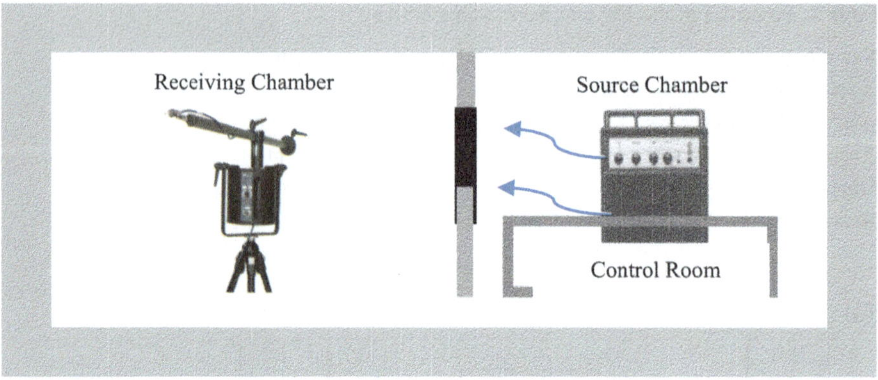

Fig. 2 Two chambers connected by a (2 m × 2 m) window

The sound source is a special B & K sound type 4224 with different sound spectra (100–4000 Hz). The sound is measured using rotating microphone boom, rotating on average rotation every 32 s, and connected to building acoustic analyzer type 4418, which is equipped with two channels, for the receiving and the source room. The sound in the receiving room is averaged by the rotating microphone and measured and stored in the analyzer.

3 Results and Analysis

The Sound Transmission Class (STC) gives an indication of the insulating property of the tested specimen. The higher the STC rating is the greater the sound insulating of the tested wall.

3.1 Results

Table 4 shows the test results for both specimens tested in wide range of octave band frequencies (100–3150 Hz). The sound level meter for 500 and 4 kHz frequencies are 87.3 and 97.0 dB consequently. Therefore, it can be noted that for the 500 Hz frequency the average STC for wall-A made of HCB was 41 dB and for Wall-B made of thermostone blocks was 43 dB.

The sound transmitted to the receiving room is 46.3 dB in case of using 100 mm HCB in the partition construction, and 44.3 dB when using 150 mm thermostone blocks.

Table 4 Average STC for both wall specimens

1/3 Octave band frequency (Hz)	Average sound transmission loss (dB)	
	Wall A	Wall B
100	23	30
125	27	32
160	31	34
200	33	36
250	35	38
315	37	40
400	39	42
500	41	43
630	42	44
800	43	45
1000	44	45
1250	45	45
1600	45	46
2000	45	46
2500	45	46
3150	45	46

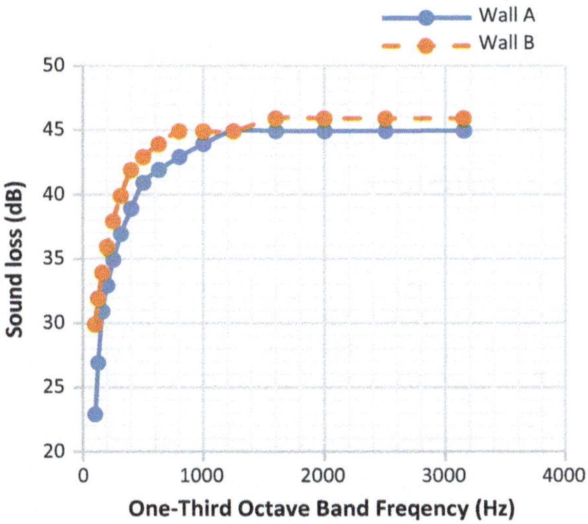

Fig. 3 Test results of sound transmission class (STC) for walls A and B

The sensitivity of noise reduction above 1.0 kHz octave band frequency is low for both walls, for example in case of using HCB the STC for 1000 Hz is 44 dB and for 3150 Hz is 45 dB. When employing thermostone partition blocks the STC for 1000 Hz is 45 dB and 46 dB for 3150 Hz octave band wave frequency as it is shown in Fig. 3.

Table 5 Cost of thermostone and HCB as of April 2017

Material	Price/m^3 (JD)	Price/block (delivery included) (JD)	Block/m^2	Price/m^2 (JD)
Themostone block	68	1.4	6.6	9.24
HCB	35	0.28	12.5	3.5

3.2 *Cost Analysis*

Cost can be classified into:

A. Short term cost: which include the direct price of all materials used to construct the partitions, once the materials are the same except for the type of blocks, the following table (Table 5) summarizes the cost of each type in the Jordanian market, for the KADDB tunnels project (1500 m^2). Cost of the HCB is 5250 JD and 13,860 JD for thermostone blocks.
B. Long term cost: taking into consideration the added value of using the thermostone blocks in reducing the energy consumption (air conditioning of the tunnels) compared to HCB. The average thermal conductivity (k) for thermostone blocks is 0.77 W/mK compared to 0.9 W/mK for HCB. The lower k is the better for thermal insulation, furthermore having better efficiency and cost reduction in energy consumption.
C. HCB occupies less space than thermostone blocks, however, thermostone blocks is easier in handling and building due to its light weight

4 Conclusions

- The proposed sandwiched walls reduced the noise to acceptable levels of exposure within 8 h working time for wide range of octave band frequencies which represents both low and high weaponry sound frequencies.
- Both partition systems reduced the transmission of sound to the receiving area by 50%.
- Using HCB is a less expensive option in the short term, however, for the long term cost, the thermostone blocks may reduce the cost of energy consumption for indoor firing range tunnels.

Acknowledgements This work was financially supported by KAFD and KADDB. We would like to thank the support and effort of Eng. Amani Nawafleh, Amer Bitro, Basil Khader, Amer Bisharat and KANKON Inc for marketing and logistics.

References

1. Li H, Peng W, Xiang Y, Wenjun Z (2016) Researches on sound environment in Futian underground railway station. Procedia Eng 165:730–739
2. Pääkkönen R, Jussila J (2011) A suppressor for indoor shooting sites. Appl Acoust 72(9): 684–687
3. Paakkonen R, Lehtomaki K (2005) Protection efficiency of hearing protectors against military noise from handheld weapons and vehicles. Noise Health 7(26):11
4. Ylikoski ME, Pekkarinen JO, Starck JP, Pääkkönen RJ, Ylikoski JS (1995) Physical characteristics of gunfire impulse noise and its attenuation by hearing protectors. Scand Audiol 24(1):3–11
5. OS (2017) Occupational safety and health standards. http://www.osha.gov Accessed 10 Apr 2017
6. Brick (2017) http://www.mgc-concrete.com/en/Brick
7. Berglund B, Thomas L, and Dietrich HS (1999) Guidlines for community noise. Geneva, World Health Organization.

Information Communication Technology (ICT) Impact on Building Construction Management Practices in the South West of Nigeria

A. Olatunji Aiyetan

1 Introduction

The computer is a system used to store and manages information and this have found significant application in the Construction Industry. Bowden and Thorpe [7] state that 65% of contractor's rework is attributed to insufficient, inappropriate, and conflicting information on construction site, in addition, sometimes late information. Communication through the internet is quick and can mitigate the issue of time delay. This is irrespective of location. There are varying types that are suitable for local and remote areas in cases of construction in these types of places. Similarly, the information that is communicated is always accurate with the exception that the information captured is incorrect. An advantage that information communication technology (ICT) has is information can be retrieved anytime as they are stored on the system.

2 Literature Review

2.1 Information Communication Technology Concept and the Construction Industry

Adriaanse and Voordijk [3] explains that ICT is a neutral provider of input for decision making. Also, that communication between construction industry participants and organizations are concerned with information exchange, dealings with

A. O. Aiyetan (✉)
Department of Construction Management & QS, Faculty of Engineering
& the Built Environment, Durban University of Technology,
Durban, Kwazulu-Natal, South Africa
e-mail: ayodejia@dut.ac.za

S. Şahin (ed.), *8th International Conference on Engineering, Project, and Product Management (EPPM 2017)*, Lecture Notes in Mechanical Engineering,
https://doi.org/10.1007/978-3-319-74123-9_19

drawings, specifications; cost data, programs plus other designs and management information. Communication can be viewed as a means along which information is transferred from one person to another. Communication can be defined in many ways as it is multidimensional and indefinable. It can have a variety of different meanings, contexts, forms and impacts; therefore, it may mean different things to different people in different situations. In the construction industry diverse communication occurs simultaneously, resulting from teams in different sections of construction undertake tasks, activities and are constantly involved in communication. The various forms of communication include, conversations, listening networking, data and information collection, mails, using different mediums ranging from electronic to manual means of getting these done [8]. Of recent internet and World Wide Web are used [12].

2.2 Challenges of Using ICT for Project Management Practice by Building Contractors

Below is enumerated and discussed challenges relative to the use of ICT in project management practices by building contractors in Nigeria.

2.2.1 Insufficient/Erratic Power Supply

Electricity supply in Nigeria is not stable; it is supplied with low voltage and often erratic. These pose a problem to the use of ICT for management practices. The other option available which is the use of generators is economical to run.

2.2.2 Virus Attacks

Virus attack is a general and common phenomenon in the computer world. Virus are written program that are harmful to the computer system. When this attacks occurs, the system may stop working. The solution to this is to install a powerful antivirus on the system and to scan any document for virus attack before opening them.

2.2.3 High Cost of Hardware/Software

The initial cost of setting up an information technology system in an organization is high. It requires a medium to large organization to be able to set it up for management practices and still be able to keep the organization afloat, these is regarding payment of wages and overhead and keeping the construction site running.

2.2.4 Capable of Making Professionals Redundant

The introduction of computer and robots into our working environment has resulted in increase in productivity. Computers as it is being said can perform the function of many individuals put together more efficiently and faster. Computers can perform the following functions: production of drawings, making structural analysis, carrying out cost and time calculations and store and manage construction information Isikdag (2006) cited in Isikdag et al. [9]

2.2.5 Lack of ICT Infrastructure/ICT Content in Construction

Service providers of telecommunication service in Nigeria still need to improve their system to provide high speed internet and access in remote areas (Isikdag 2006). Some cannot be accessed with low speed internet, such as video conference or IP camera. Further [11] states the ideal level of ICT utilization will not be attained until infrastructure is improved.

2.2.6 Lack of Management Desire and Appreciation of ICT

This may be as a result of problematic experience management has had relative to the use of software's. The lack of adequate training in the use of these software's may result in erroneous result and the resultant loss of job, cost increases that will put the firm in a disadvantage position. Secondly, this software's get outdated too quickly and may be difficult to replace. This may lead to management lack of desire for their usage.

Other factors are: low return on investment; influence of competition; financial issues [5]; lack of soft skills for professional's interaction [6], and fear of mass job losses in the industry.

2.3 Factors Influencing the Adoption of ICT Devices for Project Management Practice by Building Contractors

The factors prompting the use of ICT devices for construction management by building contractors are: supplanting; mismanagement of funds and resources; organizational and individual unethical attitude; bureaucracy, and software application [1, 4, 10]

2.4 Benefits of Information Communication Technology on Project Management Practice

There are numerous benefits that are derived from the use of ICT, it ranges from improvements in public image of contractor; coordination among construction participants and documentation of presentation. The enhancement of construction productivity speed, exchange of information among professionals, savings in operation cost, an instrument that aids transparency and accountability and reduces direct procurement cost. Its use, reduces the extent of fragmentation, facilitates decision making and makes building contractors profession job easier.

3 Research Methodology

A study on ICT impact on Building Construction Management practices in the South West of Nigeria was conducted to assess ICT impact and challenges on project manage with it. The study was conducted among contractors, registered with the state ministry of works. The sample size was one hundred and fifty (150) contractors. They were accessed through the state ministry of works of both Lagos and Ondo states. The random sampling technique was used for the selection of samples. Questionnaire survey was conducted. Questionnaires were administered via post. One hundred and fifteen (115) were returned filled representing a response rate of was 76.7% achieved. Descriptive statistics is employed for the analysis of data.

Respondents having experience of between 6 to 21 years (56%) were surveyed and medium and small firms were surveyed (having less than 81 employees).

Table 1 reveals the software packages for project management practices. The most used software package is viewpoint v6 construction management software (39.8%). The likely reason for its high frequency of usage may be its capability to solve unique problems of construction professionals and ability to schedule labour and equipment usage. Next to viewpoint v6 construction management software is none (22.2%). None implies the use of traditional method which is manual.

Table 1 Software packages for project management practices

Project management software's	Frequency	Percentage (%)
Viewpoint v6 construction management software	43	39.8
None	24	22.2
Priosoft construction management software	12	11.1
Accobuild construction software	10	9.3
Corecon construction management software	10	9.3
Aptora total office management software	9	8.3

The least used package is Aptora total office management software (8.3%). The reason for it low usage could be attributed to lack of awareness, non-availability, and lack of application knowledge.

It is noteworthy than all the factors that constitutes challenges to the use of ICT for project management hence a mean score (MS) greater than 3, which indicates moderate to a high influence regarding usage.

Table 2 presents challenges regarding the use of ICT for project management. Insufficient/erratic power supply (MS = 4.33) is rated as the most challenge faced by construction project professionals in the use of ICT for project management. ICT is electric power driven and Nigeria has a huge project of power outages/erratic power supply. In cases where there is a supply, the voltages are low and cannot power the systems. This constitutes a huge challenge to the effect use of ICT for project management. Next to insufficient/erratic power supply is high cost of hardware/software (MS = 4.17). Some organization cannot afford to purchase the hardware and the software for project management purposes as a result of their high cost. This makes communication and information sharing difficult.

Next to high cost of hardware and software is lack of skills for professionals' interaction (MS = 3.88). There are different types of people, introverts and extroverts; those that like to share information and those that will keep to themselves. These characters have an effect on the usage of ICT, either to embrace it or not. The group of people that will not want to share information may not see the need to employ ICT in this management.

The least factors relative to challenges regarding ICT usage for project management is high cost of hardware/software (MS = 3.15). The initial cost of purchasing and installation is high and require an organization that is financially buoyant to venture on. Despite this the rating according to respondent is contrary. This implies that there may be other issues relative to this. Next is fear of mass job losses in the industry (MS = 3.24). It is viewed that the introduction of ICT usage for project management will create job losses to staff as a result of the capability of

Table 2 Challenges regarding the use of ICT in project management

Challenges	Mean score	Ranking
Insufficient/erratic power supply	4.33	1
High cost of hardware/software	4.17	2
Lack of skill for professional's interaction	3.88	3
Fear of virus attack	3.70	4
Inadequate ICT content in construction	3.49	5
Low return on investment in ICT	3.48	6
Influence of competition (less)	3.47	7
Lack of management desire and appreciation of ICT	3.43	8
Fear of ICT making professional redundant	3.36	9
Fear of mass job losses in the industry	3.24	10
High cost of hardware/software	3.15	11

computer to perform task of more than two or more professional more quickly, faster and neatly, and next to fear of mass job loss is fear of ICT making professionals redundant (3.36). Organizations are looking for better ways of doing job with high productivity. ICT is a mean to perform job faster, with accuracy and neatly. The use of ICT could lead the professionals being redundant.

Based on the rating of respondents, it is of note that all factors influence the usage of ICT device for project management have mean scores greater than 3, indicating moderately high influence. Table 3 presents factors influencing the usage of ICT device for project management practices. Top among the factors rated is individual's unethical behaviour (MS = 3.64). This is about the work, bribery, falsification and so on. With ICT these are eliminated and stands as an advantage to its use. Supplanting (MS = 3.63), with the use of ICT for management of construction. Project processes, the attack, enchantment and charming relative to job prospecting will be eliminated. The least factor influencing the use of ICT for project management. Practices is bureaucracy (MS = 3.04). Bureaucracies which are bottle necks is relative to the administration work needed to be performed that are problematic to the effective delivery of project are reduced drastically with the usage of ICT for project practices.

All factors of benefits of ICT on project management practices have MSs greater than 2.5 and indicates moderate to high benefits. Table 4 reveals respondents rating of benefit of ICT on project management practices. The factor with the most rating is reduces time for data processing and communication of information (MS = 4.21) ICT is said to be fast, safe and neat in work production. The reason is for its rating. Next to reduces time for data processing and community information is enhance efficiency through improved coordination (MS = 3.85). The use of ICT enhances efficiency that is much can be done in a short space of time, also having the capability of ease coordination of task, labour and equipment. This can lead to cost savings to the organisation. Next to enhance efficiency through improved coordination is facilitate decision making (MS = 3.69). Stemming from the enormous storage of information and quick rate of retrieval of information, this assists in quick decision making without have to disrupt the speed of execution of work, thereby eliminating delays on project, coupled with the advantage of accessibility of consultant. Performance of work is fast, neat, to avoid errors depending on accuracy of data captured.

Table 3 The factors influencing the usage of ICT device for project management practices

Factor	Mean score	Ranking
Individual's unethical behaviour	3.64	1
Supplanting	3.62	2
Organization unethical attitude	3.25	3
Software application	3.16	4
Mismanagement of funds and resources	3.05	5
Bureaucracy	3.04	6

Table 4 Benefits of ICT on project management practices

Benefits	Mean score	Ranking
Reduces time for data processing & communicating information	4.21	1
Enhance efficiency through improved coordination	3.84	2
Facilitate decision making	3.69	3
Savings in operating cost	3.64	4
Enhance construction productivity	3.63	5
Improve public image of building contractors	3.62	6
Improves documentation presentation	3.54	7
Speedy exchange of vital information	3.36	8
Make professional job easier	3.31	9
It aids transparency and accountability	3.27	10
Improve communication for effective decision making and coordination	3.19	11
Reduces direct procurement costs	2.84	12

The factor rated least regarding benefits of ICT to project management practices is reduces direct procurement costs (MS = 2.84). The cost of advertisements in the dealings is eliminated. Likewise, the cost of printing and postage by contractors are eliminated, though this seems to be negligible. Next to reduces direct procurement cost is improving communication for effective decision making and coordination (MS = 3.19). It is a fact that ICT usage in any discipline improves communication for effective decision making due to access to storage of information relatively. Next factors are, it aids transparency and accountability (MS = 3.27). The ability of ICT to store information for a very long time without it being damage and its feature of accessibility makes ICT usage transparent and accountable.

4 Conclusion

The following were conclusion drawn based on data analysis:

- Insufficient and erratic power supply constitutes a major challenge to the use of ICT for project management practices.
- The unethical behaviour that prevails in the traditional method of doing work has prompt a shift to ICT in the building construction industry.
- Time for data processing and communicating information and improving efficiency and coordination are the main benefits that could be derived from the use of ICT in the construction industry.

5 Recommendations

Based on the conclusion the following recommendations were made:

- Power supply should be made readily available to afford and sustain this development by government.
- Emphasis should be placed by professional's institute and the government on the use of ICT for construction management practices to eliminate the traditional method of doing work and regulate the practice.
- Training should be given by institute to professionals on a regular basis on construction project management software's application for management practice to mitigate it non usage.
- Some aspects of documentation relative to contract award should be made compulsory to be done using software packages to ensure usage of ICT by contractors and enforced.

Acknowledgements This work is supported by the Durban University of technology. The author also gratefully acknowledges the helpful comments and suggestions of the reviewers, which have improved the presentation.

The publication of this work was supported by Durban University of Technology in South Africa.

References

1. Adeyemi AY, Oladapo AA, Akindele O (2005) Balancing globalisation, localisation and the sustainable development equation in the Nigerian Construction Industry. In: Proceedings of the 3rd postgraduate conference, Johannesburg, South Africa, pp 289–302
2. Adriaanse A, Voordijk H, Dewulf G (2010) The use of interorganisational ICT in United States construction projects. Autom Constr 19:73–83
3. Adriaanse A, Voordijk H (2005) Interorganizational communication and ICT in construction projects: a review using metatriangulation. Constr Innov 5(3):159–177
4. Alutu OE (2007) Unethical practices in Nigerian Construction Industry: prospective engineers' viewpoint. J Prof Issues Eng Educ Pract, pp 84–88
5. Ashworth A, Hogg K (2002) Willis's practice and procedure for the quantity surveyor, 11th edn. Blackwell Science, Oxford
6. Bäckblom M, Ruohtula A, Björk BC (2003) Use of document management systems—a case study of the finish construction industry. IT Con. 8:367–380. http://www.itcon.org/2003/26/
7. Bowden S, Thorpe A (2002) Mobile communications for on-site collaboration. Civ Eng 150 (2):4–14
8. Dainty A, Moore D, Murray M (2006) Communication in construction: theory and practice. Taylor & Francis, Milton Park

9. Isikdag U, Underwood J, Kuruoglu M, Goulding J, Acikalin U (2009) Construction informatics in turkey: strategic role of ICT and future research directions. J Inf Technol Constr 14(47):412–428
10. Kivrak S, Arslan G, Cagatay O (2010) Information technology usage impacts on construction projects' success
11. Ng ST, Chow L (2004) Evaluating engineering consultants' general capability during the process—a Hong Kong study. Eng Constr Architectural Manage 11(3):150–158
12. Sun M, Howard R (2004) Understanding I.T. in construction. Spon Press, London

Analysis of Changes in Perception of Organizations Quality Maturity

Anna M. Olszewska

1 Introduction

Quality maturity is a broad topic. It can be referred to, while talking about every living being, including humans, as well as social phenomena or human creations. With regard to the letter, quality maturity from the point of process, project or organisation is discussed [1]. Regardless of whether quality maturity of process, project or whole organisation is taken into consideration, it is unalterably a concept difficult to define in unequivocal terms. It can represent both the status of certain organisation and indicate its trend of changes. This view is included in the definition of maturity by Looy et al. who identify maturity as a measure, which evaluates how excellent are processes and results able to become [2]. Another modern definition describes maturity as a collection of stages, which indicate an anticipated, desirable, or logical course of action, that shapes the state from beginner level to full maturity [3].

The concept of maturity was introduced in the 1970s by Ph. B. Cosby. Then the term related to the level of organisation development evaluation (QMMG—Quality Management Maturity Grid) subsumed under six aspects: management understanding of quality, quality organization status, problem handling, cost of quality, quality improvement action and summary of company quality posture [4]. Comparison of indicated then stages has been presented in Table 1.

In the 1990s CMM (Capability Maturity Model) model was developed, which was later expanded to CMMI (Capability Maturity Model Integration). The first one included the process of software development assessment and differentiated five stages of maturity presented in Table 2.

A. M. Olszewska (✉)
Faculty of Management, Bialystok University of Technology, Bialystok, Poland
e-mail: a.olszewska@pb.edu.pl

S. Şahin (ed.), *8th International Conference on Engineering, Project, and Product Management (EPPM 2017)*, Lecture Notes in Mechanical Engineering,
https://doi.org/10.1007/978-3-319-74123-9_20

Table 1 Levels of maturity in QMMG

Stage I: uncertainty	Stage II: awaking	Stage III: enlightenment	Stage IV: wisdom	Stage V: certainty
We don't know why we have problems with quality	Is it absolutely necessary to always have problems with quality	Through management commitment and quality improvement we are identifying and resolving our problems	Defect prevention is a routine part of our operation	We know why we do not have problems with quality

Source Cosby [4], Maier et al. [1]

Table 2 Levels of maturity in CMM

Level I: initial	Level II: repeatable	Level III: defined	Level IV: managed	Level V: optimizing
Processes are not defined	Basic departmental processes are defined and are repeated more or less consistently	The organization, as a whole, knows how all their processes work together and can perform them consistently	Managers consistently capture data on their processes and use that data to keep processes on track	Managers and team members continuously work to improve their processes

Source Brocke and Rosemann [12]

Expansion of the CMM model to CMMI was interlinked with introduction of project teams' management. That model became a formula to create models of maturity in other aspects [5].

In addition to abovementioned models, in literature there can be found many other ones, including for example: Business Process Maturity Model (BPMM), Barkeley PM process maturity model, Knowledge Management Maturity (KMM), Information Process Maturity Model (IPMM), Effective Teamwork Matrix, Communication Gird Method (CGM), Process and Enterprise Maturity Model (PEMM), Design safety Capability Maturity Model (DCMM) [1, 6–8].

The view on maturity, from the beginning, has been connected with quality. Quality can be seen as 'fitness for use' [9], 'fitness for purpose' [9] or 'as conformance to requirements, not as goodness' [10]. Similar to maturity, its understanding is different depending on the context of usage. In reference to management of organisation is defined as conscious, deliberate meeting customer expectations.

Therefore, it is interesting, if Cosby's organisations maturity assessment from a quality perspective is still a field of researchers' interest and how the context of discussions in this area conducted by the authors has been changing. The identification in which context the quality maturity has mainly been developed over the last years was chosen as a purpose of this article. Years 2000–2016 were chosen for the analysed period. Obtained results have been virtualised using a tool VOSviewer. This software gives a possibility to highlight relations between multi-element data sets by creating various kind of maps, emphasizing different aspects of data presentation [11]. In this article the network visualization was used.

2 Preliminary Analysis

The first step of analysis was the collection of information regarding the number of publications relating to the issue of maturity. The results have been reduced to the data from 2000 to 2016, including four types of publications: article, conference paper, book and book chapter. In the first step, the Author used two bases: Web of Science and Scopus, however, due to the larger number of documents available in Scopus, further analyses covered Scopus only. Entering the phrase 'maturity' resulted in over 60,000 results. Even more publications were connected with the word 'quality'. In this instance, the number of publications reached nearly two million. Given such an extensive data set, reducing parameter of appearance of two phrases simultaneously has been applied.

The results of the analysis of connections of 'quality' and 'maturity' showed nearly 9000 positions. The map made on this basis (Fig. 1), presented the authors with five areas, in which given notions have been mentioned. With this information, clusters related to many fields were made, including for example human, plant and animal development. Only one of the clusters was connected with maturity and quality in the context of organisations. Therefore, the criteria of the next search were made in regard to the mentioned context. Used choice criteria have been noted in Table 3.

As the result of mentioned criteria application, the number of publications totalled 2907. The number of publications in years 2000–2016 from both analysed data bases has been presented in Fig. 2.

While analysing Fig. 2, since the 2000–2013, a gradual growth of interest in issues connecting the aspect of maturity with quality can be noted. After the year 2013, a minor drop in interest took place and plateaued at approximate level. It can be seen that this notion is still present in literature.

Further analysis was conducted based on Scopus database. The areas of education, where issues of quality and maturity were brought up, have been analysed.

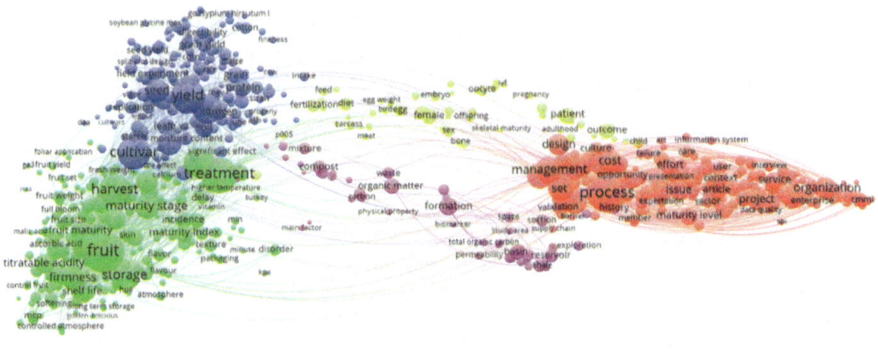

Fig. 1 Map of research trends based on co-occurrence of words 'maturity' and 'quality' in publications from Scopus database in the years 2000–2016. *Source* author's elaboration on the basis of (www.scopus.com, 18.05.2017)

Table 3 Selection criteria applied to scopus database

(TITLE-ABS-KEY (maturity OR matureness) AND TITLE-ABS-KEY (organization OR organisation OR business OR enterprise OR firm OR economy OR industry OR company OR management) AND TITLE-ABS-KEY (quality)) AND PUBYEAR >1999 AND PUBYEAR < 2017 AND (LIMIT-TO (DOCTYPE , "ar") OR LIMIT-TO (DOCTYPE , "cp") OR LIMIT-TO (DOCTYPE , "ch") OR LIMIT-TO (DOCTYPE , "bk"))

Source: author's elaboration.

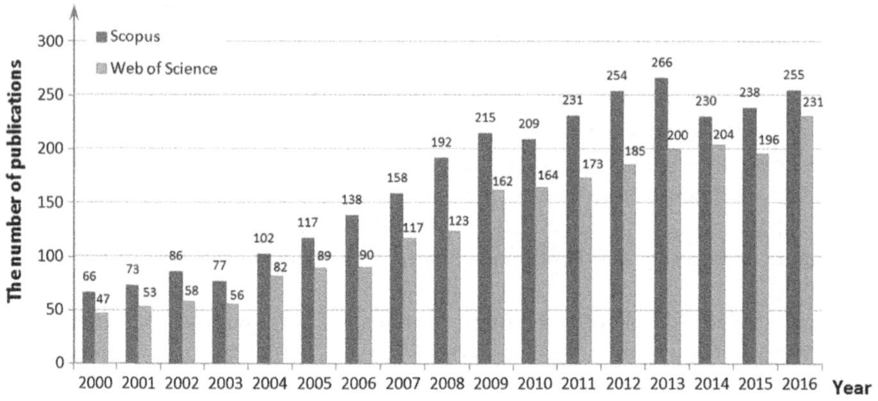

Fig. 2 The number of publications in data bases: Web of Science and Scopus connected with the notion of 'maturity' and 'quality'. *Source* author's elaboration on the basis of (webofknowledge.com, www.scopus.com, 18.05.2017)

The results have been presented in Fig. 3, where the size of the names of certain areas corresponds to an occurrence frequency (i.e. larger font—more frequent).

The main fields, in which both the notion of 'quality' and 'maturity' appeared alongside were: Computer Science, Agricultural and Biological Sciences, Engineering and Business, Management and Accounting. Figure 4 shows how the number of publications related to the issues of quality and maturity has been changing.

Fig. 3 Areas of publications connected with the notion of 'maturity' and 'quality' from Scopus database in the years 2000–2016. *Source* author's elaboration on the basis of (www.scopus.com.pl, 20.04.2017)

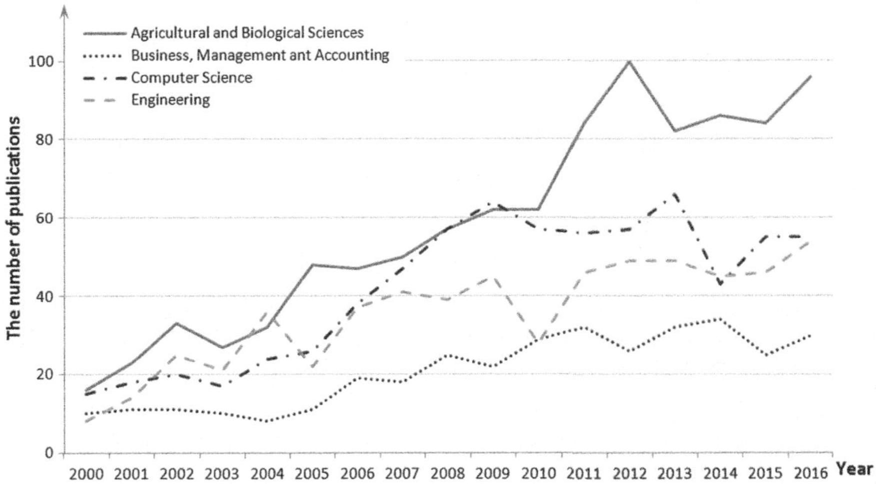

Fig. 4 The number of publications connected with the notion of 'quality' and 'maturity' in selected areas from Scopus database in the years 2000–2016. *Source* author's elaboration on the basis of (www.scopus.com, 18.05.2017)

In each of examined fields, in which documents referring to maturity and quality appear most often, graduate increase is noted, which at the end of considered period slightly abates. It should be stated that in all of analysed areas, the number of inspected publications has raised several-fold. The least significant increase appears to happen in the field of Business, Management and Accounting (total amount of publications rose by 200%, what gave an average increase of 7.1% per year). The most substantial growth in number of publications related to quality and maturity transpired in the case of the area of Engineering and it amounted to 575% (average of 12.5% per year).

3 Enhanced Analysis

For further analysis, the publications retrieved using the criteria specified in Table 3 have been used. Using this data with the help of the tool VOSviewer, an analysis of co-appearance of the keywords pointed by the authors was made. Considering the large amount of keywords, only those which have appeared at least a dozen or more times have been chosen for the construction of the map. The map is shown in Fig. 5.

Fig. 5 Map of research trends based on co-occurrence of the author' keywords with notion of 'quality' and 'maturity' in publications from Scopus database in the years 2000–2016. *Source* author's elaboration on the basis of (www.scopus.com, 18.05.2017)

At presented map (Fig. 5) four different clusters can be noted. The factsheet of the specific words belonging to certain clusters is shown in a Table 4.

First of the highlighted areas covers the issues related to the maturity in the context of development and cultivation of crops, despite the usage of the references regarding organisations. Publications of that type do not fall under the strand considered in the context of quality maturity evaluation. The other group is composed of publications related to the technical development of production processes and their modelling. This area as well, although it is much closer to quality maturity evaluation, doesn't include the issue described in this paper. The most important area in the author's assessment from the perspective of assumed goal is third cluster. Those are the issues connected with described at the beginning types of evaluation of the degree, level of process maturity, as well as related to quality assessment in the context of TQM, or knowledge management. The last cluster, on the other hand, covered publications connected with software maturity evaluation. Those issues also do not have to be analysed in the context of organisation quality maturity evaluation.

With such a selection criterion, the results of the previous step have been reduced to the publications, in which the keywords noted in third cluster appeared. That way, the number of publications, which have to be further analysed was greatly reduced. The order of publications should be determined by the number of quotations or time of development. Still it should be kept in mind that given procedure could lead to omitting worthy, from the point of view of quality maturity, publications. Therefore, while analysing given this way articles or books, the other positions, which are referenced in them, should be also taken into consideration.

Table 4 Cluster indentified through analysis

Keyword	Occurrences	Co-occurrence
Cluster 1		
Fruit quality	67	38
Firmness	47	52
Ripening	38	42
Compost	37	12
Ethylene	28	31
Storage	26	38
Postharvest	25	23
Soluble solids	19	18
Cultivar	15	15
Maturity index	15	10
Prunus avium	15	14
Cluster 2		
Yield	35	32
Model	34	33
Management	20	9
Evaluation	19	9
Growth	19	11
Nitrogen	19	14
Process	18	30
SME	18	19
Innovation	17	13
Cotton	15	8
Cluster 3		
Maturity model	89	43
CMMI	78	70
CMM	68	60
Quality management	60	30
Process improvement	33	27
Knowledge management	31	17
TQM	26	11
Information quality	19	8
Quality assurance	17	14
Maturity level	15	11
Cluster 4		
Software quality	53	61
Software process improvement	52	52
Project management	36	42
Software engineering	33	41
Software process	18	24
Risk management	17	14

Source author's elaboration on the basis of (www.scopus.com, 18.05.2017)

4 Conclusion

Currently it is not possible to conduct the analysis without including in this procedure available bibliographic databases. However, the usage of that vast information sources lead to the overflow of results. The analysis of thousands of records is not possible to conduct by a single person, but needs a whole team of people. Therefore, needed is the search for methods, which can limit, received from bases, data to the scope which is possible to analyse by the individual. An example of such an analysis is method of systematic literature overview presented above, which was used to point a bibliographic resources connected with the notion of organisations quality maturity.

It should be emphasized, that the presented procedure can greatly help researchers, interested in the issue of quality maturity, in getting to those publications which describe this matter in the context of organisations evaluation. The usage of only 'quality maturity' in the analysis of publications from the bibliographic database like Scopus or Web of Science, can lead to excessive reduction. Introduction of such a term led to indication of only 57 records in Scopus base and 43 in Web of Science. However, the usage of 'maturity' and 'quality' separately lead to too vast a collection (nearly 10,000 records). Therefore, it is necessary to greatly reduce the amount of data. Presented above procedure not only made the number of publication reduction possible, but also indicated how this issue was being developed over the last couple of years. Worth emphasizing is the fact, that it is all only a step preceding the proper analysis.

Acknowledgements The research was conducted within S/WZ/1/2017 project and was financed from Ministry of Science and Higher Education funds.

References

1. Maier AM, Moultrie J, Clarkson P J (2012) Assessing organizational capabilities: reviewing and guiding the development of maturity grids. IEEE Trans Eng Manage 59(1):138–159. https://doi.org/10.1109/tem.20; https://doi.org/10.1109/tem.20
2. Van Looy A, De Backer M, Poels G (2014) A conceptual framework and classification of capability areas for business process maturity. Enterp Inf Syst 8(2):188–224. https://doi.org/10.1080/17517575.2012.688222
3. Domingues P, Sampaio P, Arezes PM (2016) Integrated management systems assessment: a maturity model proposal. J Clean Prod 124:164–174. https://doi.org/10.1016/j.jclepro.2016.02.103
4. Cosby PB (1979) Quality is free. MacGraw-Hill, New York
5. Skrzypek E (2014) Dojrzałość jakościowa organizacji w świetle teorii i doświadczeń organizacji [Organizations' qualitative maturity in theory and practice]. Marketing i Rynek 5:579–588
6. Łukasiński W (2016) Dojrzałość organizacji zarządzanej projakościowo. Polskie Wydawnictwo Ekonomiczne, Warszawa

7. Strutt JE, Sharp JV, Terry E, Miles R (2006) Capability maturity models for offshore organisational management. Environ Int 32(8):1094–1105. https://doi.org/10.1016/j.envint.2006.06.016
8. Tarhan A, Turetken O, Reijers HA (2016) Business process maturity models: a systematic literature review. Inf Softw Technol 75:122–134. https://doi.org/10.1016/j.infsof.2016.01.010
9. Liepiņa R, Lapiņa I, Mazais J (2014) Contemporary issues of quality management: relationship between conformity assessment and quality management, contemporary issues in business, management and education 2013. Procedia Soc Behav Sci 110:627–637. https://doi.org/10.1016/j.sbspro.2013.12.907
10. Johnson K (2001) Philip B. Crosby's mark on quality. Qual Prog 34(10):25–30
11. Gudanowska AE (2015) Tworzenie mapy wiedzy opartej na tematyce projektów badawczo - rozwojowych na przykładzie województwa podlaskiego [Creating knowledge maps based on the themes of R&D projects on the example of the Podlaskie region]. Econ Manage 7(1): 257–270
12. Brocke J, Rosemann M (2010) Handbook on business process management 1: introduction, methods and information systems. Springer, Berlin. https://doi.org/10.1007/978-3-642-00416-2

Perceived Impacts of Industry 4.0 on Manufacturing Industry and Its Workforce: Case of Germany

Markus Haeffner and Kriengsak Panuwatwanich

1 Introduction

Nowadays, a large variety of goods can be ordered online, unlimited information has become available and all different kinds of communication happens through the use of the internet and social media. The development and rise of smart phones has further improved the availability of these as well as other internet services, and has facilitated the development of a whole new business segment—applications. This development opened up a whole new market, leading to the internet becoming an important part of people's everyday life.

However, everyday life is not the only area where the computer and internet revolution, also known as the third industrial revolution, has dramatically changed. It has also had a radical impact on the world of work comparable in scale to the First Industrial Revolution [9]. This has brought about a radical transformation with more IT infrastructure and networks being integrated into all areas of businesses and companies [4]. Nowadays, most of the work is carried out with the help of personal computers, communication is done via E-mail or other services enabled by the internet, whereas goods are manufactured by programmable machines in production lines that are coordinated through integrated networks.

M. Haeffner
Kuka Robotics, Augsburg, Bavaria, Germany
e-mail: markus.haeffner@kuka.com

K. Panuwatwanich (✉)
School of Civil Engineering and Technology, Sirindhorn International Institute of Technology, Thammasat University, Bangkok, Thailand
e-mail: kriengsak@siit.tu.ac.th

K. Panuwatwanich
Griffith School of Engineering, Griffith University, Gold Coast, QLD, Australia

S. Şahin (ed.), *8th International Conference on Engineering, Project, and Product Management (EPPM 2017)*, Lecture Notes in Mechanical Engineering,
https://doi.org/10.1007/978-3-319-74123-9_21

The internet still has a far-reaching potential despite having already had a significant impact on both everyday-life as well as the business world. When combining all the opportunities the internet entails with some other factors, a potential of transforming both business and private life in a way that can only be compared to the first industrial revolutions is being created [6]. What is being referred to is the potential fourth industrial revolution that can be related to with terms like "Industry 4.0", "The Internet of Things", "Industrial Internet", "The internet of things, services, data and people" or simply "The internet of everything" [5]. In general, this emerging trend can be described as "the increasing digitalisation and the interconnection of products, value chains and business models" or simply "the ubiquitous connection of people, things and machines" [12].

However, the terms "Industry 4.0" within Europe and "Industrial Internet" in the US appear in a less general but mainly manufacturing-related context. They are referred to as the digital transformation of manufacturing, driven by the development of modern information technologies, more specifically "cyber-physical production systems" (CPPS). In other words, IT will be linked with mechanical and electronic components, leading to the formation of an independent "social system" [5]. Parts of this development can already be seen in current manufacturing. However, *Industry 4.0* is meant to bring this linkage to a whole new level and form independent production systems that communicate across the entire lifecycle of the products; from the suppliers to the retailers, customers and the after-sales services [6]. Taking this into account, *Industry 4.0* entails the potential to dramatically transform the manufacturing-world as well as the skills and knowledge required from the respective workforce over the next decades, leading to a potential fourth industrial revolution.

Therefore, with an exploratory approach, the main objective of this study is to examine how *Industry 4.0* will change manufacturing and affect the skills and knowledge required from the manufacturing workforce over the next decades.

2 Background of Industry 4.0

There are many terms such as "Industry 4.0", "The Internet of Things", "Industrial Internet", "The internet of things, services, data and people" or "The internet of everything" [5, pp. 3, 12] used to refer to the new era of internet-driven business and society. These terms all describe a similar development currently underway. The terms "Industrial Internet" is more broadly used in the US whereas the term "Industry 4.0" is more widespread within Europe and especially in Germany to describe the introduction of "cyber-physical systems" (CPS) into industrial production systems, leading to "cyber-physical production systems" (CPPS). The other terms however are more generally used, also outside the manufacturing and production sector and describe the general trend towards networks and the use of the internet in everyday life [6]. Therefore, *Industry 4.0* is used in this study for describing the development within the production process, whereas the other terms are used to describe the effects of this development in a general context. As the main facilitator for the implementation of *Industry 4.0* in the manufacturing process is CPS or CPPS respectively, the following paragraphs describe what can be understood with these terms.

Generally speaking, and regardless of all the machinery and devices involved in the actual production process, it can be stated that a CPS consists of two central functional components: firstly, "the advanced connectivity that ensures real-time data acquisition from the physical world and information feedback from the cyber space" and secondly the "intelligent data management, analytics and computational capability that constructs the cyber space" [10]. Nevertheless, besides those two core functions mentioned above, Lee et al. [10] propose a five-layer architecture for the CPS, related to as "5 level CPS structure" or "5C architecture" [10, p. 19]. The functions of the five layers are described in the following.

- Level 1—Smart Connection: This level acquires reliable and accurate data from machines and components by directly measuring through sensors or obtained from "controller or enterprise manufacturing systems".
- Level 2—Data-to-information conversion: Within this level, the collected data needs to be transferred into "meaningful information" by specific algorithms that have been developed over the past years, especially for the purpose of health management and prognostics.
- Level 3—Cyber: This level represents the "central information hub" within the CPS architecture. It is being fed with all kinds of information from every machine and component within the network that is connected to the system.
- Level 4—Cognition: This level is all about presenting the gathered information and knowledge about the system to expert users in an appropriate way
- Level 5—Configuration: This final level is the supervisory control level of the whole architecture and enables the machines to be "self-adaptive" and "self-configure". It also represents the feedback from the cyber world.

This five-layer model by Lee et al. [10] provides a simple and easy to understand architecture for developing and deploying a CPS structure into the manufacturing industry, leading to a "cyber-physical production system" (CPPS). However, it is not to be seen as a model of five individual levels but rather of five levels that are strongly connected to each other through interfaces, which in turn enables them to communicate and exchange data and information. Once the machines are connected to the cyber-level infrastructure they can exchange data and information through a machine-cyber interface that can be considered similar to social networks as outlined in the introduction.

Overall, this models show that *Industry 4.0* or the introduction of CPS into manufacturing has the potential of transforming the entire structure of the production process and the whole company respectively. However, this will also lead to tremendous changes throughout the entire supply chain as CPSs from different companies will be able to communicate to each other and handle tasks independently [2]. This will lead to the fact that a lot of work will be done automatically by machines and humans will be less involved in the actual production process. Hence, besides changing the skills and knowledge required from the respective workforce, it is also likely that this development will lead to an overall reduction in labour working within the actual production process [11]. It will also enable new business opportunities and even potential new business models [14]. However, replacing humans by machines also means that the dependencies on technologies will strongly increase. By creating independent and intelligent production processes, control and responsibility is handed over to machines, networks and technologies [12], rendering minimal human involvement.

3 Methodology

A literature review pertaining to the impacts of *Industry 4.0* on the manufacturing industry and its respective workforce was firstly conducted. There were various studies that generally address this future development. However, most of them neither specifically outline what this means for manufacturing nor expound on what the impacts and changes for their respective workforce are. Particularly for the effects this development might have on the employees, there was a literature gap that needs addressing.

Given the review of literature pointed to the above limitation, semi-structured interviews with experts in the field were conducted in an exploratory manner to address such knowledge gap. The interviewees were chosen by using convenience sampling. This method refers to a sample that "is simply available to the researcher by virtue of its accessibility" [3, p. 190]. This can be considered a very convenient method; however the results most likely are not representative. Nevertheless, according to Bryman and Bell [3], this method can be considered when there is a chance of gathering data and information that represents a too good opportunity to miss. Hence, the gathered data will not allow definite findings because it cannot be generalised but it could potentially provide a springboard for further research to be conducted or links to already existing findings [3]. As in this case, the interviews will only be used to acknowledge, disprove or supplement the literature findings, this non-probability sampling method can be considered appropriate. In this study, the following four participants located in Germany (all anonymised) were contacted and interviewed:

- Interviewee A, head of corporate development, marketing, enterprise communication as well as investor relations in a large international engineering services company.
- Interviewee B, key account manager within the engineering department of a large international recruitment and HR services company.
- Interviewee C, working for the department for bio-mechatronic systems within the largest applied research organisation in Europe.

- Interviewee D, Head of the School of Engineering Management of a University of Applied Science.

These interviewees were chosen as not only had they all been following the *Industry 4.0* development over the past years but also had to deal with it in their capacity. In addition, their experience with the topic is based on multiple cases, as they work with and provide services for various companies that deal with *Industry 4.0*. To maintain their anonymity, the interviewees will be referenced as A, B, C or D respectively, throughout the reporting of the findings presented in the following section.

4 Findings

4.1 Manufacturing Process

Through the implementation of "cyber-physical systems" (CPS) into the manufacturing industry, the production process will become more integrated. On the one hand, vertically through all levels of the company, and on the other, horizontally throughout the entire supply chain. Hence, different departments, organisational levels as well as all organisations included in the value chain or supply chain respectively will be linked stronger together and communication will be partly automated and made easier (Interviewees A, B, C and D; [1]). As robot and machine support will increase or even mostly replace manual labour, they will become part of this communication process and not only communicate to each other but also with humans. Hence, human-machine and machine-machine interfaces will be required and implemented into the production process. This will enable a manufacturing that is more autonomous and independent from humans, not only within one organisation but across the entire value chain or supply chain respectively, leading to smart factories and intelligent production systems. Hence, production will move away from rigid and predefined processes and become more flexible as well as agile, enabling dynamic configurations (Interviewees A, B and C; [6, 12]). Therefore, this way of producing goods will enable manufacturing companies to react to the stronger demand for higher levels of customisation (Interviewee A; [8, 14]). Finally, this will not only enable higher degrees of customisation in a cost effective way but also lead to higher efficiency in terms of other resources such as materials, waste and energy consumption.

4.2 Manufacturing Workforce

Through the implementation of *Industry 4.0* into manufacturing, its workforce will see an increase in the required skills and knowledge (Interviewees A, B, C and D; [15]).

As most simple repetitive manual tasks will be replaced by robots and machines, the workforce will take over more management-related tasks and hence need stronger personal skills such as communication, coordination and other softs skills in order to take over more responsibility and decision-making (Interviewees A, B and C; [7, 11, 15]). In addition, software and programming skills will not only be required from the engineers but also from the typical "blue-collar" workers (Interviewees A and B; [15]). Besides having a good understanding for software and hardware, the typical manufacturing worker will need to be able to fix simple programming or coding errors (Interviewees A and B). On an engineering level, in addition to software and programming experts, more specialists with interfaces as well as data and information skills will be required. Additionally, there will be an increased demand in interdisciplinary skills, especially the combination of mechanical engineering knowledge and programming skills (Interviewees A, B, C and D; [15, 13]). It will not be enough to have one engineer and one software expert. The person for instance programming a machine or robot will need to understand what role it plays within the production process and hence need some mechanical engineering knowledge (Interviewees A, B and C). Therefore, not only the internal training will need to be adjusted but also the educational sector will need to react in order to supply the changing demand for specialists on the market (Interviewees A, B, C and D; [15]). Overall, the manufacturing workforce of the future will not only be required to provide these particular skills and knowledge, but also need to be open towards those new technologies and show trust in them (Interviewees A, B and C; [15]). Further, a willingness for continuous learning and training will be needed to keep up with the technological developments (Interviewees A and B; [15]). Hence, the *Industry 4.0* development will not only represent opportunities to the manufacturing industry, but also challenges, especially when it comes to finding the appropriate experts to make these developments happen. The above findings can be summarized in Fig. 1.

5 Discussion

5.1 Machines Replacing Humans

The *Industry 4.0* development will entail a further replacement of humans by machines, as could already be seen over the past decades. Besides the fact that work will become less tiring and more ergonomic for the factory worker, it also means that less people will be involved in the production process [12]. Hence, a reduction in the overall manufacturing workforce is likely. But will this lead to higher rates of unemployment? When looking back at the third industrial revolution, a fear of replacement of humans within the industry as well as rising levels of unemployment has been present. However, these technological advancements have not only eliminated jobs but also opened up new areas and opportunities. For instance,

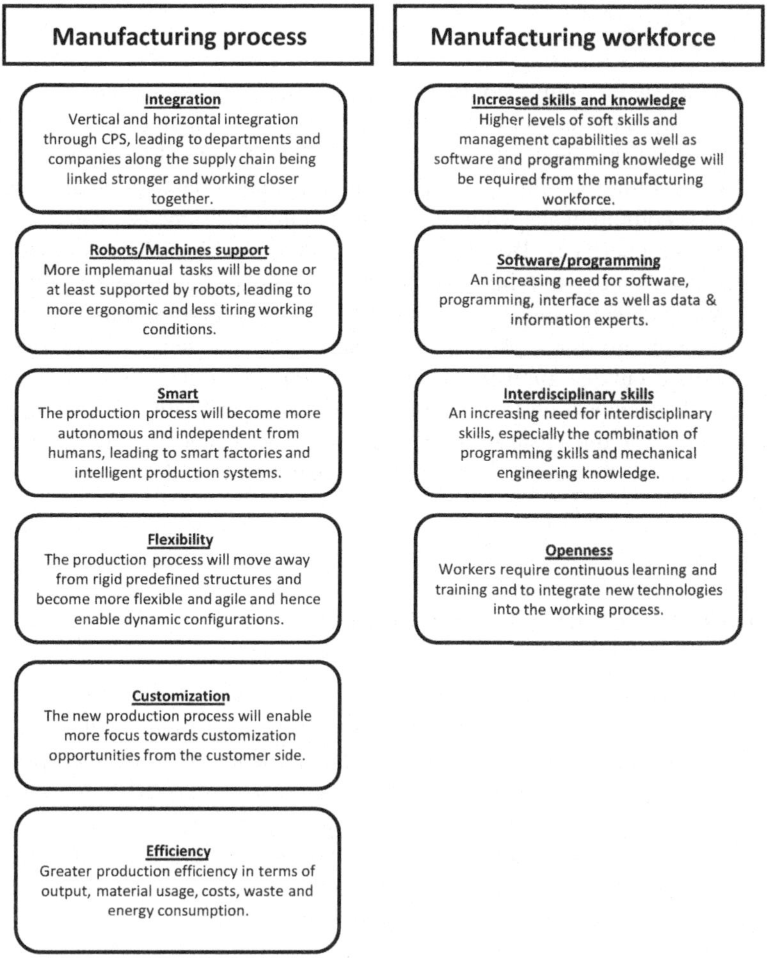

Fig. 1 Summary of perceived Industry 4.0 impacts

the introduction of the internet has seen the emergence of entire new markets such as online shops, the applications industry and many more. Although some jobs have been eliminated, many others have also been created. A similar trend might be seen with *Industry 4.0*. With this development, new kinds of business models and new ways of cooperation are foreseeable, and new kinds of services will be demanded both from end users and companies. Even though the *Industry 4.0* development is likely to result in a reduction in the overall manufacturing workforce, especially in the "blue-collar" sector, it is not guaranteed that this will lead to higher rates of unemployment. Moreover, a shift towards other areas as well as away from the secondary and tertiary sectors and more towards a fourth, information based sector, is more likely.

5.2 Social Aspects

People usually work with people in order to achieve certain goals and aims. *Industry 4.0* developments will most likely entail a change towards humans working with machines and robots. Instead of talking to other humans about issues, workers of the future would solve problems in cooperation with smart machines. They would communicate with each other through screens, voice or other human-machine interfaces. Hence, robots and other intelligent devices could even become socially integrated into the production process and potentially replace humans as co-workers. In an everyday-life context people would become even more dependent on smart phones and other high-tech devices. Not only in a working context but also in everyday-life, the *Industry 4.0* development could lead to a kind of de-socialisation that needs to be looked at critically. As *Industry 4.0* is all about digitalisation, customisation and integrated networks, this development has the potential to cause even higher dependency and addiction to these technologies.

5.3 Industry 4.0 in a Global Context

It should be noted that *Industry 4.0* is more likely to be valid for the western and so-called developed countries such as Germany, Great Britain or the US. These countries are strong enough to fund this development and bring the appropriate education for the future workforce. In addition, as *Industry 4.0* is most likely to reduce the manual labour required in the production process, it might be a reason for corporations to bring production back to their home countries such as Germany and the US. Hence, the development and implementation of *Industry 4.0* might eventually lead to an increase in unemployment in countries such as China and India due to factory resettlements.

6 Conclusion

6.1 Manufacturing Process Implications

Through the implementation of "Cyber-physical systems" (CPS) in the manufacturing industry, the production process will become more integrated. Different departments, organisational levels as well as all organisations included in the value chain will be linked stronger together and communication will be partly automated and made easier. As robots and machines support will increase or even replace manual labour, they will become part of this communication process and not only communicate to each other but also with humans. Hence, human-machine and machine-machine interfaces will be required and implemented into the production process. This will enable a manufacturing that is more autonomous and independent from humans, not only within one organisation but across the entire value chain. Dynamic configurations enabled by flexible and agile processes will enable manufacturing companies to react to the stronger demand for higher levels of customisation. Finally, this will not only enable higher degrees of customisation in a cost effective way but also lead to higher efficiency in terms of other resources such as materials, waste and energy consumption. However, this will also strongly affect the skills and knowledge required by the manufacturing workforce.

6.2 *Manufacturing Workforce Implications*

Through the implementation of *Industry 4.0* in manufacturing, its workforce will see an increase in the required skills and knowledge. As most simple repetitive manual tasks will be replaced by robots and machines, the workforce will take over more management-related tasks and hence need stronger personal skills such as communication and coordination in order to take over higher level responsibilities and make higher-level decisions. Software and programming skills will not only be required from the engineers but also from the typical blue-collar worker. At an engineering level, in addition to software and programming experts, more specialists with interfaces as well as data and information skills will be required, hence an increased demand in interdisciplinary skills. Therefore, not only the internal training will need to be adjusted but also the educational sector will need to react in order to supply the changing demand for specialists required by the market. The manufacturing workforce of the future will not only be required to provide these skills and knowledge, but also need to trust and be open towards those new technologies. A willingness for continuous learning and training will also be needed in order to keep up with the technological developments. The *Industry 4.0* development will represent not only opportunities but also challenges to the manufacturing workforce.

6.3 *Recommendations for Future Research*

Although *Industry 4.0* is a development that affects numerous areas and generally needs to be looked at from a very broad perspective, it has to be researched more specifically in regards to manufacturing companies. More specific effects of this recent development on this industry need to be further investigated. This could be done by conducting case studies, for example, on how the manufacturing process for one particular car manufacturer might change due to *Industry 4.0*. In addition, case studies like these could help to outline respective changes to the workforce requirements. Finally, future studies should focus on when exactly these changes will happen and to what degree. More research should specifically look at how the production process for various products is most likely to look like in 5, 10 or 20 years' time. In addition, the respective skill as well as knowledge requirements for the workforce could be evaluated. This would lead to more fact based and precise predictions which help guide all organisations, governments as well as institutions to best manage the *Industry 4.0* development.

References

1. Brettel M, Friederichsen N, Keller M, Rosenberg M (2014) How virtualization, decentralization and network building change the manufacturing landscape: an industry 4.0 perspective. Int J Mech Ind Sci Eng 8:37–44
2. Broy M (2010) Cyber-physical systems: innovation durch software intensive eingebettete Systeme, acatech-Deutsche Akademie der Technikwissenschaften
3. Bryman A, Bell E (2011) Business research methods, 3rd edn. Oxford University Press, Oxford
4. Communication Promoters Group of the Industry-Science Research Alliance (2013) Securing the future of German manufacturing industry: recommendation for implementing the strategic initiative INDUSTRIE 4.0. Final report of the Industrie 4.0
5. Deloitte (2015) Industry 4.0: challenges and solutions for the digital transformation and use of exponential technologies. Deloitee AG, Zurich, Switzerland
6. Drath R, Horch A (2014) Industrie 4.0: hit or hype? [industry forum]. IEEE Ind Electron M
7. Evans P, Annunziata M (2012) Industrial internet: pushing the boundaries of minds and machines. https://www.ge.com/docs/chapters/Industrial_Internet.pdf
8. Günthner WA, Ten Hompel M, Chisu R, Nettsträter A, Roidl M (2010) Internet der Dinge in der Intralogistik. Springer, Heidelberg
9. Kornwachs K (2007) Ergänzung und Verdrängung der Arbeit durch Technik—Eine Herausforder- ung für die Technikwissenschaften). In: ibid. (ed) Bedingungen und Triebkräfte technologischer Innovationen. Fraunhofer IRB Verlag, Stuttgart
10. Lee J, Bagheri B, Kao H (2015) A cyber-physical systems architecture for industry 4.0-based manufacturing systems. Manuf Lett 3:18–23
11. MHP (2014) Studie Industrie 4.0—Eine Standortbestimmung der Automobil- und Fertigungsindustrie. MHP, Reutlingen/Germany
12. PwC (2014) Industry 4.0: opportunities and challenges of the industrial internet
13. Ridgway K, Clegg CW, Williams DJ (2013) The factory of the future. The National Metals Technology Centre, University of Sheffield AMRC, Catcliffe
14. Spath D (2013) Produktionsarbeit der Zukunft-Industrie 4.0. Fraunhofer Verlag, Stuttgart
15. VDI (2015) Industry 4.0. A discussion of qualifications and skills in the factory of the future: a German and American perspective. VDI, Düsseldorf/Germany

Investigation of Roller Burnishing Process on the Mechanical Characteristics, and Micro-hardness of Al-4 wt% Cu Under Hot Work Conditions

Safwan M. A. Al-Qawabah, Nabeel Abu Shaban and Ahmad Al-Aboshi

1 Introduction

Roller Burnishing Process (RBP) considered as one of the main surface plastic deformation of metal used to enhance its surface quality and its dimensional accuracy by implementing some strain hardening. The plastic deformation resulted as the stress generated from the roller exceeds the yield stress of deformable materials, where this stress depend on certain factors namely; U.T.S, surface finish, ductility, and the radius and the geometrical shape of roll. The main reasons to use this process are its low cost, Simple Equipment, and the reduction in the surface roughness. The surface hardness that can this process used is up to 40 HRC [1]. It is essential to prepare the surface before performing this process i.e. the surface should be fine, cleaned and dimensional accurate, however this preparation is also required for operational needs such as; corrosion resistance, fritting wear resistance, fatigue life. Most of material removal processes namely; accurate turning, reaming, milling, and even grinding cannot give the required surface roughness, Therefore, to produce fine surfaces it required to use extra finishing operations like: honing, lapping, super finishing and burnishing [2, 3]. Most of the research work on roller burnishing process that already been published was investigate the effect of this process on the surface characteristics and surface hardness, where it was suggested by many investigators that an improvement in wear resistance can be achieved by burnishing process [4, 5]. As the stress applied by the roller exceeds the yield stress

S. M. A. Al-Qawabah (✉) · N. A. Shaban · A. Al-Aboshi
Mechanical Engineering Department, Al-Zaytoonah University of Jordan, P.O.Box 13720, Amman 11942, Jordan
e-mail: safwan.q@zuj.edu.jo

N. A. Shaban
e-mail: n.shaban@zuj.edu.jo

A. Al-Aboshi
e-mail: a.aboushi@zuj.edu.jo

S. Şahin (ed.), *8th International Conference on Engineering, Project, and Product Management (EPPM 2017)*, Lecture Notes in Mechanical Engineering,
https://doi.org/10.1007/978-3-319-74123-9_22

of the workpiece material the crests of the metallic surface will deform to fill the valleys [1]. Metallic parts surface properties are very important issue that it can hinder the wear and fatigue phenomenon failure [6]. Hassan investigated both number of tool passes and burnishing force on the wear resistance so to a certain limits the wear resistance of brass increases [7]. There are many advantages of the burnished surfaces compared to ground surfaces [7–10]. Other authors investigated the effect of other parameters as speed, depth of burnished layer, and burnishing force. It will established that there is an enhancement in the mechanical behavior after burnishing process [1, 11–15]. The main objective of this study is to investigate the effect of hot roller burnishing on the mechanical properties and microhardness of pure Al-4 wt% Cu.

2 Materials and Equipment and Experimental Procedures

2.1 Materials

A commercially pure Al (99.8%) of chemical composition by weight % is: 0.09 Fe, 0.05 Si, 4 Cu, 0.004 Mg, 0.004 Ti, 0.008 V, 0.005 Zn, 0.001 Mn, 0.005 Na, was used in this study.

2.2 Equipment

The main equipment that used through this study are namely; lathe machine type COLCHESTER TRIUMPH 2000, Microhardness tester model HWDM-3, Electrical furnace (Carbolite) of 1100 °C maximum temperature, and calibrated roller burnishing tool.

2.3 Experimental Procedures

A cylindrical aluminum copper alloy bars of 11.5 mm diameter and 75 mm long were used.

2.3.1 Microhardness Test

The microhardness test is performed at 300 gmf, by taking 8 reading then the average was calculated. The general microstructures of Al-4% Cu after burnishing conditions were determined by grinding, polishing and etching using an etchant

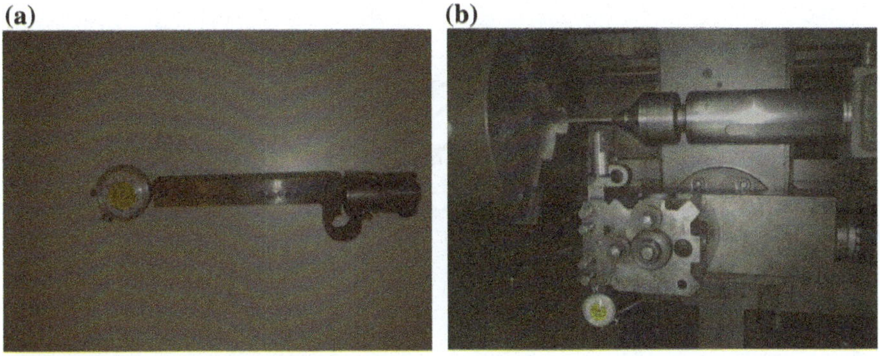

Fig. 1 Schematic photo of the **a** calibrated roller burnishing tool, **b** burnishing process

consists of 1.5% HCl acid, 2.5% HNO_3, 0.5% HF acid and 95.5% H_2O by weight, and then Photomicrographs were obtained at magnification of 200×. The hot burnishing process was performed at 200 °C. The schematic photo of the roller burnishing tool, workpiece, and the burnishing process are shown in Fig. 1.

3 Results and Discussions

In this section the effect of roller burnishing on the, microhardness and the depth of penetration are presented and discussed. It was known that after copper addition there is a reduction in the average grain size from 124 µm into 30.7 µm, while the grained transformed from columnar structure into eqi-axed one. These changes resulted in the enhancement of all mechanical properties.

3.1 Effect of Hot Roller Burnishing on the Microhardness of Al-4% Cu

It can be seen from Fig. 2 that the average microhardness start enhancing as the burnishing force increase from 20 N to 40 N for all feed rates, where the maximum is 184.4% that occurred at 0.08 mm/rev. It is also obvious from the same Fig. 2 that the microhardness start to decrease from 40 N to 80 N for all feed rates. The enhancement in the hardness values is useful for increasing the fatigue life and the wear resistance of this material.

Figure 3 obviously shows that there is an enhancement in the hardness percentage at 200 °C, that the maximum is 184.4% attained at 40 N burnishing force and 0.08 mm/rev feed rate. However the minimum enhancement is 20% that occurred at 20 N burnishing force and 0.04 mm/rev feed rate.

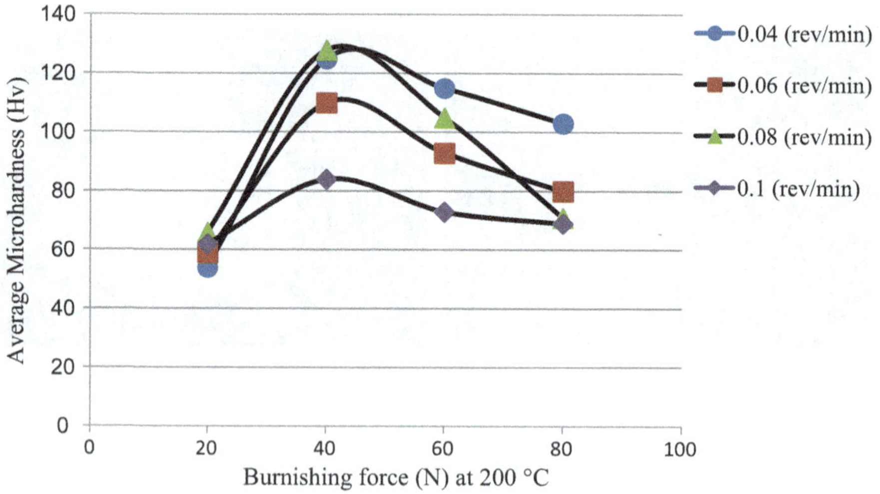

Fig. 2 Relationship between the microhardness and the roller burnishing force of Al-4% Cu alloy at 200 °C

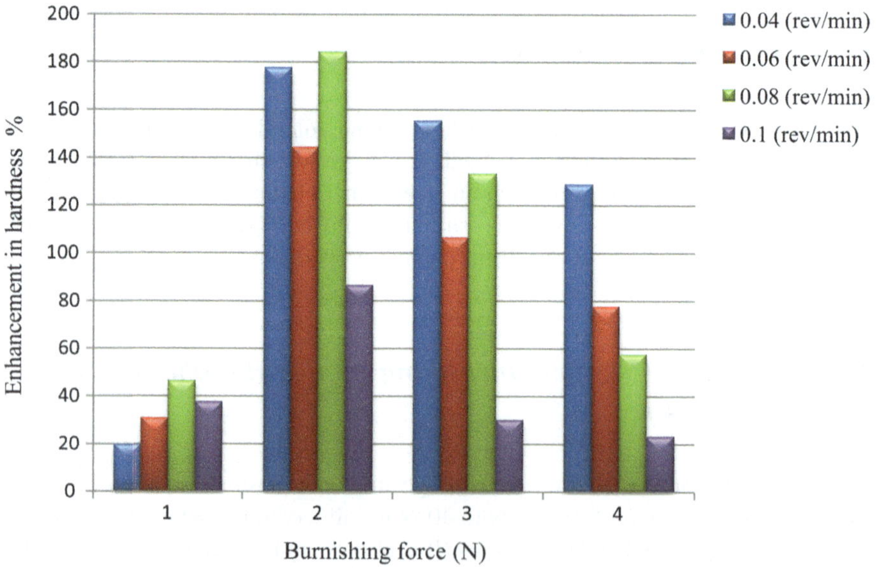

Fig. 3 Enhancement in hardness percentage after applying burnishing force

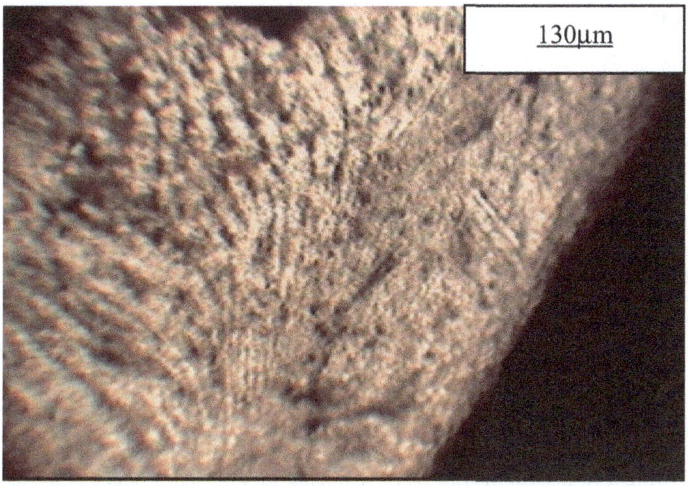

Fig. 4 Photomicroscan of burnished specimen at 200×

3.2 *Effect of Hot Roller Burnishing on the Depth of Penetration*

It recognized from the Fig. 4 that the maximum depth of penetration after roller burnishing process is 130 μm at 40 N burnishing force and 0.08 mm/rev feed. As the burnishing force applied a plastic deformation will occur which resulted in converting the tensile residual stresses into compressive stresses, so that the surface properties will be enhanced i.e. hardness, fatigue life, etc. The amount of plastic deformation for hot materials is so larger than cold work and this is the cause of high penetration mentioned above. It can be seen also from the microstructure of Fig. 4 how the grain size in the deformed zone which is so small compared to the large grains in the core of work piece.

4 Conclusions

Thus, after using the roller burnishing process it recognized that the microhardness increases by maximum of 184.4% at 40 N burnishing force and 0.08 mm/rev feed rate. On the other hand it will be established that the optimum burnishing force is 40 N at all different feed rates.

References

1. Hassan AH (1997) The effects of ball and roller burnishing on the surface roughness and hardness of some non-ferrous metals. J Mater Process Technol 72(3):385–391
2. Mahajan D, Tajane R (2013) A review on ball burnishing process. Int J Sci Res Publ 3:1–8
3. Todorovic P, Vukelic D, Tadic B, Veljkovic D, Budak I, Macuzic I et al (2014) Modeling of dynamic compliance of fixture/workpiece interface. Int J Simul Model 13:54–65
4. Murthy RL, Kotiveerachari B (1981) Burnishing of metallic surfaces—a review. Precis Eng 3:172–179
5. Fattough M, El-Khabeery MM, Serage SM (1985) Residual stress distribution in burnishing solution treated and aged 7075 aluminum alloy. Int J Mach Tools Manufact 29:153–160
6. Kragelsky IV, Alisin VV (1981) Tribology handbook. Mir, Moscow
7. Hassan AM (1997) An investigation into the surface characteristics of burnished cast Al–Cu alloys. Int J Mach Tools Manuf 37(6):813–821
8. El-Tayeb NSM (1994) Frictional behavior of burnished copper surfaces under dry contact conditions. Eng Res Bull 1:171–184 (HU, Cairo)
9. El-Tayeb NSM, Ghobrial MI (1993) The mechanical wear behavior of burnishing surfaces. In: Proceedings of the 4th international conference on production engineering and design for development, Cairo, pp 198–209
10. Hassan AM, Al-Bsharat AS (1996) Influence of burnishing process on surface roughness, hardness, and microstructure of some non-ferrous metals. Wear 199(1):1–8
11. Hassan AM, Momani AMS (2000) Further improvements in some properties of shot peened components using the burnishing process. Int J Mach Tools Manuf 40(12):1775–1786
12. Mondolfo LF (1976) Aluminum alloys structure and properties. Butterworth and Co., London
13. Hassan AM, Maqableh AM (2000) The effects of initial burnishing parameters on nonferrous components. J Mater Process Technol 102:115–121
14. Hassan AM, Al-Bsharat AS (1996) Improvements in some properties of non-ferrous metals by the application of the ball-burnishing process. J Mater Process Technol 59(3):250–256
15. El-Tayeb NSM, Low KO, Brevern PV (2006) Enhancement of surface quality and tribological properties during ball burnishing process. In: Proceedings of the ICOMAST, Melaka, Malaysia, GKH Press, pp 335–340

Systematic Review of Safety Leadership: A Fresh Perspective

Hassan M. Alidrisi and Sherif Mohamed

1 Introduction

Safety is an essential part in any organisation. Having a good safety model can assist in avoiding accidents that may lead to injuries and costs, which should have not occurred in the first place. As it is known safety threats are hidden among any work process, eliminating or containing those needs far more than one arrangement. Many studies have conducted investigations into different aspects of safety. The leadership aspect is one that has witnessed much attention in the last few years. In different industries, Leadership has been considered a key factor for improving the safety of workplace environment. Safety leadership has been positively linked to safety climate, safety culture, and safety behaviour.

Leadership is defined as a process of social influence between a leader and group of individuals to support achieving a common goal [19]. Thus, in the safety context, the goal of leadership process is creating a workplace environment where individuals work safely. To-date research studies in safety leadership are merely focused on either leadership behaviour or leadership competencies. Leadership behaviour is the practices that are adopted by a leader to influence team members [19]. Usually, these practices are grouped and composed to form leadership theories. In the safety context, the literature review shows that widely accepted leadership theories are transformational and transactional theories, which have added considerable value to improving safety performance [11]. However, following one or two theories might lead to missing other effective behaviours that are not classified into an existing leadership theory. On the other hand, leadership competen-

H. M. Alidrisi (✉) · S. Mohamed
School of Engineering, Griffith University, Gold Coast, Australia
e-mail: hassan.alidrisi@griffithuni.edu.au

S. Mohamed
e-mail: s.mohamed@griffith.edu.au

S. Şahin (ed.), *8th International Conference on Engineering, Project, and Product Management (EPPM 2017)*, Lecture Notes in Mechanical Engineering,
https://doi.org/10.1007/978-3-319-74123-9_23

cies can be recognised as the enablers that help leaders in practising their behaviours effectively [14]. Enablers can be grouped into contextual factors and/or leadership competencies. In the safety context, there is a great chance for exploring the role of leadership competencies in improving safety performance.

The aim of this research study is therefore to critically review the context of reported safety leadership research studies in the literature in order to explore other perspectives in leadership that may lead to a maximized benefit gain in the safety context.

2 Methodology

Four electronic databases were searched for peer-reviewed articles published in the last seven years (2010–June 2016): Scopus, Science Direct, PROquest, and EMERALD. The aim of exploring these databases is to gather all articles that include keywords of "Safety" and "Leadership" in their titles. The selected papers have been included in this review paper analysis upon reading their abstract and introduction. Papers that fulfilled the following eligibility criteria were selected to be on the final list of reviewed papers. First, the eligibility criteria focused on papers that demonstrate original research, thus quantitative, qualitative, theoretical, or review research studies were included; and other items such as editorial or book reviews were excluded. Second, leadership is the concept whereas safety is the context; papers that had a focus on any facet of safety leadership were included. Third, to further narrow the search, a standard for the number of citations present for each paper was established. Papers that were published in 2010 through to 2014, and have been cited five or more times, would be included within the review. As for papers published in 2015, all cited papers were considered. Papers published in the first half of 2016, cited or not, were included. It is worth noting that papers with higher citations over the study period have gone through peer-review that warranted having them as candidates for this review. Finally, papers had to have considered leadership as one of the main key interests to be included. Overall, a total of 55 research papers met the criteria and were selected to be on the final article review list for further analysis.

3 Descriptive Analysis

With an average of eight published papers per year, the highest number of published studies was found in 2013 and lowest in 2012. The second highest number of published studies was noted in 2016, indicating an ongoing interest in the field of safety leadership. On exploring the distribution of safety leadership studies by country, about 70% of studies were found distributed among 15 countries. The United States held the highest number of published studies with a total of 16 studies, accounting for 30% of total published papers. As for the distribution of studies by industry, a spread among 10 different industries was found. The health industry comprised the highest number of studies, with a total of 19 articles. The second highest number of studies is the construction industry, with seven studies. Regarding the distribution of studies by methodology type, a large number of studies, approximately 76%, followed a quantitative methodology. Qualitative studies accounted for only 9%, and the remaining 15% is spread among mixed methods, theoretical studies, and review studies. In respect to the total and average citations per journal, the 55 studies were found to be distributed among 31 journals. Out of the 55 studies, 15 papers are published by the journal of Safety Science, with a total citation of 439. However, the highest average citation per journal is 132, belonging to a paper appearing in the journal of Occupational and Organizational Psychology.

4 Discussion of Findings

4.1 Safety Leadership Competencies and Factors

Out of the 55 reviewed papers, only 10 papers have discussed safety leadership from the perspective of incitements such as competencies and factors. Their main focus was on contextual influencing factors and competencies for leadership, where the authors attempted to capture these factors and competencies, and explains it below, see Fig. 1.

Building trust between leaders and their followers is essential for an effective leadership. Deep understanding of developing trustworthiness qualities has been investigated by Conchie et al. [6]. They differentiated between developing trust and avoiding distrust. In other words, the attributes, which are needed for building trust and decreasing distrust, are different. Mediating and moderating roles of trust have been shown in safety leadership literature. Conchie et al. [7] concluded that

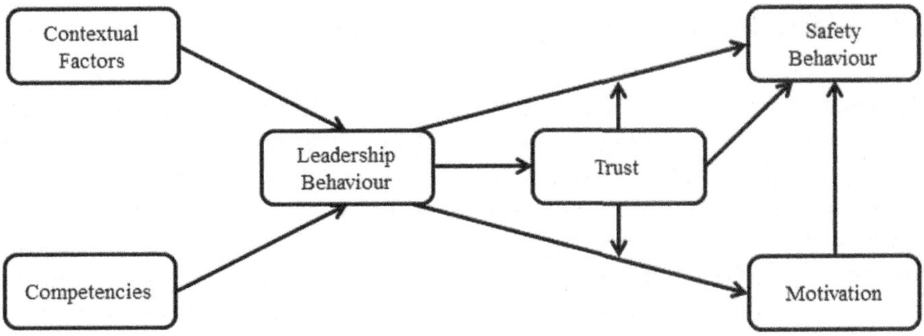

Fig. 1 Mind map: safety leadership behaviour and leadership factors

relationship between safety-specific transformational leadership and safety vice-citizenship behaviour is mediated by two facets of trust (affect-based trust and trust intentions). As a moderator, the extent to which a leader is trusted affects the relation of transformation leadership to some attributes of safety citizenship behaviour [5]. Trust also impacts the relationship between transformational leadership and intrinsic motivation that ultimately affects safety behaviours of citizenship and compliance.

Factors affecting and affected by safety leadership have been, also, discussed. Conchie [5] explored the contextual factors affecting supervisorial safety leadership behaviour, and identified role overload, production demand, workforce characteristic, and formal procedures as playing a hindering role in leadership behaviour. As for the promoting factors, social support and perceived autonomy were recognized. Newnam et al. [16] investigated the impact of leader-member relationship on safety performance. Role overload was concluded as being a moderator in the relationship between leaders and their followers. In contrast, other factors that are positively related to safety performance can be affected by safety leadership. It has been found that safety leadership style (i.e. transformational leadership) strength impacts factors of safety knowledge and knowledge-related job characteristics [12, 13].

Competencies enhance leaders' effectiveness in practising leadership behaviours [21]. In the past seven years, few studies have discussed leadership competencies in the context of safety. However, the positive effect of competency self-efficacy on safety outcomes, such as safety behaviours and safety climate, has been investigated by Chen and Chen [3], Schwarz et al. [22]. Both studies have handled self-efficacy competency as a construct that can improve safety conditions if involved with other effective elements such as safety leadership styles. Sunindijo and Zou [26] explored the interaction of emotional intelligence competency with safety leadership behaviours of transformational leadership. It was concluded that emotional intelligence competency has a positive effect over transformational leadership behaviour. Other competencies where their importance has been proven in general leadership literature, has not yet been investigated in the context of safety.

4.2 Safety Climate and Safety Leadership Behaviour

This research review shows that safety climate is affected by three groups of safety leadership behaviour (see Fig. 2). The first group encompasses the behaviours that demonstrate high level of emotional intelligence, LXM leadership and authentic leadership. Squires et al. [25] contended that behaviours such as active listening, which is associated with emotional intelligence level, has an impact on the relationship between leader and follower. This impact has been found to be positively related to safety climate. Similarly, other studies such as Nielsen et al. [17], Thompson et al. [27] have also displayed other leadership behaviours associated with safety climate. The behaviours include open communication, feedback and

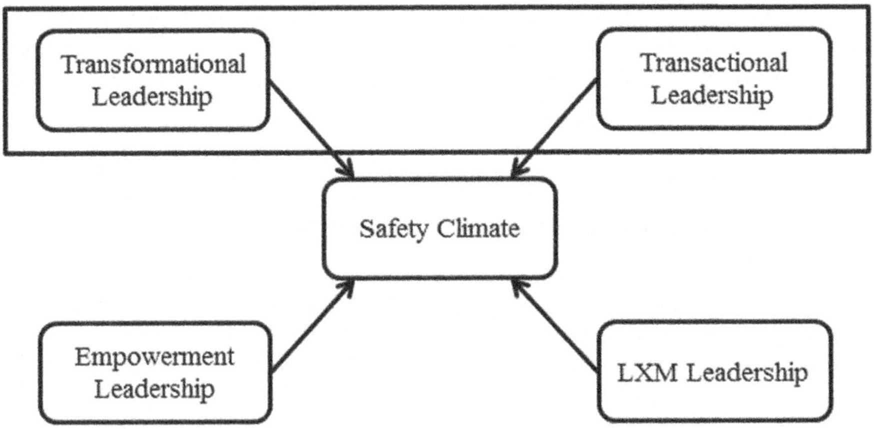

Fig. 2 Mind map: safety leadership behaviour and safety climate

communication about error, commitment to safety, non-positive response to error, and other behaviours that demonstrate transparency, self-awareness, and having a moral perspective.

As for the second group, it includes the behaviours that provide power to followers. These behaviours are listed under empowerment leadership behaviours. It includes coaching, caring, controlling, leading by example, informing, participating, and visibility at site. The effects of these behaviours on safety climate have been shown to be of positive nature with a number of studies investigating this impact and reinforcing the positive effect of empowerment leadership behaviours on safety climate.

The third group includes transformational and transactional leadership behaviours. Although these two leadership styles are different, they are combined in one group because they are usually investigated together. Transformational leadership covers four dimensions of behaviour namely individualized consideration, idealised influence, inspirational motivation, and intellectual stimulation [4]. As for Transactional leadership, in the safety context, is concerned with actions that must be taken by followers to achieve safety outcomes. Its behaviours can be shaped in one of the following styles: contingent reward, management-by-exception active, management-by-exception passive, or laissez-faire. Others, such as Du and Sun [8], have discussed transformational and transactional leadership in another perspective; Transformational leadership as representing behaviours that focus on active management and motivation towards safety and transactional leadership as representing behaviours related to monitoring safety. The relation between both leadership styles and safety climate has been explored. Safety climate is positively affected by transformational leadership, and contingent reward and management-by-exception active styles of transactional leadership. Hoffmeister et al. [11] identified safety climate as being affected by intellectual stimulation and idealized influence dimensions of transformational leadership and contingent reward style of transactional leadership. Other studies contended leaders combining transformational and contingent reward styles have higher positive impact on safety climate [22]. In contrast, management-by exception passive and laissez-faire transactional leadership styles negatively affect safety climate [18, 24].

5 A Fresh Perspective of Safety Leadership

Systems Thinking gives an interdependent view of systems [21]. In other words, it considers that an event in one part of the system could affect another. Thus, it provides the ability to understand the linear and non-linear cause-and-effect relationship and comprehend the underlying pattern of events [23]. Moreover, it emphasises a dynamic view as a means of understanding the context [21]. Therefore, the holistic view of Systems Thinking could appear as a complex process.

In order to work out and understand the whole of a system, Gharajedaghi [9] has suggested applying an iterative process. An iterative process, simply, is a relationship circle consisting of four interdependent variables namely, function, structure, process, and context. Each variable works as a co-producer for other variables; and the circle is closed once the holistic view is distinguished [9]. Moreover, Gharajedagh [9] demonstrates how seeing the whole is difficult if one fails to figure out these interdependencies. Therefore, an iterative process is important to understanding the complexity of seeing the whole system.

By applying the iterative process to the leadership definition adopted for this research study, the leadership concept can be interpreted holistically (Fig. 3). Here, Leadership is defined as "a process whereby an individual influences a group of individuals to achieve a common goal" [19]—a definition that clearly presents influencing others as the key function of leadership. The structure of leadership consists of the leader, the followers, and the environment. The process that explains how the structure generates the function of leadership is basically represented in: leaders' ability to influence followers; their ability to understand how their followers are influenced; and their ability to understand the environment for efficient influencing. By understanding the relationship between the function, structure, and process, and then placing the leadership in the appropriate context, leadership would enable the achievement of the desired goal, which is improving leaders' influence level. Thus, this holistic view adds a valuable explanation of how leaders influence their followers by synthesizing leaders' competencies along with their practices. Three types of leadership competencies have been known to assist in this holistic view of leadership namely, emotional, social, and cognitive intelligence competencies. Boyatzis [1] defined the three intelligence competencies as follows:

1. Emotional intelligence competency is "the ability to recognize, understand, and use emotional information about oneself that leads to or causes effective or superior performance".

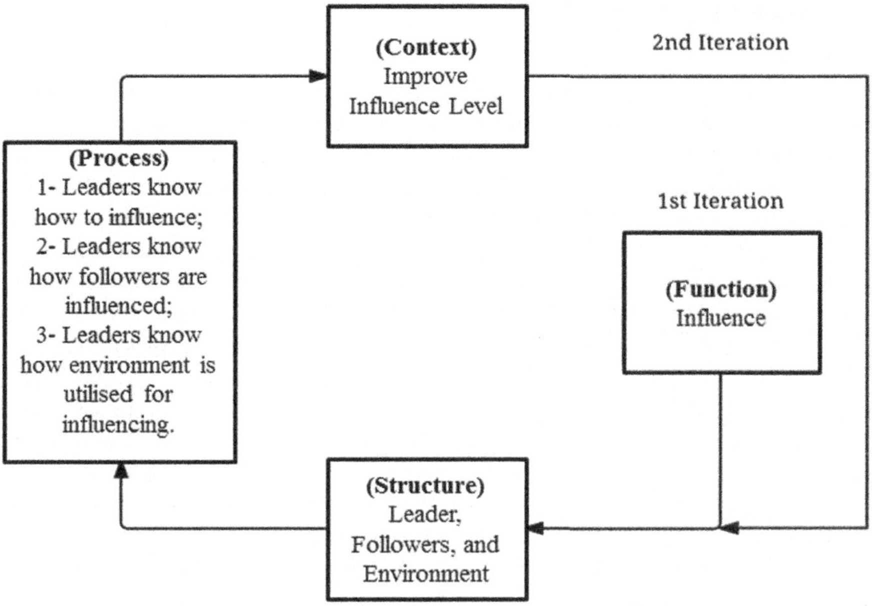

Fig. 3 Iterative process for mastering influence skills

2. Social intelligence competency is "the ability to recognize, understand, and use emotional information about others that leads to or causes effective or superior performance".
3. Cognitive intelligence competency is "the ability to think or analyse information and situations that leads to or causes effective or superior performance".

Safety Leadership, from a holistic view, is about how leaders utilise themselves, their followers and the environment when exercising influence. Leaders simply need to know how to influence, how followers are being influenced, and how the environment is utilised for influencing. This holistic view of safety leadership aligns with the general view of flexible leadership proposed by Mumford et al. [15], who argue that leadership is more than only exercising influencing behaviour. It also aligns with the general view of flexible leadership proposed by Yukl and Mahsud [28], who contend that emotional, social and cognitive intelligence competencies provide leaders with a better understanding of themselves, others, and the environment. Thus, by reinforcing leaders' emotional, social, and cognitive intelligence competencies, they become more capable of performing better in influencing their followers toward safety. Collectively, these three leadership competencies provide leaders with flexibility and a holistic view when practising their safety leadership roles.

6 Concluding Remarks

This systematic review highlights the limitations of existing literature on safety leadership. Future research should explore and investigate the aspects of these limitations to overcome. There is a chance to explore additional safety leadership behaviours and to understand their impact on safety performance. Clarke [4] argued that the combination of the transformational and transactional leadership behaviour is effective in improving safety performance. Therefore, the adoption of further leadership behaviours, in conjunction with the transformational and transactional ones, is suggested for future investigation. This will offer a flexible style of leadership that may be more appropriate in the complex environment of safety.

The result of this review has also uncovered the limited exploration of leadership competencies in the safety context. While competencies have been found valuable in experiencing outstanding success [2, 10], much attention should be directed to explore a range of other competencies and to investigate their impact on safety performance. Mumford et al. [15] argued that the role of leaders should not simply be to exercise 'influence' behaviours, but should also include deciding when, where, and how the influence is exercised to attain goals. Thus, the focus should be on the competencies, skills, and knowledge to develop effective leaders [15].

Acknowledgements The first author would like to thank King Abdulaziz University, Jeddah, Saudi Arabia, for providing the required funding and support to conduct and publish this research study.

References

1. Boyatzis R (2008) Competencies in the 21st century. J Manag Dev 27(1):5–12
2. Butler CJ, Chinowsky PS (2006) Emotional intelligence and leadership behavior in construction executives. J Manag Eng 22(3):119–125
3. Chen C, Chen S (2014) Measuring the effects of safety management system practices, morality leadership and self-efficacy on pilots' safety behaviors: safety motivation as a mediator. Saf Sci 62:376–385
4. Clarke S (2013) Safety leadership: a meta-analytic review of transformational and transactional leadership styles as antecedents of safety behaviours. J Occup Organ Psych 86(1):22–49
5. Conchie SM (2013) Transformational leadership, intrinsic motivation, and trust: a moderated-mediated model of workplace safety. J Occup Health Psych 18(2):198
6. Conchie SM, Taylor PJ, Charlton A (2011) Trust and distrust in safety leadership: mirror reflections? Saf Sci 49(8):1208–1214
7. Conchie SM, Taylor PJ, Donald IJ (2012) Promoting safety voice with safety-specific transformational leadership: the mediating role of two dimensions of trust. J Occup Health Psych 17(1):105
8. Du X, Sun W (2012) Research on the relationship between safety leadership and safety climate in coalmines. Procedia Eng 45:214–219
9. Gharajedaghi J (2011) Systems thinking: managing chaos and complexity: a platform for designing business architecture, 3rd edn. Elsevier, Burlington, MA

10. Goleman D (1995) Emotional intelligence. Bantam Books, New York
11. Hoffmeister K, Gibbons AM, Johnson SK et al (2014) The differential effects of transformational leadership facets on employee safety. Saf Sci 62:68–78
12. Jiang L, Probst TM (2016) Transformational and passive leadership as cross-level moderators of the relationships between safety knowledge, safety motivation, and safety participation. J Saf Res 57:27–32
13. Lievens I, Vlerick P (2014) Transformational leadership and safety performance among nurses: the mediating role of knowledge-related job characteristics. J Adv Nurs 70(3)
14. Müller R, Turner JR (2007) Matching the project manager's leadership style to project type. Int J Proj Manag 25(1):21
15. Mumford MD, Zaccaro SJ, Harding FD et al (2000) Leadership skills for a changing world: solving complex social problems. Leadersh Q 11(1):11–35
16. Newnam S, Lewis I, Watson B (2012) Occupational driver safety: conceptualising a leadership-based intervention to improve safe driving performance. Accid Anal Prev 45:29–38
17. Nielsen MB, Eid J, Mearns K et al (2013) Authentic leadership and its relationship with risk perception and safety climate. Leadership Org Dev J 34(4):308–325
18. Nielsen MB, Skogstad A, Matthiesen SB et al (2016) The importance of a multidimensional and temporal design in research on leadership and workplace safety. Leadersh Q 27(1): 142–155
19. Northouse PG (2015) Leadership: theory and practice, 7th edn. Sage publications Inc., Thousand Oaks
20. O'Connor S, Carlson E (2016) Safety culture and senior leadership behavior: using negative safety ratings to align clinical staff and senior leadership. J Nurs Admin 46(4)
21. Palaima T, Skaržauskiene A (2010) Systems thinking as a platform for leadership performance in a complex world. Balt J Manag 5(3):330–355
22. Schwarz UV, Hasson H, Tafvelin S et al (2016) Leadership training as an occupational health intervention: improved safety and sustained productivity. Saf Sci 81
23. Senge PM (2006) The fifth discipline: the art and practice of the learning organization. Doubleday/Currency, New York
24. Smith TD, Eldridge F, DeJoy DM (2016) Safety-specific transformational and passive leadership influences on firefighter safety climate perceptions and safety behavior outcomes. Saf Sci 86:92–97
25. Squires M, Tourangeau A, Spence Laschinger HK et al (2010) The link between leadership and safety outcomes in hospitals. J Nurs Manag 18(8):914–925
26. Sunindijo RY, Zou PX (2013) The roles of emotional intelligence, interpersonal skill, and transformational leadership on improving construction safety performance. Constr Econ Build 13(3):97–113
27. Thompson DN, Hoffman LA, Sereika SM et al (2011) A relational leadership perspective on unit-level safety climate. J Nurs Admin 41(11):479–487
28. Yukl G, Mahsud R (2010) Why flexible and adaptive leadership is essential. Consult Psychol J Pract Res 62(2):81–93

Cloud Manufacturing—The Adoption of Virtual Production Line to Soft Resources Analysis

Julia Siderska

1 Introduction

The modern manufacturing industry is experiencing thorough transformation because of the constant development of advanced information technologies. The emergence of cloud computing initiated a new paradigm of servitization, assuming changing physical product (software or hardware) into a service. This concept was adopted as a basis for delivering shared, on-demand manufacturing services—Cloud Manufacturing (CM). There exist an ongoing paradigm assuming that the modern manufacturing industry is changing to become the global manufacturing networks and supply chains allowing for commonly using of the globally distributed manufacturing systems and resources. Cloud Manufacturing, as a model integrating innovative technologies (Internet of Things, cloud computing, service-oriented technologies, virtualization, semantic web, advanced high-performance computing technologies) with advanced manufacturing, enables and supports cooperation, sharing and management of manufacturing resources. These resources and know-hows, e.g. software tools, knowledge, applications, equipment, fabrication capabilities, etc. of manufacturing companies will be inserted into the cloud and thereby accessible to presumptive consumers [1].

Cloud Manufacturing paradigm has been expeditiously growing in popularity among scientists over the past few years. The number of articles including the discussion about CM concept is increasing rapidly. The author examined three academic databases: Science Direct, Scopus and Web of Science. It is worth indicating that most of the authors taking up the CM idea in their investigations are from Chinese research units: Beihang University, Zhejiang University, Beijing Simulation Center, Ministry of Education China, Chongqing University, etc.

J. Siderska (✉)
Faculty of Engineering Management, Bialystok University of Technology,
Tarasiuka #2, 16-001 Kleosin, Poland
e-mail: j.siderska@pb.edu.pl

S. Şahin (ed.), *8th International Conference on Engineering, Project, and Product Management (EPPM 2017)*, Lecture Notes in Mechanical Engineering,
https://doi.org/10.1007/978-3-319-74123-9_24

Table 1 The number of Cloud Manufacturing papers by year

Year	Science direct	Scopus	Web of science
2010	6	7	0
2011	9	38	16
2012	13	100	20
2013	22	100	39
2014	38	142	52
2015	45	120	50
2016	63	143	57
2017	53	65	25

Moreover, the other scientists taking up in their considerations CM idea come mainly from: Sweden (University of Skövde, KTH Royal Institute of Technology); US (University of Arkansas, Georgia Institute of Technology, Old Dominion University); UK (Falmouth University, University of Nottingham) and New Zealand (University of Auckland). Most of their papers are published in International Journal of Computer Integrated Manufacturing, Robotics and Computer-Integrated Manufacturing, Journal of Manufacturing Systems and Procedia CIRP—affiliated by The International Academy for Production Engineering. Table 1 presents the number of papers considering Cloud Manufacturing concept by year from 2010 to 2016 and also published in the first four months of 2017. The literature search was carried out in mentioned academic databases using the query string title-abstract-keywords “Cloud Manufacturing”.

Majority of the authors focus on discussing the particular elements of software and hardware concerning this issue, e.g. platform technologies, ontologies, multi-task scheduling [8, 9], programming models, file systems as well as system architectures (layers) for the development of Cloud Manufacturing platforms [5]. They also often consider information and communication models and new business models referring to Cloud Manufacturing [20]. Most of the researchers indicate manufacturing resources, which should be considered as the objects of sharing between three main types of actors participating in CM: providers, consumers and operators (brokers). All of manufacturing capabilities require support from the related manufacturing resources, including soft resources (software, engineering knowledge, skills, experience, business networks, etc.) and hard resources, for instance manufacturing equipment, computational resources, monitoring resources, storage, etc. [18].

The paper is structured as follows: Section 1 demonstrates the introduction to Cloud Manufacturing concept and indicates the growing interest of this idea among scientists. Section 2 describes the background of Cloud Manufacturing as well as presents several definitions of this paradigm. Section 3 presents the concept of Virtual Production Line and provides the possibility of adopting this approach to analyze the open source manufacturing software development. The last section includes the authors’ conclusions, acknowledgements and references.

2 Background of Cloud Manufacturing

For few recent years' collaboration as well as innovative IT technologies, mainly Internet of Things and cloud, have been profoundly reshaping enterprises and creating key business technology trends worldwide. The manufacturing industry experiencing fundamental transitions enabled by these smart technologies. Cloud computing—heretofore commonly and frequently considered concept and paradigm—is nowadays being adopted to transform the possibilities of sharing and consuming the IT resources.

Cloud computing technology, providing on-demand computing services with high reliability, scalability and availability in a distributed environment, allows for achieving competitiveness in the global market. In Cloud computing, everything is treated as a service (XaaS): Infrastructure as a Service (IaaS), Software as a Service (SaaS) and Platform as a Service (PaaS). These services define a layered system structure for cloud computing. At the Infrastructure layer, processing, storage, networks, and other fundamental computing resources are defined as standardized services over the network. Besides those XaaS include also the following items: Network as a Service (NaaS), Communications as a Service (CaaS), Storage as a Service (SaaS) and Monitoring as a Service (MaaS). The idea of providing computer resources as services can be adopted in manufacturing industry, with manufacturing resources being offered as different services, i.e. Machining-as-a-Service (MCaaS), Design-as-a-Service (DaaS), etc. Examples of other contributing and enabling technologies are: Internet of Things (IoT), Semantic Web, embedded systems and virtualization technologies [1].

Many manufacturing companies build the factors, and thereby also the manufacturing resources, in different and distributed locations. It is becoming more and more essential to integrate decentralized manufacturing resources and establish a collaborative infrastructure among those units. This idea requires building the networked manufacturing environment to integrate manufacturing resources and applications [5].

Cloud Manufacturing is an emerging business model that enables manufacturing companies the virtualization of their manufacturing resources and offers them as cloud-based services via the Internet. This concept allows also the transformation of traditional, production-oriented manufacturing to an integrated and service-oriented manufacturing paradigm. CM focuses on cloud computing to simplify the management of distributed resources in a centralized way, building scale, maximizing resource utilization and increasing productivity. CM extends the idea of cloud computing to manufacturing, so that their capabilities and resources are componentized, integrated, optimized and provisioned globally [13].

Cloud Manufacturing, as a service-oriented networked manufacturing model, requires and relies on the innovative technologies, such as mainly Internet of Things (IoT), cyber-physical systems (CPS) and manufacturing data management. It also allows for extending the occurring manufacturing models and technologies, including: agile manufacturing, Web-based manufacturing and manufacturing grid as well as CM adopts the "Manufacturing-as-a-Service (MaaS)" concept.

The one and commonly accepted definition of Cloud Manufacturing has not been proposed so far, however the researchers and members of manufacturing community elaborated a comprehensive prospect of what CM would comprehend and facilitate [1]. The conducted by the author literature review allowed to identify several, often cited explanations. The first pioneering attempt to define this concept and paradigm was taken up in 2010 by Chinese researcher Li, who introduced the following definition: "*Cloud Manufacturing is service-oriented, knowledge-based smart manufacturing system with high efficiency and low energy consumption*" [6].

Xu relied his considerations and research on the assumptions of National Institute of Standards and Technology (NIST) concerning cloud computing [10]. Reflecting the nomenclature proposed by NIST Xu defined Cloud Manufacturing as *"a model for enabling ubiquitous, convenient, on-demand network access to a shared pool of configurable manufacturing resources (e.g. manufacturing equipment, manufacturing capabilities and manufacturing software tools) that can be rapidly provided and released with minimal management effort"*. This paradigm is considered as a multidisciplinary concept encompassing the following, advanced technologies: cloud computing, networked manufacturing, manufacturing grid, virtual manufacturing, agile manufacturing and Internet of Things. CM mirrors the concept of "integration of distributed resources" as well as the concept of "distribution of integrated resources". Such globally distributed resources are encapsulated into cloud and should be managed in a centralized way [21].

Several authors stress the essential role of human factor and collaboration when adopting cloud computing paradigms within the manufacturing context, underlining the importance of fostering trust within cloud networks. The conducted literature review allowed identifying three groups of actors involved in manufacturing in the clouds: consumers (requesting and using cloud manufacturing processes), application providers (providing the software to enable the cloud manufacturing) and service providers (providing, owning and operating the manufacturing services [4]. Tao considered then CM as a relationship between the consumer and production services and matching service providers to product and manufacturing processes [14].

Although the proposed definitions of Cloud Manufacturing have slightly different assumptions of the system structures, almost all of them share some similar elements (i.e. cloud service composition or resource virtualization) [13]. According to Ren manufacturing resource is an object supporting the activity involved in the life cycle of a product. There are two basic types of manufacturing resource: hard resource and soft resource. A hard resource could be considered as a manufacturing cell as well as IT hardware, while a soft resource includes software, information, data or other intellectual units [12]. In author's opinion it is necessary to add to this list the social capital of the workers and of the whole company. The knowledge, skills, experience, talent, etc. of workers are the most important and valuable resources of all enterprises operating in the manufacturing industry. That is why they should be properly managed. The next section introduces the Virtual

Production Line as a concept of analyzing those intangible, soft resources of the manufacturing company.

3 Virtual Production Line

At the beginning of the twentieth century cars were manufactured in production circles. Workers were moving around the car in the center of the production circle. In such a production process the division of labor was very flexible and they could easily substitute for one another. An American industrialist and founder of Ford Motor Company—Henry Ford in 1913 introduced the following observation: if we partition the complex car manufacturing process into a fixed number of simple operations, then the productivity increases and thereby solve the problem of limited number of craftsmen. This idea completely transformed the manufacturing industry and it is considered as one of the most important achievements in management science and economics. The idea of the assembly line was then implicated in many manufacturing and service processes. Many production lines, manned by people or robots in a factory, were then combined into one production line, called the Classical Production Line (CPL). However, this concept was a routine work and as the automotive industry has been developing very rapidly and the appropriate process organization appeared to be one of the most essential priorities in the manufacturing company, the CPL concept was no longer up-to-date. Nowadays manufacturing processes are very sophisticated, requiring extensive knowledge and advanced IT technology. Thus such problems need to be solved creatively and should be analyzed with the use of modern, complex and holistic methods [17]. Also Cloud Manufacturing require more innovative approach.

Let's consider a Virtual Production Line (VPL) pictured in Fig. 1. Number of experts and engineers are solving a manufacturing problem of the company within a

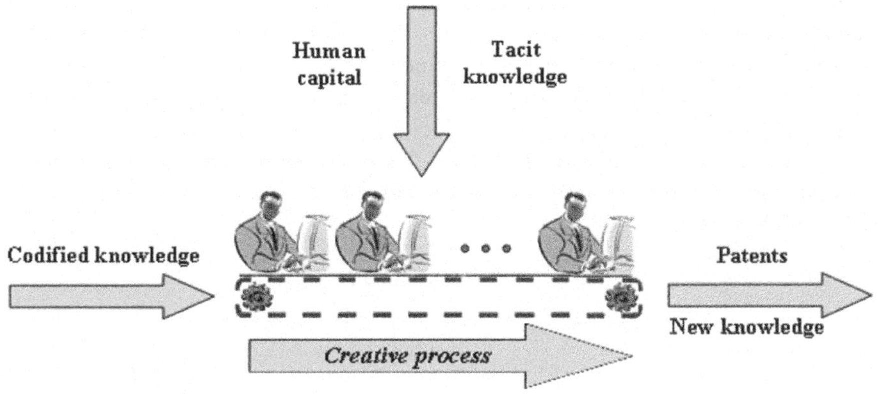

Fig. 1 The concept of Virtual Production Line (Walukiewicz 2006)

creative process equipped with their computers, software, databases, etc. There is no material representation of the VPL and the experts are usually globally distributed (denoted with a dotted line).

Specialists working on a VPL combine their human capital (mostly their tacit knowledge) with codified knowledge to solve a problem in a creative process. At the beginning, the problem seems to be unclear and unintelligibly defined, but due to the efforts of the experts, it will become more and more unmistakable and marked. It is necessary to introduce the definition of the Virtual Production Line:

> **Virtual Production Line (VPL)** is a division of a complex creative process into more or less precisely described tasks (jobs) combined with modern IT. The division into tasks, as well as the number of tasks, may be changed throughout the process by actions of experts involved in it. Such a modification is called a self-organization of the VPL.

Virtual Production Line can be considered as an instrument (a virtual conveyor belt) that experts use to combine codified knowledge with their tacit knowledge, competence, experience, etc. to introduce improvements in products, services, technology and management, and contribute to the world's stock of knowledge.

It worth underlining that Virtual Production Line allows for sharing resources and flexible division of labor, while CPL based on a rigid (stiff) partition of labor. This unique feature of Virtual Production Line, according to definitions discussed in the Sect. 1, perfectly fit in the Cloud Manufacturing concept. They also exemplify that the resources in manufacturing industries to be shared are not only hard: (machines, hardware, etc.) but they also contain human factor: knowledge, skills, abilities, talent, experience, etc.

The very illustrative example of products developed on the Virtual Production Line is manufacturing software, including Open ERP or Open Bravo. They are available as a service (SaaS) and allow for manufacturing resource planning, production management, scheduling production tasks and orders, etc. They are disseminated under the open source license and developed by volunteers, production management specialists from all over the world. Open applications are currently the most rapidly growing software in the world. Free and open software is created as a result of the work of many enthusiasts, working simultaneously, in parallel. The source code of such applications is open—the developers all over the world are involved in its creation, modification, refinement and distribution. Such idea fits perfectly into Cloud Manufacturing concept as it proves the possibility of sharing the globally distributed soft resources, including know-hows, skills, experiences etc. The essential task for companies is to elaborate the proper methods of management and of inserting into the cloud and thereby make them accessible to presumptive consumers.

4 Conclusion

The conducted literature review highlighted that Cloud Manufacturing is nowadays one of the most important and rapidly prospering directions of development in manufacturing industry. Service-oriented manufacturing, deriving from the cloud computing paradigm to integrate manufacturing resources, is becoming very promising manufacturing and business models. The idea of Cloud Manufacturing rises as an innovative paradigm contributing significantly to the success of the Industry 4.0. It should also be acknowledged as a networked manufacturing model enabling on-demand access distributed manufacturing resources to create temporary and reconfigurable cyber-physical production lines. The fundamental particularities of CM include networked manufacturing, ubiquitous access and virtualization, agility, IoT, big data and everything-as-a-service (e.g., software-as-a-service, platform-as-a-service, infrastructure-as-a-service, and hardware-as-a-service) [16].

The author presented the Virtual Production Line and conducted on it the analysis of the software development. Those concepts were the basis to pointing out that human factor, relations and collaboration among production workers should also be considered as an essential manufacturing resource and could also be shared in the cloud as well as componentized, integrated, optimized and provisioned globally.

Although cloud-based manufacturing is the consequence and evolution of existing technologies, this concept is widely accepted to be also the future-oriented, innovative paradigm that will reform the manufacturing industry. However, there exist some unexplored dilemmas and areas. The researchers are to investigate whether all types of manufacturing services are suitable to move to the cloud or to identify the possibilities of adopting Cloud Manufacturing by the manufacturing companies [20]. Moreover, knowledge workers typically are paid more than others, so increasing their productivity is critical. Improving these workers' efficiency and effectiveness is essential and should be achieved by introducing proper collaboration technologies.

Cloud computing, considering as the ability to buy software, infrastructure and platforms on a pay-per-use basis, has transformed the way organizations think about their technology investments. Nowadays a new type of service providers, basing on the Cloud Manufacturing paradigm, should be developed. These provide skilled people as a managed service—mirroring the options available for technology. Beside manufacturing resources being offered as different services, i.e. Design-as-a-Service, Machining-as-a-Service, the model Skills-as-a-service is inevitable in the manufacturing industry. Workers owing transferable skills would make themselves available through value-added skills providers and those unique competencies could be shared and provided when they are needed. This concept force manufacturing companies to adapt quickly, to determine new ways of consuming and providing data and flexibly adjust their business models.

The author's future research will concern the identification of manufacturing resources as well as services appropriate for promoting to the cloud environment.

The planned investigations will also include the diagnosis of the necessary and sufficient conditions for manufacturing resources to be adopted in the cloud and thereby accessible to presumptive consumers. Finally, it is necessary to develop a set of viable strategies and business models for service providers and consumers.

Acknowledgements The research was conducted within S/WZ/1/2014 project and was financed from Ministry of Science and Higher Education funds.

References

1. Adamson G, Lihui W, Holm M, Moore P (2017) Cloud manufacturing—a critical review of recent development and future trends. Int J Comput Integr Manuf 30(4–5):347–380. https://doi.org/10.1080/0951192X.2015.1031704
2. Bughin J, Chui M, Manyika J (2010) Clouds, big data, and smart assets: Ten tech-enabled business trends to watch. McKinsey Quarterly, McKinsey Global Institution
3. Corbet J, Kroah-Hartman G (2016) Linux Kernel development. How fast it is going, who is doing it, what they are doing and who is sponsoring the work. The Linux Foundation
4. Golightly D, Sharples S, Patel H, Ratchev S (2016) Manufacturing in the cloud: a human factors perspective. Int J Ind Ergon 55:12–21. https://doi.org/10.1016/j.ergon.2016.05.011
5. He W, Xu L (2015) A state-of-the-art survey of cloud manufacturing. Int J Comput Integr Manuf 28(3):239–250. https://doi.org/10.1080/0951192X.2013.874595
6. Li BH, Zhang L, Wang S, Tao F, Cao J, Jiang X, Song X et al (2010) Cloud manufacturing: a new service-oriented networked manufacturing model. Comput Integr Manuf Syst CIMS 16 (1):1–7
7. Liu Y, Xu X (2017) Industry 4.0 and cloud manufacturing: a comparative analysis. J Manuf Sci Eng 139. https://doi.org/10.1115/1.4034667
8. Liu Y, Xu X, Zhnag L, Wang L, Zhong RY (2017) Workload-based multi-task scheduling in cloud manufacturing. Rob Comput Integr Manuf 35:3–20
9. Liu ZZ, Song Ch, Chu DH, Hou ZW, Peng WP (2017) An approach for multipath cloud manufacturing services dynamic composition. Int J Intell Syst 32(4):371–393. https://doi.org/10.1002/int.21865
10. Mell P, Grance T (2009) Perspectives on cloud computing and standard. National Institute of standards and technology (NIST). Information Technology Laboratory
11. Qin J, Liu Y, Grosvenor R (2016) A categorical framework of manufacturing for industry 4.0 and beyond. Changeable, Agile, Reconfigurable & Virtual Production) Procedia CIRP 52:173–178
12. Ren L, Zhang L, Tao F, Zhao Ch, Chai X, Zhao X (2015) Cloud manufacturing: from concept to practice. Enterp Inf Syst 9(2):186–209. https://doi.org/10.1080/17517575.2013.839055
13. Siderska J, Jadaan KS (2018) Cloud manufacturing: a service-oriented manufacturing paradigm. A review paper, Engineering Management in Production and Services 10(1): 22–31; https://doi.org/10.1515/emj-2018-0002
14. Tao F, Zhang L, Venkatesh VC, Luo Y, Cheng Y (2011) Cloud manufacturing: a computing and service-oriented manufacturing model. Proc Inst Mech Eng B 225(10). https://doi.org/10.1177/0954405411405575
15. Tao F, Zhang L, Liu Y, Cheng Y, Wang L, Xu X (2015) Manufacturing service management in cloud manufacturing: overview and future research directions. J Manuf Sci Eng 137 (4):040912. https://doi.org/10.1115/1.4030510
16. Thames L, Schaefer D (2016) Software-defined cloud manufacturing for industry 4.0. Changeable, Agile, Reconfigurable & Virtual Production, Procedia CIRP 52:12–17. https://doi.org/10.1016/j.procir.2016.07.041

17. Walukiewicz S, Wiktorzak AA (2013) Measuring human capital in education. Oper Res Decisions 23(3):57–69. https://doi.org/10.5277/ord130305
18. Wang XV, Xu X (2013) An interoperable solution for Cloud manufacturing. Robot Comput Integr Manuf 29(4):232–247. https://doi.org/10.1016/j.rcim.2013.01.005
19. Wu D, Greer MJ, Rosen DW, Schaefer D (2013) Cloud manufacturing: strategic vision and state-of-the-art. J Manuf Syst 32:564–579
20. Wu D, Rosen DW, Wang L, Schaefer D (2014) Cloud-based manufacturing: old wine in new bottles? variety management in manufacturing. In: Proceedings of the 47th CIRP conference on manufacturing systems, vol 17, pp 94–99
21. Xu X (2012) From cloud computing to cloud manufacturing. Robot Comput Integr Manuf 28:75–86. https://doi.org/10.1016/j.rcim.2011.07.002
22. Yu Ch, Xu X, Lu Y (2015) Computer-integrated manufacturing, cyber-physical systems and cloud manufacturing—concepts and relationships. Manuf Lett 6:5–9
23. Yuan M, Deng K, Chaovalitwongse WA (2017) Manufacturing resource modeling for cloud manufacturing. Int J Intell Syst 32(4):414–436. https://doi.org/10.1002/int.21867
24. Zhang L, Luo Y, Tao F, Li BH, Ren L, Zhang X, Guo H, Cheng Y, Hu A, Liu Y (2014) Cloud manufacturing: a new manufacturing paradigm. Enterp Inf Syst 8(2):167–187. https://doi.org/10.1080/17517575.2012

The Effectiveness of Health and Safety Training and Its Impact on Construction Workers' Attitudes, and Perceptions

Tafadzwa Mushayi, Claire Deacon and John Smallwood

1 Introduction

Accidents that occur on construction sites can be categorized as defects of a construction project and defects of the H&S system. The reality is that the construction industry in the Republic of South Africa (RSA) experiences high injury and fatality rates CIDB [7]. In relation to previous research conducted, there seems to be a lack of correlation between the impact of effective H&S training and its influence on worker H&S attitudes and H&S behaviors. H&S training goals and objectives should be designed to influence construction workers' understanding of the specific hazards in their work environment; including the accident and injury risks posed by exposure to these hazards. Certain benefits can be attributed to the effective delivery of H&S training. These include enhanced H&S knowledge and H&S motivation among construction workers. Increased H&S knowledge and H&S motivation in turn, result in construction workers being empowered in terms of taking ownership of their own individual H&S. Empowerment of construction workers can result in a further decrease in accident and injury rates on a construction site.

T. Mushayi (✉) · C. Deacon · J. Smallwood
Nelson Mandela University, P. O. Box 77000, Port Elizabeth 6031, South Africa
e-mail: tafadzwa.mushayi@mota-engil.co.za

C. Deacon
e-mail: claire@occumed.co.za

J. Smallwood
e-mail: john.smallwood@mandela.ac.za

S. Şahin (ed.), *8th International Conference on Engineering, Project, and Product Management (EPPM 2017)*, Lecture Notes in Mechanical Engineering,
https://doi.org/10.1007/978-3-319-74123-9_25

2 Review of the Literature

2.1 H&S Culture and Climate

Wilkins [14] purports that employees in an organization are relatively governed by similar set of values. He further highlights that organizational culture is a pattern of beliefs and assumptions shared within an organization, and operates 'unconsciously', or without thought. Furthermore, Wilkins [14] states that H&S culture is a subset of organizational culture, where the beliefs and values refer specifically to matters of H&S. A good H&S culture exists when all members of the organization share a high H&S ethic. H&S culture can also be further defined as a concept used to understand a range of aspects such as values, attitudes, social norms, and practices among groups [14]. Attitudes and values can be found in policies, H&S training approaches, procedures and formal communications. Some examples include, when construction workers can demonstrate teamwork, H&S compliance, H&S commitment, and accountability towards H&S. To influence organizational culture, organizations often seek personality traits such as honesty, integrity, responsibility, a strong work ethic, professionalism, and self-motivation when recruiting new employees.

Dingsdag et al. [9] found evidence that organizational climate predicts H&S climate, and along the same line, that H&S climate influences both H&S knowledge and H&S motivation among construction workers. The evidence also suggested that H&S knowledge and H&S motivation are predictors of H&S compliance and H&S acceptance and participation among construction workers. A positive H&S climate in turn results in greater, visible H&S compliance and participation in H&S related activities. Burke et al. [2] purport that the day-to-day activities and behaviors of organizational leaders and management form the foundation of an organizations' H&S culture. Management should lead from the front with a shared vision of H&S excellence and demonstrate exemplary leadership styles. Management should implement practices to drive the desired culture change, such as encouraging employee ownership of H&S. Burke et al. [2] state that managers' own H&S behavior can negatively influence construction workers' perceptions of their support for H&S.

Management should clearly define work procedures and construction workers must fully understand these work procedures. Organizations usually implement different measures to assist construction workers understand safe working procedures, including outlining their duties and responsibilities expected from them. One of these measures is to offer H&S training to construction workers prior to starting construction work. [1] concur with Burke et al. [3] in stating that H&S related problems in organizations are often training related or training relevant. The authors also proved that well targeted H&S training interventions do not only have an impact upon construction worker practices, but also have an impact on attitudes related to H&S, or H&S culture within organizations [1].

2.2 Hazard Identification

One of the key fundamental issues identified to reduce accidents and injuries on a project or in an organization is hazard identification. Enshassi et al. [10] concur with Carter and Smith (2003) in stating that fundamental to construction H&S is hazard identification. They further state that a CW ability to deal with hazards is based upon their hazard awareness, knowledge, experience, and H&S training. For this reason, Enshassi et al. [10] state that the task-hazard relationships must be accurately defined in the design stages regarding tasks. H&S training can be designed based upon the identified hazards. Unidentified hazards present the most challenges to construction workers as not all construction workers can identify a hazard and evaluate risk.

2.3 H&S Training

Bahn and Barratt-Pugh [1] concur with Tannenbaum and Yukl (1992), in stating that effective H&S training involves changing attitudes, behaviors, and consequently the way people conduct and undertake construction tasks. Clarke and Flitcroft [6] further define H&S training as the acquisition and development of the knowledge, skills, and attitudes required by employees to adequately perform a task or job. This definition suggests that H&S training comprises of activities that have the specific goal of developing new knowledge, skills, and attitudes, which can be applied in the workplace. Weinstock and Slatin [13] argue that H&S training should not raise awareness with respect to hazards and risks, but also strengthen construction workers' knowledge and skills to manage risky situations effectively. Such skills should include communications and interpersonal skills.

H&S training offers direct benefits to construction workers regarding H&S. Some of these benefits include the improvement of H&S knowledge and skills, improved problem-solving and analytical skills, and enhanced hazard awareness [6]. Moreover, research studies have demonstrated that trained construction workers follow H&S rules and regulations more closely than those who have not been trained. Another key observation noted is trained employees are less likely to take H&S risks. Clarke and Flitcroft [6] state that several factors influence training, including the design and delivery of training, trainees' traits and motivation, and the work environment. Bahn and Barratt-Pugh [1] in their study concluded that H&S training is successful when it is delivered well.

Two key types of H&S training include H&S induction, and toolbox talks. Grobler et al. (2011) indicate that an induction process integrates new employees into the organization and briefs them regarding the details and requirements of the job. Induction entails providing employees with the necessary information of what is expected of them, to enhance a smooth path to settling in Wilkins [14] states that toolbox talks are less formal and shorter than H&S meetings and training sessions, and are designed to reinforce H&S training and information regarding a topic. Wilkins [14] further states that toolbox talks are a great refresher and keep employees abreast of changes in H&S regulations, job assignments, and responsibilities. It is also effective in terms of restoring workers' knowledge, and toolbox talks facilitate the exchange of information with the construction workers.

2.4 H&S Leadership, Education, and Training

The Occupational Health and Safety Act (OHSA) (Republic of South Africa (RSA) [11] states that every employer should provide and maintain a working environment that is safe and is without risk to the health of employees. The OHSA further states that a safe environment can be ensured through providing information, instructions, H&S training, and supervision to employees. A common approach adopted by most firms in South Africa to improve an organization's H&S culture is to provide management and leadership training to key staff from senior management to site management. Training and educational support offered to managers and leaders assists the managers and leaders in acquiring the relevant H&S knowledge and technical expertise to identify and adequately manage hazards and risks. Bailey (1997) cited by Bust et al. [4] suggests that leadership has the authority and ability to influence the desired behavior and H&S values required on a project. Employees' perceptions of management's commitment to H&S, participation in H&S, and the effectiveness of the H&S training offered, have a significant impact on their H&S behavior. Dėjus and Antuchevičienė [8] indicate that trained managers and leaders are in a better position to assess, evaluate and monitor the effectiveness of the H&S programmers on offer. Furthermore, trained managers are better equipped to appoint competent trainers and employees.

2.5 H&S Meetings

According to the OHSA [11] when two or more H&S Representatives have been appointed for a workplace, at least one H&S committee must be established. The purpose of an H&S committee is to initiate, develop, promote, maintain, and review measures to constantly ensure the H&S of employees at work. The OHSA indicates that the H&S committees should hold meetings as often as may be necessary, but at least once every three months. Most organizations in South Africa have H&S committees in place to maintain good H&S standards. Cameron and Duff [5] state that for H&S committees to be effective, H&S meetings need to be conducted frequently, preferably weekly. H&S committees and meetings usually discuss behavioral H&S issues observed, and assist in developing solutions. H&S meetings also assist in the allocation of resources and support, and discussions with respect to

HIRAs. A further significant issue discussed during H&S meetings is that of H&S induction.

According to Cameron and Duff [5] recording of H&S committee meetings' minutes is vital. Records assist in terms of identifying problem areas, and there after present the H&S committees with opportunities to brainstorm, and evolve solutions with respect to the identified problem. Records of H&S induction, toolbox talks, and HIRAs should also be noted and recorded.

2.6 *Manager H&S Inspections*

Manager H&S inspections and actions are also fundamental H&S related issues. Cameron and Duff [5] state that regular H&S inspections positively influence H&S practices as inspections monitor the construction process, and provide a platform to recommend corrective action. An inspection also provides a platform for interaction between the workers and management, which enables management to listen to the concerns of workers and supervisors, and develop an enhanced understanding of activities and tasks.

2.7 *H&S Signs and Symbols*

According to Tam et al. [12], signs and symbols have been used extensively on construction sites for H&S awareness purposes. Such signs convey different H&S messages to construction personnel. Warnings in the forms of labels and symbols have been recognized as one of the effective tools to influence H&S behavior and improve the risk perceptions of construction workers [12]. The major function of these signs and symbols is to draw observers' attention, and to convey an immediate warning with respect to the level of danger.

3 Research

The targeted population included grade 4–9 GCs registered with the CIDB in the Eastern Cape and Gauteng provinces. 25 GCs were contacted, and 11 questionnaires were returned resulting in a response rate of 44%. 3 responses were required from each GC; from a site agent, H&S Officer, and supervisor. A total of 33 responses were received within the allocated time.

All 33 responses were included in the analysis of the data. The composition of the sample consisted of 12 (36.4%) females and 21 (63.6%) males. The response rate of H&S officers (14/25) was 56%. The response rate of management supervisors (10/25) against site agents (9/25) was 40% and 36% respectively.

9 (27.3%) were holders of a bachelor's degree, 6 (18.2%) had an honors degree, 10 (30.3%) were diplomates, 7 (21.2%) had a postgraduate diploma, and lastly, 1 (3.0%) had a master's degree.

The questionnaire consisted of eleven questions, one of which was demographic related, three were open-ended, and the remaining seven questions were five- point Likert scale type questions. A measure of central tendency in the form of a mean score (MS) between 1.00 and 5.00 was computed based upon the percentage responses to the points on the respective scales to enable interpretation of the responses, and to rank factors where necessary. The responses were weighted as per the figures recorded within parentheses: never/minor extent (1); rarely/near minor extent (2); sometimes/some extent (3); often/near major extent (4), and always/major extent (5).

Respondents were required to indicate the importance of addressing the H&S training of construction workers before CWs undertake a task. The resultant MS of 4.89 is >4.20 to $\leq$5.00, indicates that the importance is between a near major extent to major/major extent.

Respondents were then asked to identify the factors that contribute to worker H&S attitudes and H&S perceptions. The MSs of the factors ranked from first to fifth are >4.20 to $\leq$5.00, and include management involvement in H&S (4.51), worker negligence (4.49), management participation (4.46), H&S culture and climate (4.34), and first line supervision (4.23). The MSs indicate that all these factors contribute to worker H&S attitudes and H&S perceptions between a near major extent to major/major extent.

Table 1 indicates the extent to which factors contribute to the causes of accidents on projects, in terms of percentage responses to a scale of 1 (minor) and 5 (major),

Table 1 Extent to which factors contribute to the causes of accidents on projects

Factor	Response (%)						MS	R
	U	Minor–Major						
		1	2	3	4	5		
Lack of hazard identification and awareness on site	0.0	5.7	2.9	5.7	31.4	54.3	4.26	1
Poor H&S culture and H&S climate	0.0	2.9	2.9	11.4	34.3	48.6	4.23	2
Lack of supervision	0.0	2.9	2.9	14.3	37.1	42.9	4.14	3=
Inadequate management commitment	0.0	2.9	8.6	17.1	14.3	57.1	4.14	3=
Inadequate H&S training	0.0	2.9	8.6	11.4	34.3	42.9	4.06	5
Inadequate management participation	0.0	5.7	8.6	14.3	20.0	51.4	4.03	6
Lack of analytical skills	0.0	2.9	8.6	28.6	25.7	34.3	3.80	7
Low knowledge retention	0.0	5.7	8.6	25.7	28.6	31.4	3.71	8
Safe working procedures	2.9	2.9	8.6	31.4	17.1	37.1	3.69	9
Inadequate toolbox talks	0.0	5.7	8.6	34.3	31.4	20.0	3.51	10
Lack of feedback platforms (H&S suggestion box)	0.0	11.4	17.1	11.4	31.4	28.6	3.49	11=
Inadequate H&S induction	0.0	8.6	11.4	28.6	25.7	25.7	3.49	11=
Frequency of the H&S training	0.0	8.6	8.6	28.6	42.9	11.4	3.40	13
Duration of training	0.0	17.1	20.0	34.3	25.7	2.9	2.77	14

and MSs. 13/14 (92.9%) factors have MSs >3.00, which indicates that the factors contribute between a major extent as opposed to minor extent to the causes of accidents, as in the case of the duration of H&S training which had a MS $\leq$3.00. Lack of hazard identification and awareness was ranked first with a MS of 4.26, followed by poor H&S culture and H&S climate with a MS of 4.23. Given that the MSs are >4.20 to $\leq$5.00, they contribute to the causes of accidents between a near major extent to major/major extent.

10/14 (71.4%) MSs >3.40 to $\leq$4.20, which indicates that the factors contribute between some extent to a near major/near major extent to the causes of accidents. Factors include lack of supervision and inadequate management commitment ranked joint third, followed by inadequate H&S training, inadequate management participation, lack of analytical skills, low knowledge retention, safe working procedures, inadequate toolbox talks, lack of feedback platforms (H&S suggestion box), and inadequate H&S induction.

Table 2 in turn indicates the extent to which factors contribute to the successful delivery of an H&S training program. All the factors have MSs >3.00, which indicates that the factors contribute a major extent as opposed to minor extent to the successful delivery of an H&S training program. The factors (45.5%) ranked joint first to fifth have MSs >4.20 to $\leq$5.00, which indicates they contribute to the successful delivery of an H&S training program between a near major extent to major/major extent—employee participation, job specific knowledge shared in the H&S training program, skills and abilities gained, competency of H&S trainers, and relevance of H&S training to worker tasks. Given that the factors (45.5%) ranked from sixth to tenth have MSs >3.40 to $\leq$4.20, they contribute to the successful

Table 2 Extent to which factors contribute to the successful delivery of an H&S training program

Factor	Response (%)						MS	R
	U	Minor–Major						
		1	2	3	4	5		
Employee participation	0.0	0.0	5.7	2.9	42.9	48.6	4.34	1=
Job specific knowledge shared in the H&S training programme	0.0	0.0	2.9	8.6	40.0	48.6	4.34	1=
Skills and abilities gained	0.0	0.0	0.0	11.4	42.9	45.7	4.34	1=
Competency of H&S trainers	0.0	2.9	0.0	8.6	37.1	51.4	4.34	1=
Relevance of H&S training to worker tasks	0.0	2.9	2.9	8.6	37.1	48.6	4.26	5
Involving workers in the development of the H&S programme	0.0	2.9	2.9	11.4	45.7	37.1	4.11	6
Additional materials used in training (videos, handouts)	0.0	8.6	5.7	17.1	22.9	45.7	3.91	7
Inadequate supervision	2.9	5.7	5.7	17.1	34.3	34.3	3.77	8
Size of group being trained	2.9	5.7	11.4	25.7	28.6	25.7	3.49	9
Duration spent on H&S training	2.9	8.6	8.6	34.3	14.3	31.4	3.43	10
Rest break during H&S training	0.0	14.3	20.0	25.7	20.0	20.0	3.11	11

delivery of training programs between some extent to a near major/near major extent—involving workers in the development of the H&S program, additional materials used in training (videos, handouts), inadequate supervision, size of group being trained, and duration spent on H&S training.

Respondents were also requested to make comments in general regarding H&S training in terms of its delivery and effectiveness. Some of the remarks include "Incompetent trainers without practical experience and thus just reading what is on the presentation." and "The effectiveness of the training is largely influenced by the enthusiasm and knowledge level of the trainer. If a trainer is boring or disinterested in the subject matter the employees quickly lose interest and turn off."

Another open-ended question requested respondents to comment on any other issues relating to H&S not addressed by the questionnaire. One respondent remarked: "In order for Health and Safety to improve in the construction industry, it must be incorporated in the curriculum of students who are studying towards any major construction related field in tertiary institutions. Management of companies need to take health and safety very seriously, also they must have knowledge of health and safety. They must be committed and be involved and lead by example."

Another respondent stated that: "Largest influence on a construction site with regards to safety is the first line supervision. He carries the legal responsibility and the accountability for his crew. Big concern is with regards to 'the best' artisan that is promoted as 'supervision' in the industry, but do not have the necessary supervision skills."

4 Conclusions

The research determined that it is essential for construction workers to receive relevant, appropriate, and adequate H&S training to gain the relevant knowledge and skills related to undertake a task. H&S training should further transmit knowledge and skills to identify hazards, change attitudes, and engender safe work behavior. H&S training should be effective in such a way that it involves changing, perceptions, attitudes, behaviors and consequently the way construction workers conduct themselves at work. If construction workers receive adequate and relevant H&S training they are empowered to effectively deal with site hazards when encountered. Another key observation is the need for the inclusion and participation of construction workers in any H&S related programs such as H&S induction training. Involvement and participation will indicate to construction workers that they are included and a part of the process, which in turn will motivate and encourage them to change their attitude and behavior.

The research also determined that H&S culture and climate have a major influence on the H&S attitudes of construction workers, and organizations should invest in building a positive H&S culture and climate. Management involvement in H&S and the presence of first line supervision are fundamental interventions in terms of developing a positive H&S culture and climate. Top management commitment is crucial in terms of achieving H&S standards for any organization, as top management is usually the driver and initiator of any organization's H&S vision.

5 Recommendations

Hazards need to be addressed from the design stages of a project to mitigate their occurrence on site. The identification of residual hazards will enable the construction team to propose and implement measures to empower construction workers to deal with such hazards when encountered.

Organizations should offer H&S training programs that are relevant to construction workers. The H&S training should transfer the relevant H&S knowledge and skills to construction workers, which will enable them to effectively deal with the identified site environment and related H&S hazards.

Organizations should invest significantly towards building a positive H&S culture and climate. Organizations should clearly define their values, and ideally, new employees should only be recruited if their values align with those of the organization. General contractors should influence and educate construction workers to empower them to take ownership of their own H&S.

Competence from management and trainers has a major influence on the success of achieving the organizations H&S goals and objectives. Management should be clearly aware of their responsibility towards implementing and monitoring H&S compliance from Construction Workers. Supervision of Construction Workers and their tasks should be regularly conducted to monitor the work procedures. Roles of management should be clearly defined and understood by each member. Competent managers should be well abreast of H&S training programs required by Construction Workers and should appoint competent trainers. Trainers together with management should design appropriate H&S training programs that are of relevance to worker tasks. These H&S training programs should be effectively delivered to Construction Workers ensuring the sharing of knowledge and analytical skills to effectively deal with hazards.

References

1. Bahn S, Barratt-Pugh L (2014) H&S training evaluation: the case of construction induction training and the impact on work-related injuries in the Western Australian construction sector. Int J Training Res 12(2):148–157
2. Burke MJ, Chan-Serafin S, Salvador R, Smith A, Sarpy SA (2008) The role of national culture and organisational climate in H&S training effectiveness. Eur J Work Organ Psychol 17(1):133–152
3. Burke MJ, Chan-Serafin S, Salvador RO, Smith-Crowe K (2011) The dread factor: how hazards and safety training influence learning and perfomance. J Appl Psychol 96(1):46–70

4. Bust PD, Gibb AGF, Pink S (2008) Managing construction H&S: migrant workers and communicating H&S messages. H&S Sci 46:585–602
5. Cameron I, Duff R (2007) Use of performance measurement and goal setting to improve construction managers' focus on H&S. Constr Manage Econ 25:869–881
6. Clarke S, Flitcroft C (2013) The effectiveness of training in promoting a positive OSH culture. The effectiveness of training in promoting a positive OSH culture. Institute of Occupational Safety & Health, Wigston
7. Construction Industry Development Board (CIDB) (2009) Construction health and safety in South Africa status and recommendations. CIDB, Pretoria
8. Dėjus T, Antuchevičienė J (2013) Survey on the H&S training (learning) for construction workers: assessment of H&S solutions at a construction site. J Civ Eng Manage 19(5): 728–737
9. Dingsdag DP, Biggs HC, Sheahan VL (2008) Understanding and defining OH&S competency for construction site positions: worker perceptions. H&S Sci 46:619–633
10. Enshassi A, El-masri F, Mayer PE, Mohamed S (2007) Perception of construction managers towards safety in Palestine. Int J Constr Manage 7(2):41–51
11. Republic of South Africa (1993) Occupational health and safety act, No. 85. Pretoria
12. Tam CM, Fung IWH, Yeung TCL, Tung KCF (2003) Relationship between construction safety signs and symbols recognition and characteristics of construction personnel. Constr Manage Econ 21(7):745–753
13. Weinstock D, Slatin C (2012) Learning to take action: The goals of health and safety training. New Solutions 22(3):255–267
14. Wilkins JR (2011) Construction workers' perceptions of health and safety training programmes. Constr Manage Econ 29:1017–1026

Graduate Employment: Introducing Construction Management Graduates to the Workplace in South Africa

Mafa Maraqana and John Smallwood

1 Introduction

The rapid changes in the construction business environment, fierce global competition, detailed client requirements, and the complexity of construction projects amplifies the need for highly educated and competent construction management graduates [17].

In the final report emanating from an internship baseline study, Koyana [13] claims that the biggest problem in the labour market is the inability to support young people in making the 'school-to-work' transition. This is arguably due to the work experience required at work only encompassing a low percentage of the invaluable theoretical knowledge acquired at HEIs. Although Lowden et al. [14] suggest that these institutions instill theoretical knowledge that is vital to any employment, and are a stepping-stone to obtaining good employment, employers and graduates perceive that more work can be undertaken to: increase the number of students graduating; develop students' wider skills and attributes, and ease the difficult transition to the workplace environment. Skills are of importance because, according to Zakariaa et al. [25], to be a competent employee, a person needs to possess two types of skills; specific skills, and employability skills.

Evidently, graduates require training and development to become effective in their roles, and to contribute to organisations. Prior to this, and depending on the circumstances, graduates may have to be absorbed in 'entry-level' graduate employment.

The primary objective of the study reported on is to investigate the integration of construction management graduates into the workplace, and the secondary

M. Maraqana (✉) · J. Smallwood
Nelson Mandela University, PO Box 77000, Port Elizabeth 6031, South Africa
e-mail: m.maraqana@gmail.com

J. Smallwood
e-mail: john.smallwood@mandela.ac.za

S. Şahin (ed.), *8th International Conference on Engineering, Project, and Product Management (EPPM 2017)*, Lecture Notes in Mechanical Engineering,
https://doi.org/10.1007/978-3-319-74123-9_26

objectives are to determine: the contribution of HEIs to students' work readiness; challenges faced by graduate employees at work, and the contribution of GDPs and internships to their integration into the workplace.

2 Review of the Literature

2.1 Transitioning into the Work Environment

Gallagher [8] states that the increasing focus on the working environment, and the fluctuating graduate labour market has raised many queries with respect to formal learning, and the capabilities and employability of graduates, despite being the driving force for employability.

The role of HEIs lies in the disseminating of knowledge, and awarding of qualifications for the main purpose of job entry and earning potential [23] and [11]. However, Othman [17] states that it appears that students and professionals within the industry perceive that HEIs are not doing enough to ensure the preparation of students and graduates for the workplace, and to ensure adequate development of these individuals. As a result, the transition to the workplace can be particularly challenging to a graduate, and there can be a significant mismatch between employer and graduate expectations, says Gallagher [8]. Rochlen and Wendlandt [21] identify three themes regarding the challenges of transitioning into work: are a change in culture associated with the transition between the different HEI and work environments; inaccurate expectations with respect to work life, and the lack of experience and skills required by employers.

Olson [15] states that this period of transition is complex, individualized, and difficult to predict. Graduates often feel unprepared for the new work settings and new modes of learning. Many recent graduates experience tension as they move into work settings with professional expectations different from those of parents or family networks. Furthermore, according to Oxenbridge and Evesson [18], research indicates that newly employed graduates consider themselves vulnerable and unprepared, or lacking in confidence when starting work as the work appeared intimidating, daunting, and provoked anxiety. These factors also contribute to the difficulty of integrating into an organization. Additionally, the graduates perceived that employers had unrealistic expectations of the skills and abilities they possessed, albeit they lacked experience.

2.2 Employability Skills

Das and Emuze [6] warn that there is a critical case for universities to evaluate their curricula and determine how the desirable skills could be included in the modules.

Affandi et al. [1] state that construction management graduates' competencies generally meet employers' expectation; however, there are skills where graduates fall below the expectation of contracting organizations: practical building knowledge; interpersonal skills; time management, and the ability to exercise professional judgment.

Mirrored to this in the study conducted by the Council for the Built Environment [4], broadly to include built environment graduates, is the notion that construction management graduates fail to meet expectations. The findings also unveil concerns that graduates lack skills and attributes that are required by built environment employers from the onset, and therefore job remedial training is necessary. The report suggests that the graduates have acquired theoretical knowledge, but are not ready for the workplace. Yusof et al. [24] suggest that employers are dissatisfied with the quality of graduates entering the work environment, and that the main culprit points to the educational programmes not being responsive to the changing realities, and practices in the construction industry.

There is ambiguity as to the skills most desired by employers. Heaton et al. [9] state that literature presents contradictions with respect to employers' requirements of graduates, with some reports suggesting that some attributes and skills are more valued than others, or that employers are satisfied (or dissatisfied) with one attribute or skill versus another. Oluwajodu et al. [16] claims that there are main categories of skills and attributes that employers seek when recruiting graduates. Basic skills, intellectual ability, workplace skills, applied knowledge, and interactive skills are most important. On the other hand, Yusof et al. [24] suggest that employers appear to be seeking a more flexible and adaptable workforce so that these graduate employees can enable their employers to become more flexible and adaptable to changing market needs.

Poon [19] notes that employers are particularly dissatisfied by (in descending order of importance): lack of commercial awareness; analysis and decision-making skills; communication skills; literacy (proficient writing skills); passion; relevant work experience; planning and organizational skills; confidence, and personal development skills.

To address the skill deficiencies of graduates, Jackling and Natoli [12] recommend increased calls for the participation of students in experiential work to enhance their work readiness. It is interesting to note, according to Crawford et al. [5], that students and graduates with work experience are more likely to achieve success in their final year of study, and earlier in their careers (both in terms of progression and compensation) in comparison to their student and graduate counterparts who do not have work experience.

Work experience also assists in introducing students and graduates to the organisational setting and culture, as graduates may find it difficult to integration into an organisation [2].

2.3 Graduate Employment

Assessing the difference between graduate and non-graduate employment, the Department of Business Innovation & Skills [7] explains that graduate occupations are so defined as those providing scope for the utilization of 'high degree-level' skills, and where the possession of a degree is a long-established, or relatively new entry requirement for the occupation. By contrast, the opposite applies to non-graduate occupations. Vansa [22] and Raftopoulos [20] outline some of the programmes available to a graduate or student upon entry into the organization as a new hire: internships; apprenticeships; learnerships; work readiness programs, and GDPs.

Connor and Shaw [3] further explain that the proportion of graduates entering the workplace via GDPs has fluctuated over the decades and argue that, in recent years, GDPs have not only grown in popularity, but that graduates on these programs have noticeably more accelerated career progression than graduates in non-graduate occupations.

Hegarty and Johnston [10] explain that organisations adopt formal GDPs to raise an organisation's intellectual capital, and to tackle the issue of the lack of skills and competencies of graduates. The design of the programmes is aimed at attracting and retaining graduates. The program endeavors to encourage graduates to take ownership of their own learning; to provide management with the opportunity to develop mentoring capabilities, and to organize graduates into cross-functional learning groups with mentors. Hegarty and Johnston [10] believe that high salary and benefits, an opportunity for development, and advancement, a stimulating and challenging work, and mentoring is most valued by prospective GDP participants. When adopting GDPs, Raftopoulos [20] advises organizations to:

- Assign mentors (mentoring program) to graduates who will guide, and coach graduates to stimulate their growth and development;
- Develop the program whereby the graduates will have the opportunity and environment to learn and grow;
- Establish a business case for the GDP and involve senior management and the directors from the start;
- Include a comprehensive induction program in the GDP;
- Make sure that the three essential R's of graduate development—recruiting, retaining, and return on investment, are fully integrated into the program, and
- Measure success by measuring the graduates' contribution to the profit, and measure their potential by rating their progress on the program.

According to Koyana [13], an internship is a means by which organizations provide an opportunity to individuals to apply the knowledge they have acquired in the classroom to the workplace. The primary purpose is to provide hands-on experience that enhances an individual's learning or understanding of issues relevant to an area of study. The programs assist the intern to bridge the gap between the academic learning process and the practical reality. Internships are often temporary or contractual in nature.

3 Research

A questionnaire (soft and hard copy) was created to capture the responses of individuals at the academic level and in industry. The soft copies were e-mailed to potential respondents and hardcopies physically distributed. An online (electronic format) questionnaire/survey was also created, to which a website link was sent via e-mail and LinkedIn (social network). A total of 243 e-mails were sent, 19 LinkedIn messages, and 13 questionnaires were physically distributed. This totalled 275 potential responses. 29 Responses were received, which represents a response rate of 10.5%.

Responses where received from construction management students and graduate employees, and employers (or representatives thereof). The responses were as follows: students (11); graduate employees (5), and employers and representatives (13).

Three similar, but not alike questionnaires were compiled, which included predominantly five-point Likert scale type questions. Given the latter, a measure of central tendency in the form of a mean score (MS) between 1.00 and 5.00 was computed based upon the percentage responses to the points on the scale to enable interpretation of the responses, and to rank factors where necessary. The responses were weighted as per the figures recorded within parentheses: no/minor extent (1); minor/near minor extent (2); some extent/some extent (3); major/near major extent (4), and extensive/major extent (5).

Table 1 indicates the contribution of HEIs to student's work readiness relative to thirteen factors, in terms of percentage responses to a scale of 1 (no) and 5 (extensive), and MSs between 1.00 and 5.00. 11/13 (84.6%) factors have MSs >3.00, which indicates that the factors contribution is extensive as opposed to minor. It is notable that no MSs are >4.20 to $\leq$5.00—between a major to extensive/extensive contribution.

6/13 (46.2%) MSs >3.40 to $\leq$4.20, which indicates that the HEIs contributed between some extents to a major/major extent to student's work readiness: knowledge (science & technology); enhancing career development; providing academic support; knowledge: CM discipline related, equipping students with employability skills, and knowledge: economics. The MSs of the remaining seven (53.8%) factors are >2.60 to $\leq$3.40 which indicates that the HEIs contribute between a near minor to some extent/some extent to student's work readiness: knowledge: business and management; knowledge: accounting and finance; providing information regarding construction industry trends and developments; knowledge: other e.g. general; providing counselling to students; promoting self-development, and preparing students for the work environment. Clearly the emphasis is on academia as opposed to all round development, especially personal development, and preparing students for the work environment.

Table 2 indicates the extent to which graduate employees face fifteen challenges at work, in terms of percentage responses to a scale of 1 (no) and 5 (extensive), and

Table 1 The contribution of HEIs to students' work readiness

Factor	Response (%)						MS	R
	U	No				Extensive		
		1	2	3	4	5		
Knowledge: science & technology	0.0	0.0	17.2	20.7	31.0	31.0	3.76	1
Enhancing career development	3.5	3.5	10.3	3.5	62.1	17.2	3.72	2
Providing academic support	0.0	0.0	13.8	27.6	34.5	24.1	3.69	3
Knowledge: CM discipline related	3.5	0.0	10.3	41.4	17.2	27.6	3.55	4
Equipping students with employability skills	0.0	3.5	10.3	34.5	37.9	13.8	3.48	5
Knowledge: economics	6.9	6.9	10.3	20.7	31.0	24.1	3.41	6
Knowledge: business and Management	10.3	0.0	20.7	17.2	27.6	24.1	3.34	7
Knowledge: accounting and Finance	10.3	6.9	6.9	27.6	24.1	24.1	3.31	8
Providing information regarding construction industry trends and developments	3.5	3.5	17.2	41.4	13.8	20.7	3.24	9
Knowledge: other e.g. general	3.5	3.5	17.2	41.4	17.2	17.2	3.21	10
Providing counselling to students	3.5	6.9	20.7	24.1	34.5	10.3	3.14	11
Promoting self-development	3.5	3.5	27.6	37.9	17.2	10.3	2.97	12
Preparing students for the work environment	0.0	10.3	31.0	24.1	27.6	6.9	2.90	13

MSs between 1.00 and 5.00. 6/15 (40%) MSs >3.00, which indicates that the extent to which graduate employees face the challenges is extensive as opposed to minor. It is notable that no MSs are >4.20 to ≤5.00—the extent is between major to extensive/extensive.

3/15 (20%) MSs >3.40 to ≤4.20, which indicates that the extent is between some extent to a major extent/major extent: feeling vulnerable, and lacking in confidence; having inaccurate expectations about work life, and the lack of experience and skills required by employers.

A further five (33.3%) MSs are >2.60 to ≤3.40, which indicates that the extent is between a minor extent to some extent/some extent: adapting to a change in culture from HEIs to the work environment; generational difference (age gaps) issues with older employees; the lack of planning and preparation for commencing work; coping with difficult and complex tasks, and inadequate mentoring.

The remaining seven (46.7%) MSs are >1.80 to ≤2.60, which indicates that the extent is between a minor extent to some extent/some extent: difficulty in understanding the degree of work responsibility; difficulty in understanding organisational social relations and hierarchy; could not relate to the organisational culture; cultural, religious and ethnicity issues; difficulty in understanding the norms, policies, and rules; discrimination against race, and discrimination against gender.

Table 2 Challenges faced by graduate employees at work

Challenge	Response (%)						MS	R
	U	No				Extensive		
		1	2	3	4	5		
Feeling vulnerable, and lacking in confidence	0.0	5.6	5.6	33.3	38.9	16.7	3.56	1=
Having inaccurate expectations about work life	0.0	0.0	5.6	44.4	38.9	11.1	3.56	1=
The lack of experience and skills required by employers	0.0	0.0	11.1	38.9	44.4	5.6	3.44	3
Adapting to a change in culture from HEIs to the work environment	0.0	11.1	16.7	16.7	50.0	5.6	3.22	4=
Generational difference (age gaps) issues with older employees	0.0	5.6	16.7	33.3	38.9	5.6	3.22	4=
The lack of planning and preparation for commencing work	0.0	0.0	16.7	50.0	33.3	0.0	3.17	6
Coping with difficult and complex tasks	0.0	11.1	16.7	38.9	33.3	0.0	2.94	7
Inadequate mentoring	0.0	27.8	5.6	38.9	22.2	5.6	2.72	8
Difficulty in understanding the degree of work responsibility	5.6	27.8	16.7	16.7	27.8	5.6	2.56	9
Difficulty in understanding organisational social relations and hierarchy	0.0	27.8	22.2	33.3	11.1	5.6	2.44	10
Could not relate to the organisational culture	0.0	27.8	27.8	22.2	22.2	0.0	2.39	11
Cultural, religious and ethnicity issues	0.0	33.3	16.7	38.9	11.1	0.0	2.28	12
Difficulty in understanding the norms, policies, and rules	0.0	33.3	22.2	33.3	11.1	0.0	2.22	13
Discrimination against race	0.0	33.3	27.8	33.3	5.6	0.0	2.11	14
Discrimination against gender	5.6	38.9	22.2	27.8	5.6	0.0	1.94	15

The challenges faced are possibly attributable to the emphasis on academia as opposed to all round development, especially personal development, and preparing students for the work environment as reflected in Table 1.

Table 3 indicates the contribution of GDPs and internships relative to ten factors, in terms of percentage responses to a scale of 1 (minor) and 5 (major), and MSs between 1.00 and 5.00. All ten factors have MSs >3.00, which indicates that the factors contribution is major as opposed to minor. It is notable that no MSs are >4.20 to ≤ 5.00—between a major to extensive/extensive contribution.

All ten are MSs >3.40 to ≤ 4.20, which indicates that the factors contributed between some extent to a near major/near major extent. It is also notable that the

Table 3 Contribution of GDPs and internships

Factor	Response (%)						MS	R
	U	Minor				Major		
		1	2	3	4	5		
Induction to the job and organisation	0.0	0.0	5.6	11.1	44.4	38.9	4.17	1
Enhancing job knowledge	0.0	0.0	5.6	11.1	55.6	27.8	4.06	2=
Enhancing skills	0.0	0.0	5.6	11.1	55.6	27.8	4.06	2=
Developing graduates	0.0	0.0	11.1	11.1	38.9	38.9	4.06	2=
Registration to a professional body (SACPCMP)	0.0	5.6	0.0	22.2	44.4	27.8	3.89	5
Job training	0.0	5.6	5.6	11.1	55.6	22.2	3.83	6=
Mitigating the challenges faced in transitioning into the workplace	0.0	0.0	0.0	27.8	61.1	11.1	3.83	6=
Effective mentoring	5.6	5.6	0.0	16.7	50.0	22.2	3.72	8=
Enhancing self-confidence	0.0	5.6	5.6	22.2	44.4	22.2	3.72	8=
Mitigating the challenges faced in integrating into an organisation	0.0	0.0	5.6	33.3	50.0	11.1	3.67	10

MS of the first ranked factor, namely 'induction to the job and organisation' is marginally outside the upper range. It is notable that the surface competencies in the form of knowledge and skills are ranked joint second with developing graduates, the latter being the primary objective of such GDPs and internships. Fifth ranked 'registration with a professional body (SACPCMP)' should be encouraged as the related candidacy period requires mentorship, documentation of development, and reporting thereon. Joint sixth ranked 'job training' is also a secondary objective of such GDPs and internships. The second joint sixth ranked 'mitigating the challenges faced in transitioning into the work-place' and tenth ranked 'mitigating the challenges faced in integrating into an organisation' are also key secondary objectives of such GDPs and internships. Joint eighth ranked 'effective mentoring' is promoted by formal GDPs and internships, and 'enhancing self-confidence' is an expected and very necessary output.

Clearly the emphasis is on academia as opposed to all round development, especially personal development, and preparing students for the work environment.

4 Conclusions

There is sufficient evidence from the literature review and the empirical findings that suggests that students and graduates lack employability skills that are highly sought after by employers. The lack of these skills emanates partly because of students and graduates not engaging in work experience, and HEIs not doing enough to equip students with skills. It is noted, however, that most respondents and the literature state that HEIs, and work experience, play a crucial role in preparing students for the workplace, albeit there is debate.

This is of importance, as graduates experience challenges when they transition into the working environment, and are adopted into organizations. The empirical findings suggest that perhaps some challenges are experienced less extensively than others.

GDPs and internships are specifically meant to assist in the socialization, and development of graduates. The literature, and empirical findings attest to this by indicating that GDPs and internships: induct graduates to the job and organization; enhance job knowledge; enhance skills; develop graduates; assist in the registration of graduate employees to a professional body; provide job training; attempt to mitigate the challenges faced by graduates in transitioning into the workplace; attempt to institute an effective mentoring program; enhance the self-confidence of graduates, and attempt to mitigate the challenges faced by graduates in integrating into an organization.

References

1. Affandi HM, Hassan F, Ismail Z, Kamal MF (2012) Soft skills implementation in construction management program: a comparative study of Lecturers and students' perspective. Universiti Teknologi MARA, Sabah, Malaysia
2. Bauer TN (2010) Onboarding new employees: maximizing success. SHRM Foundation, Alexandria
3. Connor H, Shaw S (2008) Graduate training and development: current trends and issues. Educ Train 50(5):357–365
4. Council for the Built Environment (2014) Is there a gap between graduate competencies and the expectations of the built environment industry in South Africa. Council for the Built Environment, Pretoria, RSA
5. Crawford I, Wang Z, Andrews G (2016) Exploring the influence of individual and academic differences on the placement participation rate among international students. Educ Train 58 (4):342–357
6. Das DK, Emuze FA (2015) Contextualising urban engineering education for future cities. J Constr Proj Manag Innov 5(2):1139–1151
7. Department of Business Innovation & Skills (2015) Understanding employers' graduate recruitment and selection practices. Institute for Employment Studies (IES), Brighton, UK
8. Gallagher P (2015) Graduate transition into work: the bridging role of graduate placement programmes in the small- and medium-sized enterprise workplace. J Educ Work 28(5): 461–480
9. Heaton N, McCracken M, Harrison J (2008) Graduate recruitment and development. Educ Train 50(4):276–288
10. Hegarty C, Johnston J (2008) Graduate training: evidence from FUSION projects in Ireland. Educ Train 50(5):391–405
11. Hendricks KB (2014) Graduates' perspectives on transition from college graduation to the workplace: a qualitative approach. University of Kentucky, Kentucky
12. Jackling B, Natoli R (2015) Employability skills of international accounting graduates. Educ Train 57(7):757–773

13. Koyana S (2013) Report on the internship baseline study 2013, South Africa: The National Youth Development Agency (NYDA) and The South African Graduates Development Association (SAGDA)
14. Lowden K, Hall S, Elliot D, Lewin J (2011) Employers' perceptions of the employability skills of new graduates. Edge Foundation, London, UK
15. Olson JS (2016) Chasing a passion: first-generation college graduates at work. Education 58 (4):358–371
16. Oluwajodu F, Blaauw D, Greyling L, Kleynhans EPJ (2015) Graduate unemployment in South Africa: perspectives from the banking sector. J Hum Resour 13(1):1–9
17. Othman AAE (2014) A study of the competency requirements of property development graduates to meet the qualifications of the South African construction industry. J Eng Design Technol 12(1):2–28
18. Oxenbridge S, Evesson J (2012) Young people entering work: a review of the research. Acas Research and Evaluation programme, UK
19. Poon J (2012) Real estate graduates' employability skills. Prop Manag 30(5):416–434
20. Raftopoulos M (2006) Work readiness and graduate recruitment in the fasset sector. University of South Africa, Pretoria, RSA
21. Rochlen AB, Wendlandt NM (2008) Addressing the college-to-work transition: implications for university career counselors. J Career Dev 35(2):151–165
22. Vansa (2015) Internship toolkit. Vansa and Africalia, Johannesburg, RSA
23. Yang J, Schneller C, Roche S (2015) The role of higher education in promoting life long learning. UNESCO Institute for Life Long Learning, Hamburg, Germany
24. Yusof N, Fauzi SNFM, Abidin NZ, Awang H (2013) Improving graduates' employability skills through industrial training: Suggestions from employers. J Educ Pract 4(4):1–7
25. Zakariaa MH, Yatima B, Ismail S (2014) A new approach in measuring graduate employability skills. AIP Publishing LLC, Malaysia

A New Method to Tackle the Duration Risks of a Construction Project

Wen-der Yu, Hsien-kuan Chang and Shao-tsai Cheng

1 Introduction

Accurate estimation of the activity durations is the most essential element for the effective management of a construction project. Due to the complex conditions existing in the construction environment, there are many influential risk factors associated with an activity. It makes the accurate estimation of activity durations extremely difficult. The most well-known method to deal with such risks has been the Project Evaluation and Review Technique (PERT). In PERT, the different influential risk factors associated with an activity is considered simultaneously by the scheduler to come up with a set of three point estimates (*i.e.*, most-optimistic, most-likely, and most-pessimistic) for the duration of an activity [4]. Although PERT provides a convenient approach to tackle the risks associated with activity duration, it has been criticized to be overly simplified and inaccurate. Using the three point estimates mentioned above, the expected time (t_e) for the duration of an activity can be calculated. The t_e implies the average time the task requires if the task were repeated over an extended period of time. There are several drawbacks associated with the PERT method, such as: historical records are usually not available; most schedulers are not familiar with statistic distribution;

W. Yu (✉)
Department of Construction Engineering, Chaoyang University of Technology, Taichung, Taiwan
e-mail: wenderyu@cyut.edu.tw

H. Chang
Department of Civil Engineering, Chung Hua University, Hsinchu, Taiwan
e-mail: d10304006@chu.edu.tw

S. Cheng
Department of Construction Management, Chung Hua University, Hsinchu, Taiwan
e-mail: shaotsai@chu.edu.tw

S. Şahin (ed.), *8th International Conference on Engineering, Project, and Product Management (EPPM 2017)*, Lecture Notes in Mechanical Engineering,
https://doi.org/10.1007/978-3-319-74123-9_27

Traditional PERT provides only schedule planning but no schedule controlling method, it is difficult to control the risk factors effectively.

Very little improvement was achieved regarding the above drawbacks since PERT was proposed in 1958. As a result, the PERT method was rarely implemented in practice. The present research aims at developing a systematic approach to tackle the different influential risk factors that affect the activity durations of a construction project, so that a more accurate planned schedule can be attained.

2 Previous Works on the Integration of Risk Management and Scheduling

Risk is the probability of events that may cause negative impacts on the objectives. Since the essential nature of a risk is uncertainty, the primary objective of risk analysis is to evaluate whether we should undertake or avoid such risks [4]. The risk category characterizes the risks with their essential characteristics; it facilitates the process of risk management. The structure of the category usually contains a major category that can be furtherly derived into layers of subcategories by the nature of risk events. Finally, it will form a Risk breakdown structure (BRS) [4].

In order to better tackle the risk in a construction activity, previous researchers Mol [9], Yi and Langford [13] proposed an equation to express all risks involved in a construction activity as bellow:

$$Total\,Risk = P \times H \times T \times E \tag{1}$$

where, P means 'Process Risk'; H means 'Human Risk'; T means 'Technology Risk', E means 'Physical Environment Risk'; and the symbol '$\times$' implies that the impacts of the risk factors to the schedule is a product of the impacts caused by the individual risk factors.

The above risk breakdown concept is very useful for analysis of the risk factors affecting construction schedule. However, no systematic approach was proposed yet to implement the above-mentioned factors in construction scheduling process. Other literature related to the integration between risk management and scheduling includes: (1) Application of risk analysis to float utilization and optimization [2, 3] (2) Application of risk evaluation to scheduling optimization [10, 14]; (3) Risk-based scheduling and safety programming [13]; (4) Algorithm of risk reduction [6]; (5) Integration of risk management and scheduling technique [11]; (6) Case study of risk-based scheduling application [15]; (7) Quantitative method for risk analysis [5, 7]; (8) Integration of schedule control and risk information [12].

3 Proposed Risk Critical Path Scheduling Method (R-CPSM)

3.1 Definition of the Seven Risk-Levels of the Activity Duration

Activity duration forms the foundation for the schedule of a construction project, but estimating the duration of an activity is not easy, since it is effected by different risk factors. In this paper, such risk factors are defined as 'constraints' of performing the activity. The spatial restriction of the construction site is the most essential constraint to perform an activity [9, 13]. The duration for a construction activity considering the spatial constraint is called 'base duration'. The second constraint for the duration of completing an activity is the 'physical environment risk' [9, 13]. The physical environment risk considering all factors of the surrounding environment on the construction site, e.g., weather, local social and cultural factors. As a result, while taking the physical environment risk into account, the activity duration is called 'empirical duration'. Estimation of the empirical duration can be obtained by adding an extra duration to the base duration, thus it is longer than the base duration.

Beside the physical environment risk, there are five categories (*i.e.*, the 5 Ms of construction management) of resource constraints that can be altered by management schemes including: Man, Machine, Material, Method, and Money. Considering all of the above risk factors, seven risk-levels are defined:

- Level-0 duration—no risk event and no physical environment constraint occurs to the activity, this is also considered as the base and shortest duration. The Level-0 duration is denoted as RD_0 and calculated as follows:

$$RD_0 = \frac{Quantity}{Rate} \tag{2}$$

where, 'Quantity' is the quantity of product to be produced by the activity; 'Rate' is the productivity rate that can be referred to any cost reference, *e.g.*, RS Means Book [8] or Dodge Estimating Guide [1]; RD_0 is the 'base duration'.

- Level-1 duration—only the 'physical environment risk' occurs, so the extra duration caused by the physical environment risk estimated by the experienced engineer is added to the base duration, resulting in the 'empirical duration'. The Level-1 duration is denoted as RD_1 and calculated as follows:

$$RD_1 = RD_0 + (Environment\ Effect) \tag{3}$$

where, RD_0 is 'base duration' of the activity; 'Environment Effect' is the extra duration caused by the physical environment risk and is estimated by the experienced engineer. RD_1 is the 'empirical duration'.

- Level-2 duration—both the physical environment risk and one resource risk occur, denoted as RD_2.
- Level-3 duration—not only the physical environment risk but also two resource risks occur, denoted as RD_3.
- Level-4 duration—not only the physical environment risk but also three resource risks occur, denoted as RD_4.
- Level-5 duration—not only the physical environment risk but also four resource risks occur, denoted as RD_5.
- Level-6 duration—not only the physical environment risk but also all five resource risks occur. This is considered the worst case scenario of the activity duration, denoted as RD_6.

The most significant resource risk factor for each of the five resource types is identified as M_x using the following equation:

$$M_x = Max(M_{x1}, M_{x2}, M_{x3}, M_{x4}, \ldots, M_{xn}) \tag{4}$$

where, M_x represents the most significant resource risk factor in a specific resource type (*e.g.*, M_1 means the most significant human resource risk, M_2 means the most significant machine resource risk, and so forth.); M_{x1}, M_{x2}, …, M_{xn} are all possible risk factors for a specific resource type (*e.g.*, M_{23} means the third type of machine resource).

The most dominating resource type is determined according to their contribution to the activity duration, and is calculated by the following equation:

$$R = \{R_2, R_3, R_4, R_5, R_6\} = Sort(M_1, M_2, M_3, M_4, M_5) \tag{5}$$

where, M_x represents the most significant resource risk factor for each of the five resource types; R_2 is the risk of duration for the Level-2 Risk; R_3 is the risk of duration for the Level-3 Risk; and so forth; 'Sort(…)' is sorting function (using descending ordering).

The activity duration for Risk Level-2–Level-6 can be calculated as follows:

$$RD_x = RD_{x-1} + R_x \tag{6}$$

where, RD_x is the activity duration for Risk Level-*x* (*e.g.*, RD_2 is the activity duration for Risk Level-2), *s.t.* $2 \leq x \leq 6$; the 'Round Law' is adopted in calculating RD_x in order to obtain the most conservative estimate of risk duration for the activity.

3.2 Simulation Procedure of the Proposed R-CPSM

The simulation procedure of the proposed R-CPSM is depicted in Fig. 1, which includes: (1) Activity risk probability (P) and effect duration (D); (2) Random number (R) generation; (3) Check if there is any ($R_x \leqq P_x$)—if ($R_x > P_x$), if the risk occurs and the activity duration is increased; (4) Update activity duration;

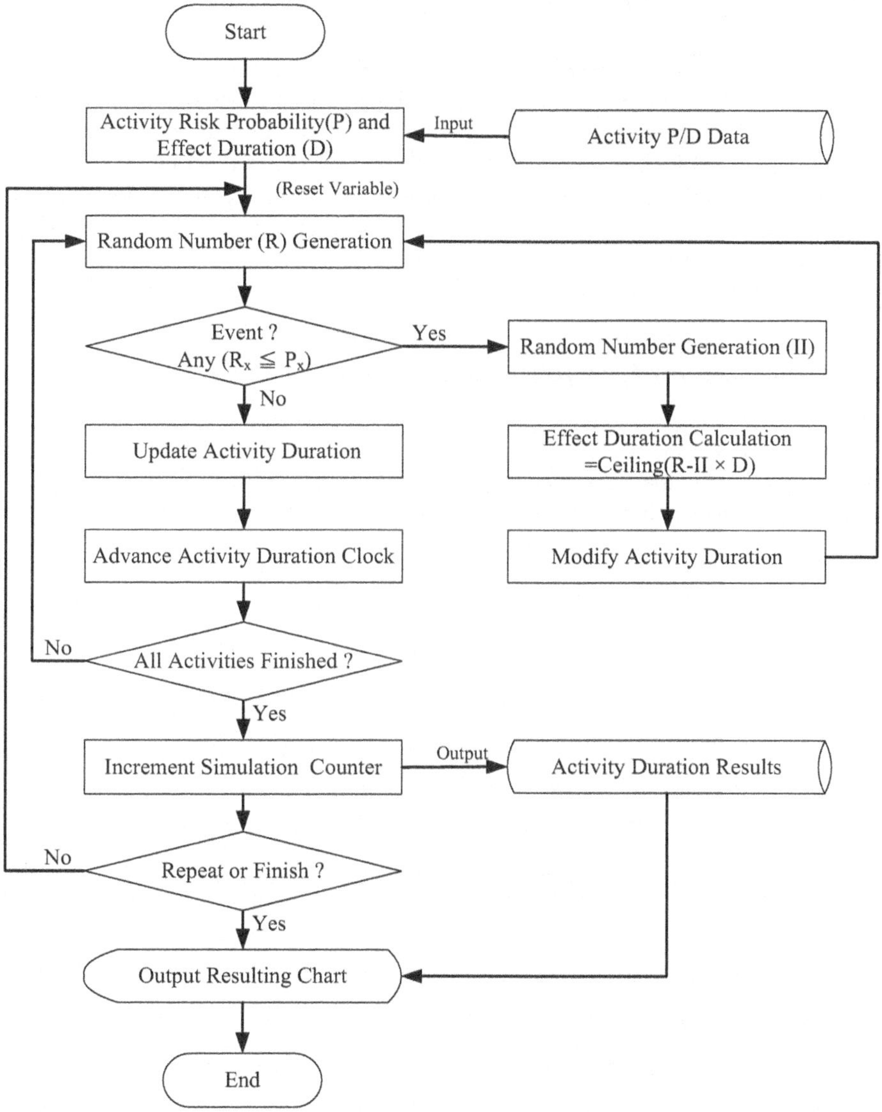

Fig. 1 Operation procedure of R-CPSM

(5) Advance activity duration clock; (6) Check if all activities are finished? If not, go back to Step (2); (7) Increment simulation counter; (8) Repeat or Finish, if not finished go back to Step (2); (9) Output resulting chart.

4 Demonstration Case

4.1 Information of Demonstration Case

The critical path network diagram of the demonstration case is depicted in Fig. 2. It consists of 12 activities, including: (1) Mobilization, (2) Materials Stocking, (3) Clean & Grubbing, (4) Grading, (5) Grade Finishing, (6) Bleacher Prefabrication, (7) Landscaping, (8) Road Paving, (9) Court Placing, (10) Bleacher Erection, (11) Curbing, and (11) Final Finishing. From Fig. 1, the project duration using 'base durations' for all activities is 32 days. The Activity durations of the 7-risk-levels in R-CPSM is shown in Table 1.

4.2 Simulation of Project Risk Durations

Using the activity durations of the 7-risk-levels in Table 1 and following the calculation procedure, the associated project durations of 7 risk-levels can be obtained, as shown in Fig. 3, where the probability of 50-day project duration is 79.1%, Risk

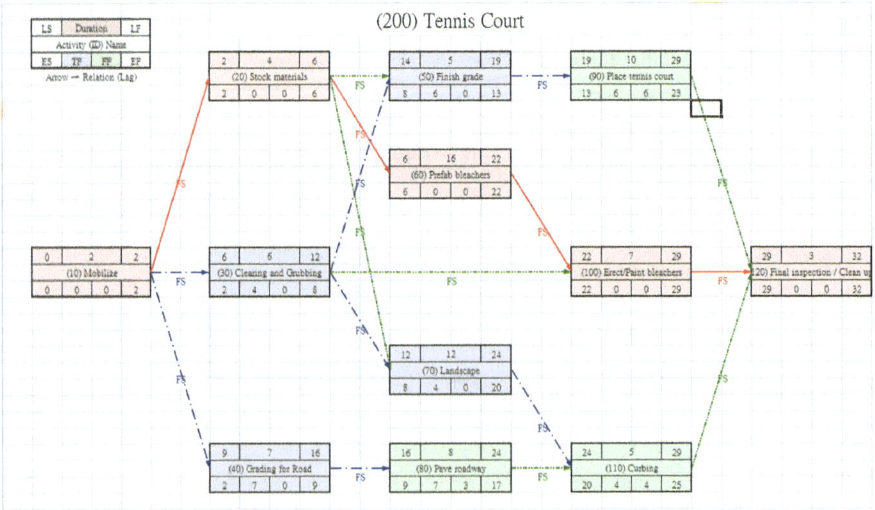

Fig. 2 Critical path network diagram of the demonstration case

Table 1 Activity durations of the 7-risk-levels

AID	ACode	AName	Level 0	Level 1	Level 2	Level 3	Level 4	Level 5	Level 6	Max
10	M	Mobilize	2	3	4	5	6	7	8	8
20	SM	Stock materials	4	5	6	7	8	9	10	10
30	CG	Clearing and grubbing	6	7	8	9	10	11	12	12
40	GD	Grading for road	7	8	9	10	11	12	13	13
50	FG	Finish grade	5	6	7	8	9	10	11	11
60	PB	Prefab bleachers	16	18	19	20	21	22	23	23
70	L	Landscape	12	14	15	16	17	18	19	19
80	PW	Pave roadway	8	9	10	11	12	13	14	14
90	PTC	Place tennis court	10	11	12	13	14	15	16	16
100	EPB	Erect/paint bleachers	7	8	9	10	11	12	13	13
110	C	Curbing	5	6	7	8	9	10	11	11
120	F	Final inspection/ clean up	3	4	5	6	7	8	9	9
		Project duration	32	38	43	48	53	58	63	63

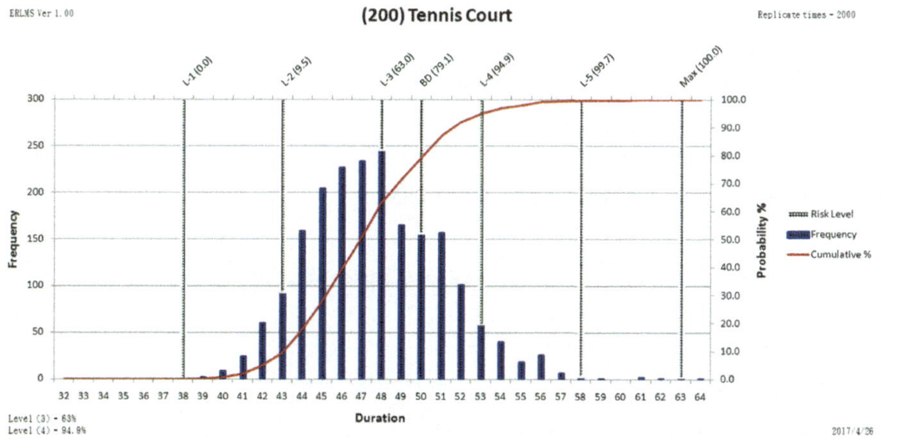

Fig. 3 Project duration distribution of risk level-3 of the demonstration case

Level-3 (63.0%) and Risk Level-4 (94.9%). As a result, planning the project duration as 50 days under Risk Level-3 is reasonable.

4.3 Project Planning and Scheduling with R-CPSM

Using the previous case for project planning and scheduling with R-CPSM, the result is shown in Fig. 4. The project is scheduled for 48 days (near 50 days).

4.4 Project Controlling with R-CPSM

During project controlling phase, the project progress status data need to be collected in the field. On the 15th day after project execution, the status data are collected for all activities, as shown in Fig. 5. Based on the progress status data, the predicted project completion date is simulated and shown in Fig. 6.

From Fig. 6, it is found that the predicted completion date for 50 days is only 39.4% (compared with 79.1% in planning). It means that the de facto execution of the project encounters an unexpected much worse situation than it was planned. However, the predicted completion date for 55 days is 90.4%. Since the 55-day schedule is acceptable by the owner. No controlling activities are needed.

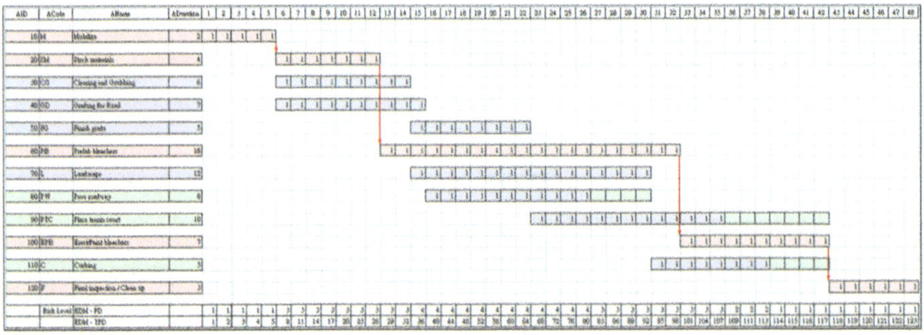

Fig. 4 Project planning and scheduling with R-CPSM

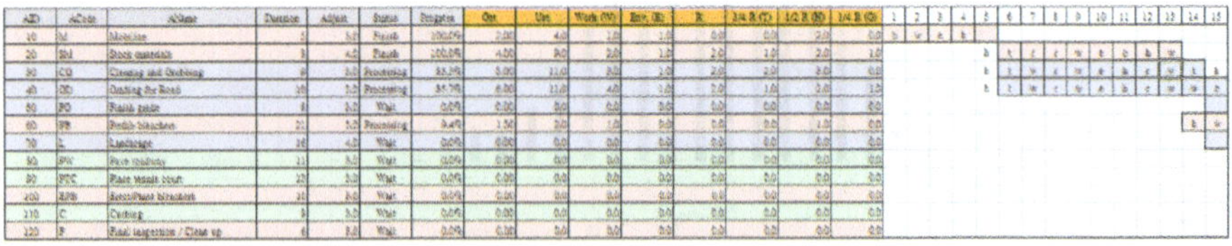

Fig. 5 Project progress status data of the 15th day after execution

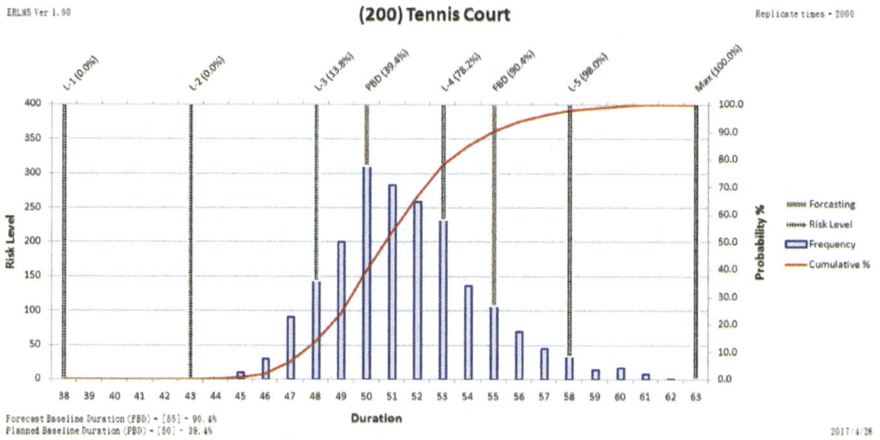

Fig. 6 Predicted project completion date

5 Conclusion

This paper proposes a new approach to tackle the risks with the activity duration by evaluating the constraints associated with different construction resources, including spatial, physical environment, man, machine, material, method, and money constraints. A CPM-based risk critical path scheduling method, namely R-CPSM, which takes into account the seven risk levels of activity duration is proposed for scheduling of a construction project. A tennis court construction project is selected as a case for implementation of the proposed R-CPSM to demonstrate its applicability. It is found that the proposed R-CPSM tackles the risks of activity duration more appropriately; since it replaces the lump-sum uncertain duration estimation in the traditional PERT with a systematic method that considers the 7-level resource constraints. Moreover, the proposed R-CPSM provides functionalities for real-time risk updating and controlling that is lacked in the traditional PERT. As a result, it is concluded that the proposed R-CPSM has a potential to improve the management of risks in construction schedule management.

Acknowledgements This research project was funded by the Ministry of Science and Technology, Taiwan, under project No. MOST 105-2221-E-324-025-MY2. Sincere appreciations are given to the sponsor by the authors.

References

1. Dodge (2017) Dodge estimating guide. https://construction.com. Visited June 2017
2. Gong D (1997) Optimization of float use in risk analysis-based network scheduling. Int J Proj Manag 15(3):187–192

3. Gong D, Rowings JE Jr (1995) Calculation of safe float use in risk-analysis-oriented network scheduling. Int J Proj Manag 13(3):187–194
4. Hsu K-H (2006) Project management: a view of the body of knowledge. Hua Tai Publishing, Taipei (in Chinese)
5. Jun DH, El-Rayes K (2011) Fast and accurate risk evaluation for scheduling large-scale construction projects. J Comput Civil Eng 25(5):407–417
6. Kılıç M, Ulusoy G, Şerifoğlu FS (2008) A bi-objective genetic algorithm approach to risk mitigation in project scheduling. Int J Prod Econ 112(1):202–216
7. Luu VT, Kim SY, Tuan NV, Ogunlana SO (2009) Quantifying schedule risk in construction projects using Bayesian belief networks. Int J Proj Manag 27(1):39–50
8. Means (2017) RS means book. https://www.rsmeans.com. Visited June 2017
9. Mol T (2003) Productive safety management. Butterworth-Heinemann, Oxford, UK
10. Moon H, Kim H, Kamat VR, Kang L (2015) BIM-based construction scheduling method using optimization theory for reducing activity overlaps. J Comput Civil Eng ASCE 29(3). https://doi.org/10.1061/(asce)cp.1943-5487.0000342
11. Schatteman D, Herroelen W, Van de Vonder S, Boone A (2008) Methodology for integrated risk and proactive scheduling of construction projects. J Constr Eng Manag 134(11):885–893
12. Vanhoucke M (2011) On the dynamic use of project performance and schedule risk information during project tracking. Omega 39(4):416–426
13. Yi KJ, Langford D (2006) Scheduling-based risk estimation and safety planning for construction projects. J Constr Eng Manag 132(6):626–635
14. Zafra-Cabeza A, Ridao MA, Camacho EF (2004) An algorithm for optimal scheduling and risk assessment of projects. Control Eng Pract 12(10):1329–1338
15. Zafra-Cabeza A, Ridao MA, Camacho EF (2008) Using a risk-based approach to project scheduling: a case illustration from semiconductor manufacturing. Eur J Oper Res 190 (3):708–723

Assessment of the Effect of Alligator Cracking on Pavement Condition Using WSN-Image Processing

Turki I. Al-Suleiman (Obaidat), Zoubir M. Hamici, Subhi M. Bazlamit and Hesham S. Ahmad

1 Introduction

Computer vision has become the digital eye of artificial intelligence. In fact, the development of high definition digital cameras along with the advancements of smart algorithms operating on high performance processors have pushed computer vision and its applications into new dimensions that were not accessible before. The ingathering of these technological revolution tools in the service of human welfare and safety reveals the real weight represented by new technologies, where man can only play a limited role. The maintenance of road pavements and the management of this maintenance was and remains a difficult task, in terms of monitoring and diagnosis of the distresses of roads and the establishment of timetables and priorities for intervention to address them. Moreover, the introduction of digital image processing with wireless sensor networks technology increases the efficiency of analysis-repair service and spreads such added value, so that tasks are accomplished effectively and at closer intervals of time compared to what was done without intervening of these techniques. The last decade has given us an unprecedented technological tools allowing us to segment, understand, measure, quantify, recognize image patterns in a widespread domain of application fields [1]. The new challenge is therefore, the development of new algorithms that take advantage of these tools and exploit the rich technological resources of sensors, artificial

T. I. Al-Suleiman (Obaidat) · S. M. Bazlamit · H. S. Ahmad
Department of Civil and Infrastructure Engineering,
Al-Zaytoonah University of Jordan, Amman, Jordan

Z. M. Hamici (✉)
Department of Electrical Engineering, Al-Zaytoonah University of Jordan,
Amman, Jordan
e-mail: zoubir.h@zuj.edu.jo

S. Şahin (ed.), *8th International Conference on Engineering, Project, and Product Management (EPPM 2017)*, Lecture Notes in Mechanical Engineering,
https://doi.org/10.1007/978-3-319-74123-9_28

intelligence, computer vision, signal processors and wireless sensor networks that represent the building blocks of modern intelligent equipment serving a wide range of fields.

Pavement management systems are effective tools for management of large pavement networks. The effectiveness of such tools is based on the how the distress of pavements is quantified, and how data is collected. Traditional pavement distress identification relays on the visual inspection of pavement sections. Pavement distresses can be caused by surface fatigue, consolidation or shear developing in the subgrade, sub base, base or surface layers. The earlier signs of pavement distress take the form of interconnected lateral and longitudinal cracks. These cracks are referred to as alligator cracking which are caused by the excessive movement of the supporting layers or the fatigue of the surface layer. These cracks may or may not be progressive. Weakness in the base layer resulting from improper compaction may lead to the development of this surface condition [2]. Management of pavements has become necessary as pavements continue to age and deteriorate with increasing funding limitations. The objective of a pavement maintenance management system is to provide a systematic maintenance process for decision making. Many papers have introduced approaches to identify thin entities in textured images, like in medical imagery, for the detection of blood vessels [3]. Others used satellite imagery, for roadway network detection and representation [4]. In addition, algorithms have been proposed for the detection of road cracks using road imagery and image processing. These algorithms are dependent on parameters such as acquisition, storage and image processing [5–17].

The following contribution is divided into four sections. Section 2 introduces a novel architecture of a pavement management system based on wireless sensor networks, data fusion and image processing. Section 3, presents a novel image processing algorithm used to extract alligator cracks and convert the distress density into pavement condition index. Section 4, presents results of the alligator cracks detection algorithm and shows the effectiveness of the histogram filter for cracks filtering. Finally, Sect. 5 concludes about the novel wireless sensor networks pavement management system and alligator cracks parametrization algorithm.

2 WSN Pavement Management System

The advancement of electronics, wireless networks and web-based software motivate the raise and growth of a new generation of project management systems, with both quantitative and qualitative objectives. With such novel architecture a pavement network should be managed efficiently and its performance should be improved. Pavement Management Systems (PMS) have been in use around the world by many highway agencies. In general, these systems are comprised of key components that allow these systems to yield the desired benefits to agencies. There is a component that entails the development of a pavement inventory database which contains records of pavement condition information, construction, maintenance, and rehabilitation. It also allows the decision makers to evaluate the consequences of various funding levels on pavement conditions. Agencies also may improve their scheduling of pavement repairs; assisting in optimizing rehabilitation and maintenance alternatives by taking advantage of the information needed to compare the cost-effectiveness of treatments.

Pavement systems deteriorate over time primarily due to fatigue. This deterioration increases with time and is highly influenced by repeated axle loadings of the vehicular traffic. Generally, early pavement deterioration contains four types of cracking, namely: transverse, longitudinal, block and alligator cracking. In order to manage pavements effectively, one need to monitor the development of these cracks and maintaining safety of personnel responsible for inspection of pavements. Figure 1 shows a proposed 3D visualization system. One method is based on dividing the pavement into cells and displaying 3D hexagonal cell with the height proportional to the pavement condition index. The second visualization method is based on displaying colored hexagonal cells with three colors identifying severity level of a cell. Information about cells are provided by a mobile pavement management system equipped with the Global Positioning System (GPS)/Global Navigation Satellite System (GLONASS) GPS/GLONASS geo-location receivers as explained in the next section.

Data fusion is defined as the process of integrating multiple sensors data and knowledge representing the same analyzed object into a consistent, accurate, and useful representation. We propose a system to enhance decision making. Such system should use multi-sensors data fusion; data from multiples sensors of different types; a full system mounted on a vehicle should be equipped with cameras, vibration sensors on each side of the vehicle, laser finder for depth analysis and a high accuracy high speed GPS/Glonass satellites antenna to acquire vehicle speed, time and position. The information extracted from still images are to be merged with vibration signals in order to decide about the level of severity of a pavement. Concerning Acquired images; they should be taken in a uniform condition of

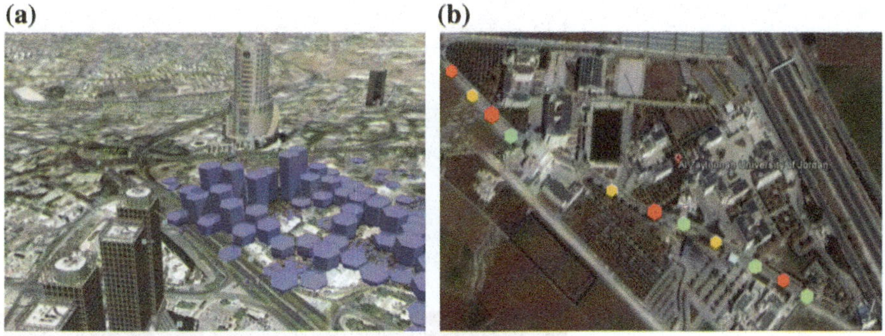

Fig. 1 3D global visual categorization of a pavement management system. PCI data is mapped into a satellite image with geographical information system. **a** 3D cells PCI indexing. **b** Cells coloring

lighting. To achieve such a goal and eliminate the effects of shadows or illumination artifacts and non-uniform illumination, appropriate illumination is to be used and image acquisition taken during night.

GPS or GLONASS are space-based radio-navigation systems, that provide geolocation along with atomic clock time information to a GPS/GLONASS receiver anywhere on Earth where there is an unobstructed line of sight to four or more satellites (four satellites needed to estimate four variables x,y,z,t). From the data provided by the satellites using a microcontroller ready NEO-7 module receiver, an accurate estimate of the vehicle speed can be computed. This latter metrics, together with the vibration information received from two sensors, each one placed on a car side (wheel path), are merged with the PCI metric and transmitted in real-time to a remote central pavement management system. An internet connection is performed through a wireless connection integrated into the device to a mobile hotspot network. The data is then conveyed to a remote server for analytics, storage, classification and management. Many vehicles can perform the task in different areas at the same time. The system is relatively cost-effective compared to other systems.

3 Image Processing Algorithms

The complexity of pavement textures, lighting diversity, shadows, renders image features extraction a big challenge for computer vision algorithms due to nature of complex-varying noise affecting such images. Algorithms are then designed for special purpose based on the nature of image and the different natural conditions affecting such images. Figure 2 shows the structure of the image processing algorithm. It starts by acquiring a color image which is then converted to grayscale and passed to histogram equalization. The next stage consists of pre-filtering with denoising. Edge detection is performed before the result is fed into morphological

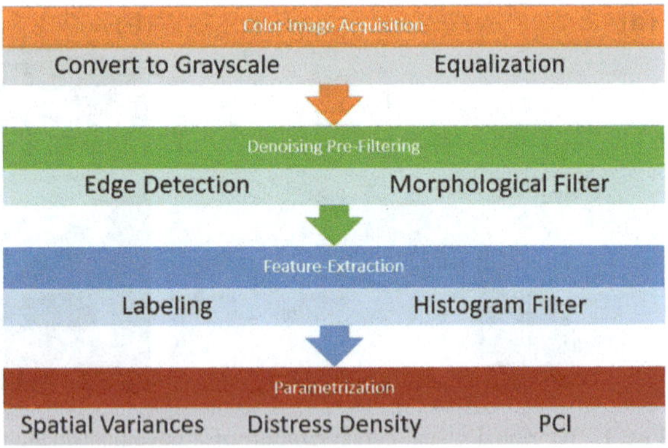

Fig. 2 Flowchart for image processing algorithm. Cracks extraction, parametrization and PCI estimation

filter with dilation and erosion sequences using a squared structuring elements of size 4 and 3 respectively, in order to remove low level noise. The third stage is the identification of connected components, also known as labeling, representing the cracks. The number of cracks is normalized to 255 and then fed into a histogram filter which removes all edges that are below a predefined threshold. The resulting image is then ready for final parametrization process. In order to localize the concentration of edges inside the image, a spatial analysis is performed in both horizontal and vertical dimensions. The horizontal and vertical concentration signals are analyzed as random signals and the spatial variances (or spatial spreads) estimates (σx, σy).

The spatial spreads show the spread of cracks across the image in one dimension. These parameters are then used to define the pavement distress density. The pavement distress density is converted the pavement condition index using a cubic spline model. The relationship between distress density and condition index is categorized into three classes; namely: low severity, moderate severity and high severity. Standard data points are used to interpolate the three curves.

The distress density is then estimated using (1) given by:

$$\mathrm{Ditstress}_{\%} = \mu_1 \frac{\mathrm{S_{Cracks}}}{\mathrm{MxN}} + \mu_2 \frac{\sigma_x}{\mathrm{N}} + \mu_3 \frac{\sigma_y}{\mathrm{M}} \tag{1}$$

$$\mathrm{S_{Cracks}} = \sum_{i=1}^{N} \sum_{j=1}^{M} \mathrm{Cr(i,j)} \tag{2}$$

where $\mathrm{S_{Cracks}}$ is the area of alligator cracks, M, N, σx, σy are the image width, image height, horizontal spatial spread, vertical spatial spread, respectively. Spatial spread measures the flatness or spread of cracks. μ1,μ2, μ3 are normalization control coefficients estimated based on image set in order to scale the distress density to the range [0–100]. If μ2, μ3 are set to zero only the density of cracks represents the distress density regardless of the spread of cracks inside the image.

We are introducing the spatial spread of cracks in both horizontal and vertical dimensions of the image. They are given by Eqs. (3) and (4) where xPeak and yPeak are the peak values in the spatial distributions and SCracks/N, SCracks/M are the average levels in x, y directions respectively. The spatial spread may be also defined as spatial flatness in contrast to spectral flatness for filters in frequency domain.

$$\sigma_x = N\left(1 - \frac{x_{Peak} - \frac{S_{Cracks}}{N}}{S_{Cracks} - \frac{S_{Cracks}}{N}}\right) \tag{3}$$

$$\sigma_y = M\left(1 - \frac{y_{Peak} - \frac{S_{Cracks}}{M}}{S_{Cracks} - \frac{S_{Cracks}}{M}}\right) \tag{4}$$

Table 1 Pavement condition index categorization

Overall Condition Index	Qualitative condition	Description
85–100	Very good	Very good surface
75–85	Good	Low severityalligator cracking
40–75	Fair	Moderate severityalligator cracking
0–40	Poor/very poor	High severity alligator cracking

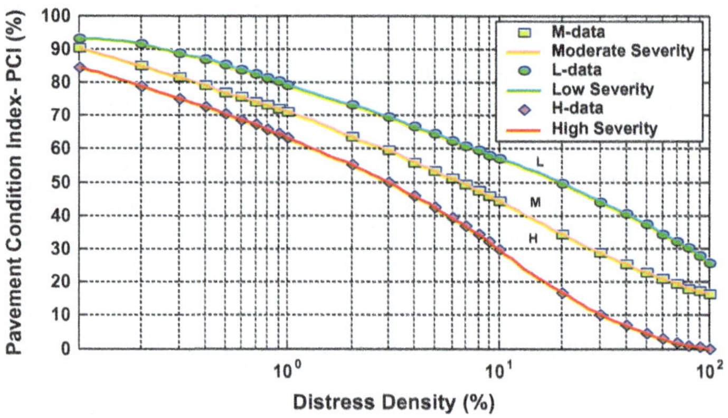

Fig. 3 Pavement condition index versus pavement distress density. PCI versus the logarithmic scale distress density

Table 1 shows the overall pavement condition index grouped by qualitative condition. The three categories of distress severity are also shown. The same coloring is used on satellite maps in Fig. 1.

The functions of PCI versus distress density are modeled using splines. Cubic spline interpolation is a well-known polynomial interpolation used in order to avoid the Runge's effect of polynomial oscillation at the edges intervals of equispaced data points, yet it provides a smooth curve fitting for nonlinear curves modeling with a minimized error relative to other polynomials approximations.

Figure 3 illustrates the PCI for the three categories of severity; low, moderate and high. The data points are standard points used in research literature, while the curves are the result of spline interpolation. These functions allow the conversion of distress density extracted using image processing into pavement condition index used as a key parameter for decision making about maintenance, repair or rehabilitation. In this paper, it was assumed that the pavement sections do not include other distresses.

4 Results and Discussion

This section presents results obtained by the novel cracks histogram filtering algorithm described in Sect. 3. Figure 4 shows the case of processing a high severity alligator cracks image. The classification of images in three categories is bases on the ratio of cracks surface relative to the total area represented by the image. Experimental results classify the low severity for a ratio below 4%. A ratio between 4 and 8% is considered as moderate severity, while a ratio above 8% is considered as high severity. These values depend on the area covered by of the image. Figure 4a shows the grayscale image with intensity correction based on histogram equalization. Figure 4b is the result of edge detection with morphological pre-filtering to remove residual edge noise. Figure 4c show an advanced filtering using a histogram filter. Figure 6 shows the histogram of edges (cracks) obtained by searching connected components. With a noise floor of 25 pixels, the histogram filter reduces the number of connected components (cracks) from 369 to 175 representing 12.1%, while with a noise floor of 100 pixels; the histogram filter reduces the number of connected components to 46 representing 10.8% of the surface of the image. The dotted rectangle in Fig. 4d depicts the concentration of cracks as expected by spatial spreads. Figure 4d is the result of labeling, it shows connected cracks. The surface of cracks represents about 11% of the total image surface and hence it is considered as high severity cracks image.

Figure 5 shows the histogram of connected components (cracks). The noise floor indicates small isolated cracks that can be removed (filtered) for a more robust estimation. The image of the cracks is labeled for connected components, and normalized to 255 as maximum label corresponding to maximum greyscale value in a histogram. A full scan of components is then performed with estimation of the size of each crack. Cracks that present a density less than a predefined threshold

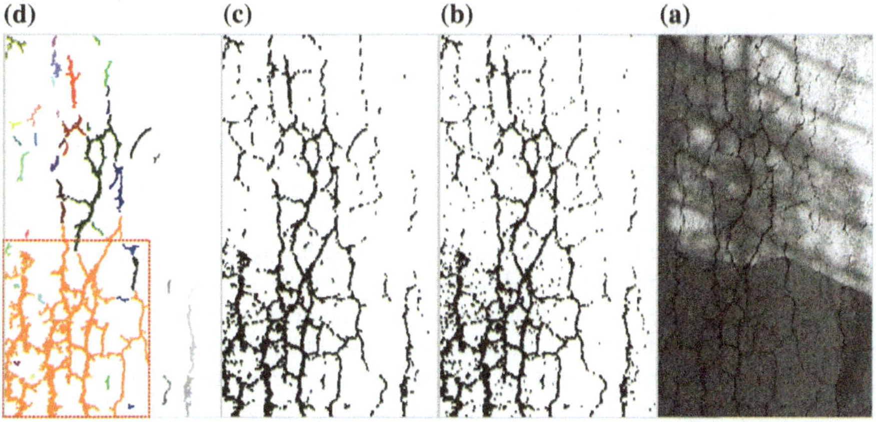

Fig. 4 Cracks detection and labeling as connected components. Further filtering is based on crack size removal based on Histogram filtering. For printing purpose, the colors of **b**, **c** and **d** are inverted

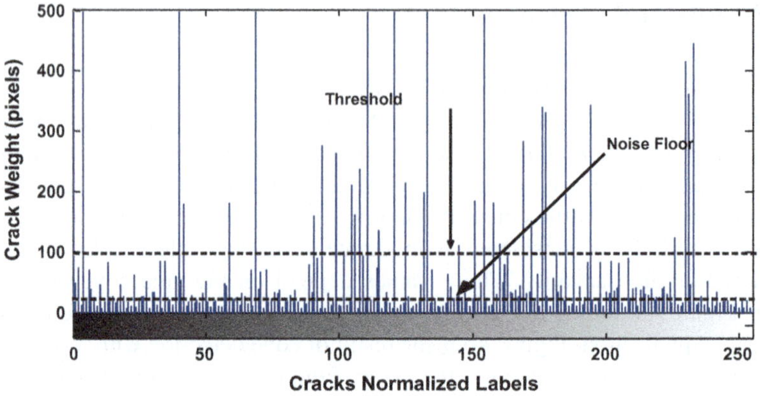

Fig. 5 Histogram filtering. Each bin indicates a separate crack. The intensity of the bin represents the size of the cracks in pixels

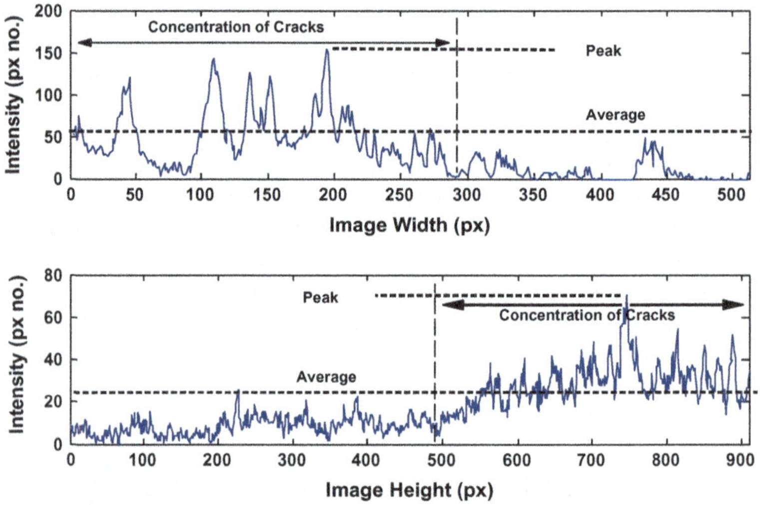

Fig. 6 Spatial distribution analysis and localization of cracks. The spatial variances (or spatial spreads) are used for distress density calculation

representing the noise floor in pixels are removed by histogram filtering. The image of cracks is than reconstructed and the number of cracks is re-estimated.

In order to analyze the two dimensional distribution of cracks inside an image, a spatial analysis is performed. Both horizontal and vertical distribution (of pixels) of cracks are extracted then a statistical analysis of the variance of the two dimensional cracks is performed, which permits the localization of the severe deterioration area inside the cracks image. The equations of both distributions as a spatial spread are given by Eqs. (3) and (4). Figure 6 shows the spatial distribution of cracks. One for horizontal distribution and the other for vertical distribution.

5 Conclusion

This paper presents a novel architecture of a pavement network management system based on wireless sensor networks architecture. A network node is mounted onboard on a PMS vehicle and enables real time data transmission to a central web-based management system. The wireless sensor network is a scalable network that permits introduction of new vehicle-nodes with high flexibility. Data analytics, storage are then centralized which allows an optimum use of resources in term of prioritization of the maintenance and rehabilitation. The geographical size of the management system is unconstrained since it is web-based system. Concerning pavement images processing, it is performed in situ. The system analyses captured images and performs a full image processing sequence in order to extract the pavement distress density which is then converted to pavement condition index. The results of the image processing algorithm called cracks histogram filter show the robustness of the algorithm in term of noise removal, classification of pavements based on severity level of the alligator cracks and quantification of the pavement distress using spatial variances along the two dimensions of the image.

References

1. Hamici Z (2006) Real-time pattern recognition using circular cross-correlation: a robot vision system. Int J Robot Autom 21(3):174–183
2. Yoder E, Witczak MW (2008) Principles of pavement design, Wiley & Sons, 1975, 2nd edn. Wiley
3. Chaudhuri S, Chatterjee S, Katz N, Nelson M, Goldbaum M (1989) Detection of blood vessels in retinal images using two-dimensional matched filters. IEEE Trans Med Imaging 8(3):263–269
4. Geman D, Jedynak B (1996) An active testing model for tracking roads in satellite images. IEEE Trans Pattern Anal Mach Intell 18(1):1–14
5. Scheffy C, Diaz E (1999) Asphalt concrete fatigue crack monitoring and analysis using digital image analysis techniques. In Accelerated pavement testing international conference, 1999
6. Chambon S, Moliard JM (2011) Automatic road pavement assessment with image processing: review and comparison. Int J Geophys 2011:1–20
7. Georgopoulos A, Loizos A, Flouda A (2006) Digital image processing as a tool for pavement distress evaluation. ISPRS J Photogram Remote Sens 50(1):23–33
8. Corso D, Fioravanti R, Fioravanti S (1995) Morphological analysis of textured images for identification of thin structures. In: Proceedings of the 20th international conference on acoustics, speech, and signal processing, vol 4, pp 2359–2362
9. Tanaka N, Uematsu K (1998) A crack detection method in road surface images using morphology. In: Proceedings of the workshop on machine vision applications, pp 154–157
10. Rughooputh HC, Rughooputh SD, Kinser JM (2000) Automatic inspection of road surfaces. In: Proceedings of the machine vision applications in industrial inspection VIII, SPIE, Ed., vol 3966, pp 349–356
11. Mahler DS, Kharoufa ZB, Wong EK, Shaw LG (1991) Pavement distress analysis using image processing techniques. Microcomput Civ Eng 6(1):1–14
12. Iyer S, Sinha SK (2005) A robust approach for automatic detection and segmentation of cracks in underground pipeline images. Image Vis Comput 23(10):921–933

13. Oliveira H, Correia PL (2009) Automatic road crack segmentation using entropy and image dynamic thresholding. In: Proceedings of the European signal processing conference
14. Hsu CJ, Chen CF, Lee C, Huang SM (2001) Airport pavement distress image classification using moment invariant neural network. In Proceedings of the Asian conference on remote sensing, vol 1, pp 216–220, 2001
15. Augereau B, Tremblais B, Khoudeir M, Legeay V (2001) A differential approach for fissures detection on road surface images. In: Proceedings of the international conference on quality control by artificial vision
16. Chambon S (2011) Detection of points of interest for geodesic contours: application on road images for crack detection. In: Proceedings of the international conference on computer vision theory and applications (VISAPP '11), Algarve, Portugal
17. Coudray N, Karathanou A, Chambon S (2010) Multi-resolution approach for fine structure extraction—application and validation on road images. In: Proceedings of the 5th international conference on computer vision theory and applications

System Dynamics Simulator of Inventory Management as a Learning Tool to Improve Undergraduate's Decision Making

Raed M. Alqirem and Khaled S. Al Omoush

1 Introduction

This paper presents a model development of inventory management for undergraduate business student. The model was developed based on system dynamics methodology in order to simulate different scenarios in managing an inventory level in an organization and to measure the effectiveness of their decisions. It depicts variables, their relationships and feedback loops in a computer software as an interactive learning tools to support the students in their decision making process, and to allow them to explore the effect of their decisions.

2 Literature Review

The literature review focuses on two areas that provides the methodology behind this paper: System Dynamics methodology and learning with interactive learning tool.

R. M. Alqirem (✉)
Management Information Systems Department,
Al-Zaytoonah University of Jordan, Amman, Jordan
e-mail: Drraed@zuj.edu.jo

K. S. Al Omoush
Business Administration Department, Al-Zaytoonah University
of Jordan, Amman, Jordan

S. Şahin (ed.), *8th International Conference on Engineering, Project, and Product Management (EPPM 2017)*, Lecture Notes in Mechanical Engineering,
https://doi.org/10.1007/978-3-319-74123-9_29

2.1 System Dynamics

System Dynamics is a computer-aided approach for analyzing and solving complex problems with a focus on policy analysis and design. Initially called 'Industrial Dynamics' [2], the field developed from the work of Jay W. Forrester at the Massachusetts Institute of Technology. System Dynamics has its origins in control engineering and management; the approach uses a perspective based on information feedback and delays to understand the dynamic behavior of complex physical, biological, and social systems.

System Dynamics is a well formulated learning and teaching method for understanding a system that include cause and effect relationships and their underlying logic and mathematics, time delays, and feedback loops. After the successful policy changes in lots of fields such as management, engineering, economy, chemistry and physics, the system dynamics researchers gone towards applying system dynamics in the educational fields too [1, 6, 10].

SD methodology concentrates on the importance of conceptualization, formulation and simulation [9]. It has two stages, qualitative and quantitative analysis. Modelers study system description and its environment for problem development and formulate a qualitative analysis, and the relationships between the system variables are formulated in conceptual model for qualified analysis using simulation techniques. The use of simulation for learning and teaching is becoming increasingly popular as shown in the next section.

2.2 Learning with Interactive Learning Environments (ILEs)

System Dynamics Interactive learning environments SD-ILEs is a term sufficiently general to include system dynamics model constructed in Microworld, DSS, and management flight simulator. In which represents a computer simulation environment.

The basic idea is to facilitate learning in complex and dynamics environment such as in this paper managing the inventory in a organization [3, 4]. An ILE is considered a significant tool and its more popular for teaching and learning, and they often used to enhance decision making in the context of the dynamic complexity of business interactions, this is done by facilitating user learning [5, 7].

It is consisting of four main components (i) a computer simulation model to adequately represents the issue or problem on hand, which allows decision makers to experience real world-like responses, and this component could be improved by using System Dynamics tools, (ii) a user friendly interface that allows bidirectional communication between the user and the machine, and thus gives the decision maker the opportunity to test and explore the effect and consequences of their decisions. Furthermore, it provides them with access to the feedback on interactive basis, (iii) database or knowledge component that contains varying and related data and knowledge that emerges within or outside the organization to be analyzed.

(iiii) the teacher or facilitator to control and facilitate the learning process through the simulation and to observe many simulation runs that could happen [8, 11].

3 The Research Approach

The aim of this study is to explore the use of system dynamics simulator which considered an effective and interactive learning tool as a decision support system (DSS) to improve the undergraduates' business student's decision making.

A simple model represents an inventory management in a company is constructed, based of system dynamics methodology tools such as causal loop diagram (CLD) and SD model.

This dynamic model is capable of addressing the relevant problem, and then the problem is defined to the students over an appropriate time horizon. Finally, the policies and scenario's analysis stage seeks to transfer the insights and understanding gained from the simulator to the students. It is important to note that testing and going through these scenarios is iterative, and the steps proceed in a recursive manner, which provide insights and may lead to revision and refinements in the way they perceive or react in their decisions. In this study, POWERSIM software is used to construct a SD simulation. POWERSIM is software designed for modeling one or more variables and quantities that change over time.

3.1 System Dynamics Model

Before constructing the SD model, a Causal Loop diagram is drawn to presents the different relations among the inventory variable.

For the purpose of this study, we have created a simple causal loop diagram that represents some variable surrounding the inventory in the company and the relations between these variable, and of course this brief description was translated to students and with their help to construct a simple example for teaching purposes. For instant, if the company wants to pay 100% of its purchases in cash, the cash will decrease and then the firm's purchases from the supplier will be less that the case in which the firm have to pay nothing to receive its material purchases. This is the loop (B7). Loop (B5) represents the relation between sales and the inventory in stores. The more products in the firm's store the more completed sales. Delivering the sales to the buyers will decrease the inventory in store which creates an inventory gap between the quantity of products are available in the firm's store and the desired inventory level (B6). These loops were constructed to enable students to create scenario on their own, and to think about the consequences of their decision deeply, and there are shown in Fig. 1.

The above causal loops are the foundation of creating the SD model. These loops were translated into a mathematical model using POWERSIM software which are shown in Fig. 2.

The students observed drawing the causal loop and we tried to simplify the process as we can, then they were taught how to build a model that behaves like the real system in order to understand the relations among the variable and to be ready to make decisions on their own using the software interface.

3.2 SD-ILEs Interface

Below is the interface that students interact with. Variables can be changed through sliding the bars in the interface, and then moving to see the consequences of their decision of the company's financial performance.

Figure 3 shows the window in the Interface which represents the parameters which the students can change to test their inventory control decisions. Examples are adjustment times for finished and material inventories, coverage ratios for finished and materials inventory, and existing production capacity.

Fig. 1 Causal loop diagram

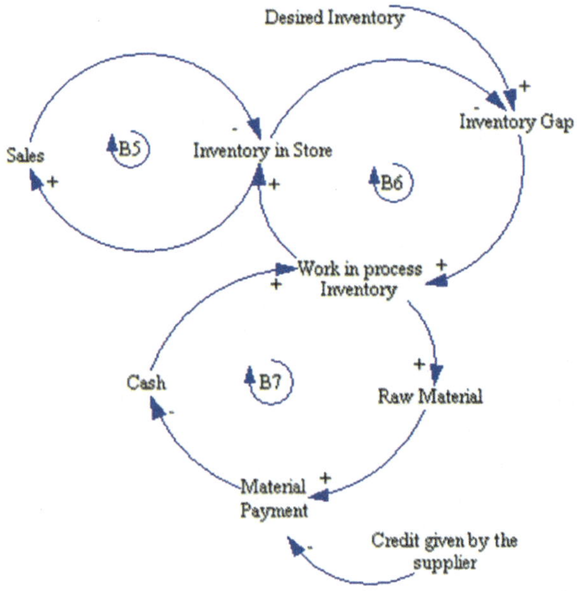

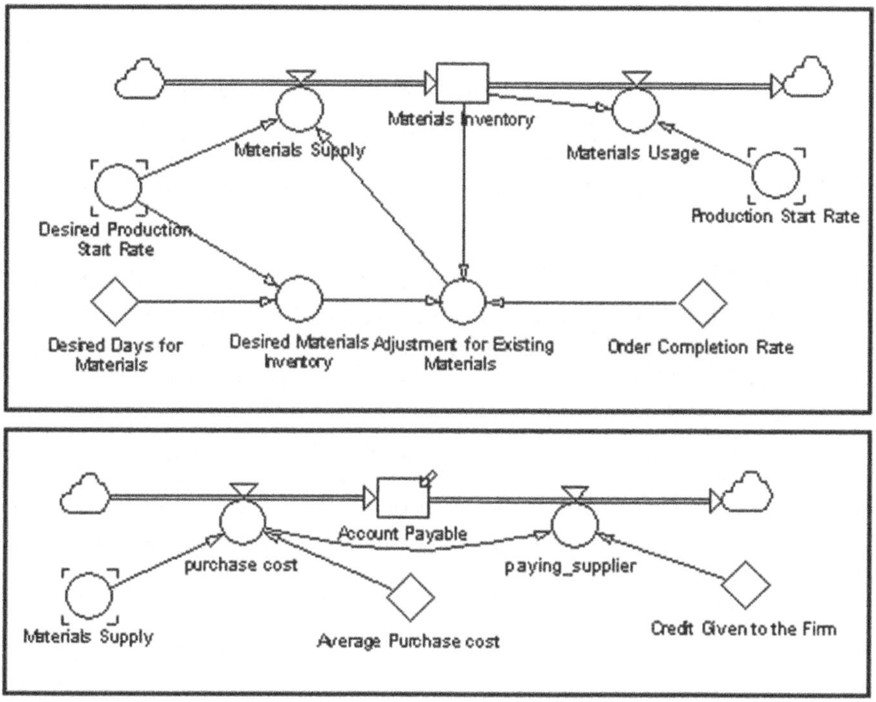

Fig. 2 System dynamics models is also known as (stock and flows)

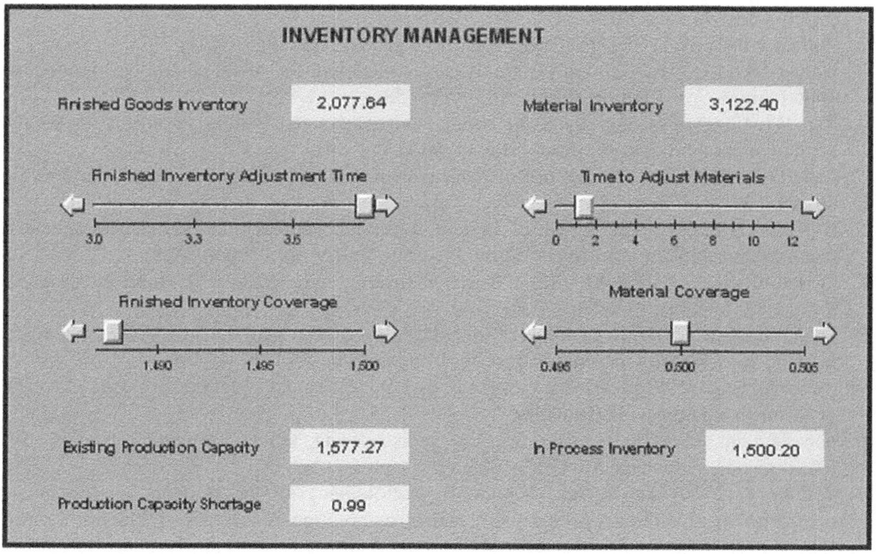

Fig. 3 The interface window for managing inventory

4 Conclusion

We have presented a brief and general learning procedure that provide to business students the required knowledge on inventory and production issues through testing their decisions, and gaining deep insight of how to manage delays, production change, inventory level, and raw materials in the organization in a dynamic and interactive environment.

We have demonstrated in this study the application of the simulator as a tow steps process (1) a brief presentation of modeling theory using system dynamics that constructed with a simple mode, and the application of SD methodology in a real time, (2) the construction of a computer simulator as an interactive learning environment that help the students to simulate, examine, and solve different problems.

The procedure can be easily modified to be use a teaching tool in different business fields such as finance, marketing, planning, and other business areas depending on the characteristics and need of the case, specially that the attempt done in this study showed the increase of awareness and knowledge gaining that students had and eventually leaded to enhance their decision making and solving problems skills.

References

1. Enos J (2010) Learning with loops: applying feedback to teaching system dynamics to undergraduates. In: Proceedings of the 28th international conference of the system dynamics society, Boston, July 2010
2. Forrester JW (1961) Industrial dynamics. MIT press, Cambridge, MA
3. Homer JB, Hirsch GB (2006) System dynamics modeling for public health: background and opportunities. Am J Public Health 96(3):452–458
4. Kriz WC (2003) Creating effective learning environment and learning organizations through gaming simulation design. Simul Gaming 34(4):569–591
5. Lane DC (1995) On a resurgence of management simulations and games. J Oper Res Soc 46:604–625
6. Nuhoglu H (2010) The effect of system dynamics approach on understanding causal relationships skills in science education. Proc Soc Behav Sci 2:3614–3618
7. Qudrat-ullah H, Karkul M (2007) Decision making in interactive learning environment: towards an integrated model. J Decis Syst 16(1):79–99
8. Qudrat-ullah H (2010) Perceptions of the effectiveness of system dynamics-based interactive learning environment: an empirical study. Comput Educ 55:1277–1286
9. Richardson GP (1991) Feedback thought in social sciences and system theory. PA: University of Pennsylvania press. Philadelphia
10. Schaffernicht M (2010) Learning from rediscovering system dynamics models. Syst d'information et Manage 14(4):87–106
11. Zydney J (2010) The effect of multiple scaffolding tools on students' understanding, consideration of different perspectives, and misconception of complex problems. Comput Educ 54:360–370

Managing the Digitisation of Filing System Project at Al-Zaytoonah University of Jordan

Esra'a S. Al-Khatib, Mohammed M. Yassin and Ala'a S. Alkhatib

1 Introduction

At the beginning of the 21st century, the higher education institutions (HEIs) worldwide face unprecedented challenges result from the convergent impacts of globalization [19]. In response, countries have undertaken significant transformations of their higher education systems, including changes in patterns of financing and governance, growing institutional differentiation, the creation of evaluation and accreditation mechanisms and technological innovation. Holm-Nielson [9] concluded that this transformation has been uneven as most developing countries face several difficulties towards the developments of their educational systems such as insufficient financial resources to improve quality, rigid governance models and management practices [9]. These issues are preventing the HEIs from embracing change and launching reforms and innovation in these countries including Jordan.

Both the Ministry of Higher Education and Scientific Research (MoHE) and the Higher Education Accreditation Commission (HEAC) place the matter of developing standards for the assessment of quality as a top priority. [1] HEAC has recently adopted eight quality assurance standards at institutional level for the external quality assurance of HEIs [8]. These standards are: (1) strategic planning, (2) governance, (3) academic programs, (4) scientific research, scholarship and

[1]The "accreditation" refers to a process of assessment and review which enables a higher education course or institution to be recognized or certified as meeting specified standards. The term "quality assurance" in higher education refers to systematic management and assessment procedures adopted to monitor performance and achievements and to ensure achievement of specified quality or improved quality [19, p. 4].

E. S. Al-Khatib (✉) · M. M. Yassin · A. S. Alkhatib
Al-Zaytoonah University of Jordan, P. O. Box 130, Amman 11733, Jordan
e-mail: e.alkhatib@zuj.edu.jo

S. Şahin (ed.), *8th International Conference on Engineering, Project, and Product Management (EPPM 2017)*, Lecture Notes in Mechanical Engineering,
https://doi.org/10.1007/978-3-319-74123-9_30

innovation, (5) financial, physical and human resources, (6) student services, (7) community service and foreign relations and (8) quality assurance [8].

Although Jordan took the lead in the Arab world, in adopting rules and regulations concerning accreditation in particular, quality assurance and accreditation of higher education institutions and programs are new and are still under the control of the Jordanian government [19]. During the last two decades the laws have been issued to govern higher education in Jordan are Higher Education Act number (28) for 1985, Public Jordanian Universities Act number (29) for 1987, Private universities Act number (19) for 1989, and Higher Education Provisional Law number (41) for 2001. El-Gharaybeh [7] argued that the vast number of laws and legislation issued on higher education represents a major problem to universities and concerned bodies. The HEIs are required to show their compliance with laws and issued legislation [19].

As the HEIs in Jordan are subject to the laws and regulations of the MoHE and according to these laws, they are required now to modify their modes of operation and delivery and take advantage of the opportunities offered by the new information and communication technologies. On the other hand, having to file all academic and managerial documents in accordance with the quality assurance standards is cumbersome on private universities and their faculties. Using paper-based filing or other electronic learning systems cannot drive out the duplicated data and unnecessary descriptions of files prepared by a particular university. In essence, the higher education policy stresses on the need to promote use of Information and Communication Technology (ICT) for teaching, learning and networking of universities in order to ease the pressure on its limited available resources. It seems that no prior study has been carried out on managing the digitisation of the filing system in Jordanian private universities.

Jordan has a total of 29 universities: 10 public and 19 private universities. Among all Jordanian private universities, Petra University was the first academic institution which has been awarded Quality Assurance Certificate in 2015. Since 2015, Al-Zaytoonah University of Jordan (ZUJ) through the accreditation and quality assurance office started keeping all accreditation and quality assurance documents as required by the Accreditation and Quality Assurance commission for higher education institutions. ZUJ was established after receiving its license and general accreditation by Decision No. 848 on March 6, 1993. ZUJ has seven faculties, five centres and three departments. since 2015.

In this paper, we propose digital filing system (DFS) that is enabled by an eXtensible Mark-up Language (XML) technology as a new mechanism for digitising the filing system at ZUJ. The development of digital filing system (DFS) aims at reducing the burden on the filers of the information at the private universities. The number of implementation projects of XML is growing rapidly around the world and similar initiatives to DFS which have been developed at businesses and banks such as Standard Business Reporting (SBR) that have been successfully rolled out in the Netherlands, Australia and Finland to reduce the administrative burdens on businesses [6, 14]. The remainder of the paper is organised as follows: Sect. 2 outlines the development of paper filing systems at HEIs; Sect. 3 introduces DFS and provides an overview of XML technologies and discusses their importance and the scenarios for its implementations, followed by the expected benefits and costs of DFS. The final section draws conclusions and discusses future research directions.

2 Methodology

We performed our exhaustive search for the purpose of this study. We based on some key words on the most relevant databases such 'costs and benefits of XML technologies'; 'project management', 'digital reporting', 'standard business reporting' to obtain the relevant academic articles. The overall literature covered 124 studies dated from 1999. We evaluate the studies of similar initiatives and their possible scenarios implemented in the Netherlands, Australia, the USA and the UK. The major themes after reviewing and analyzing the literature are discussed below.

3 Paper Filing Systems at HEIs

The paper-based filing has been used by the majority of the higher education institutions (HEIs) to file, store and send the information to different users. It was criticised as slow, labour intensive, costly, error prone and inefficient [12, 15]. The emergence of some technologies such as CD-ROM, Portable Document Format (PDF) then began substituting for the use of paper medium from the mid-1980s through the 1990s [15]. Lymer [15] found that these electronic media contained weaknesses or gave little extra advantage to filers and users compared to paper-based filing. For example, the use of CD-ROM for organisations still had to be distributed by physical means, making it clumsy. In addition, the distribution of large files and reports on Adobe Acrobat files has been an unsatisfying experience for users as the files are slow and difficult to download [15].

Numerous electronic educational systems (e.g. e-Learning, e-library, e-forms, e-gate and e-book) were introduced by the HEIs in Jordan for less than a decade. These tools enable the faculty members to manage their courses electronically [1]. Nevertheless, these projects faced a number of challenges that have also affected their implementation. Some challenges like the difficulty of developing a fully online course from scratch, cost and time consuming, technological uncertainty and some students and faculty members do not like change from paper to electronic technology [1]. Although the faculty documents are filed in paper format, their formats are still technically unstructured based on a particular taxonomy. The filers of these documents are required to re-key data or spend time trying to access relevant information multiple times [5]. As a result, these educational tools could

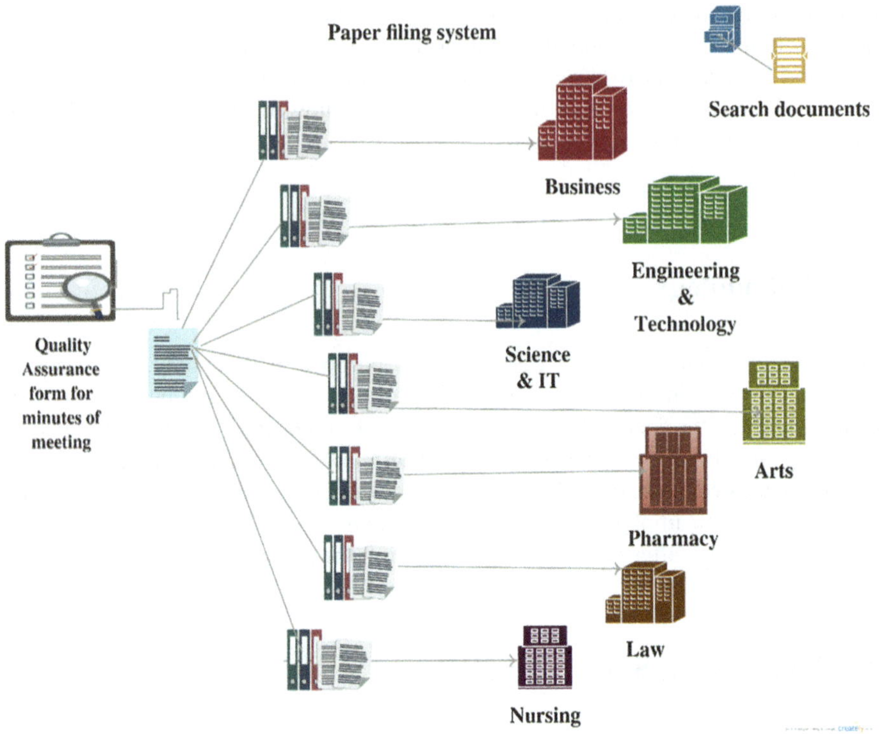

Fig. 1 Simplified pre-DFS architecture

not drive out the duplicated data and unnecessary descriptions as the architecture for the current filing system depicted in Fig. 1.

Figure 1 shows the fragmentation of paper filing system to file and store the department minutes of the meeting agenda form by seven faculties within ZUJ. For the same piece of information, the faculties have different requirements and descriptions. The different lines from the filer to the users represent this situation, demonstrating the extent to which the faculty has to produce unique reporting capabilities to fulfil its reporting responsibilities to different departments and external users. The users group include for example the faculty dean, heads of departments, quality assurance office, the departments' secretaries, the academic members within each department and the representatives of HEAC. The information models, business rules, process designs, controls, etc. are redundantly embedded within each department documents. Moreover, data validation and analysis need to be conducted by the users and manual manipulation of information from disparate sources is needed to create custom reports. In the pre-DFS situation, the filers identify what piece of information that is within their files and folders and then file information multiple times for different folders. The move to open standard and format could remove these limitations which also might enhance usability of information by Internet users' language [15]. Achieving this goal requires a major transformation of the current architecture for the faculty filing system that is discussed below.

4 Proposed Digital Filing System via XML

The development of DFS that enabled by XML takes the Internet filing a step further by using a technology to develop a standard for the information [5, 11]. XML is simply a standard for the electronic exchange of data between different parties across multiple technological platforms on the Internet. It enables the filers to report one set of information instead of reporting and filing it repeatedly in different forms to different users for different purposes by developing a taxonomy [14]. The taxonomy is essentially a data dictionary that allows the data to be tagged (labelled) based on standard [6]. This drives out duplicated data and unnecessary descriptions [6], thus reducing the burden of reporting [13]. Furthermore, the value of XML technologies lies in its reusability [6], enabling different users to reuse the data efficiently in the filing chain. XML allows the users to automatically extract and efficiently analyze the data [18].

Figure 2 gives an overview of the DFS landscape. On the left, a filer (secretary, heads of departments, quality assurance office etc.), who is supported by

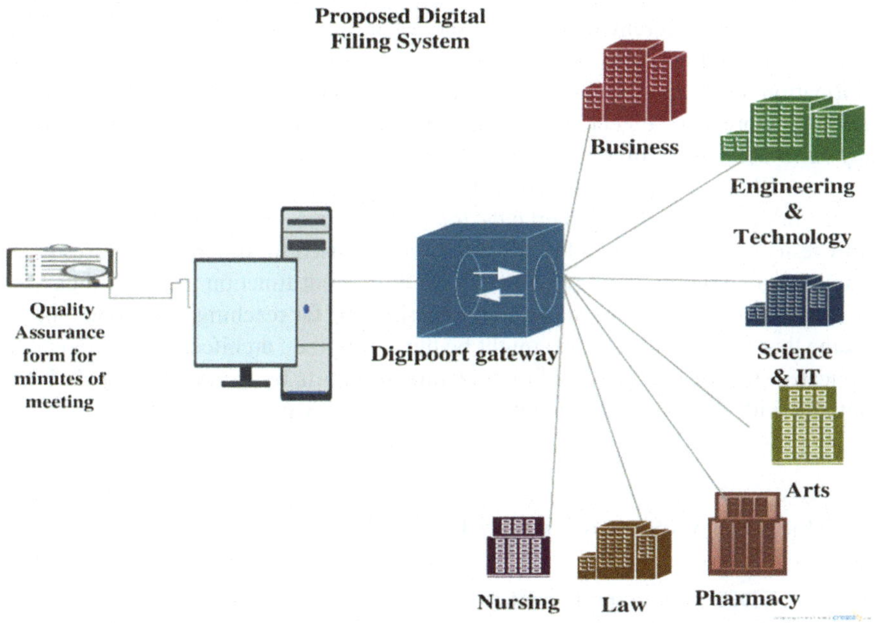

Fig. 2 Proposed digital filing system architecture

computer-based filing. In the middle, the appropriate taxonomy is chosen and developed based on the procedures and rules issued by the quality assurance office for the different filing streams, and the gateways. Seven faculties within ZUJ demanding reports are shown on the right.

As shown in Fig. 2, when it comes to DFS, the digital service (or Logius) is required to manage the gateway through which reports will be routed and which carries out authentication of the reporting organization. DFS will operate much like a post office; simply moving electronic messages from sender' system to the right receiver (user), and returning an electronic receipt. The gateway may also perform other services (e.g., authentication, logging, validation, enrichment). In particular, it could be used to allow for a single submission of faculty document which the digipoort gateway "disaggregates" sending relevant information to each user as appropriate. Digipoort represents the electronic post office which provides a safe and reliable connection for various reporting obligations and processes the messaging data between filers and users of information [3]. While the main goal is the development of a single set of reporting definitions in a single language, a further goal is for the information to be sent directly and electronically from the sender's system to the users, and remain in the control of the faculty files and documents within ZUJ. Therefore, this study suggests the DFS within ZUJ should be enabled by XML technologies.

Previous studies provided two possible scenarios to reduce the administrative burden on the filers based on XML [3, 10]. The first scenario is "one stop shop" which refers to a single point of access to electronic services and information offered by different public agencies [20]. It requires all systems to be interconnected that enable the user to access online services by a single point even if these services are provided by different filers. The users of information therefore can reuse the data. The second scenario is "store once and report many" in which the filers of information need to store the data once in a data representation format. Thus, multiple reports can be generated. Despite the data definitions and the infrastructure may be re-used over different reporting chains; the specific report will be addressed to a user. Bharosa et al. [3] found that the second scenario is more appropriate for similar initiatives in the Netherlands for two reasons. First, legislation does not allow re-using data collected for one purpose to be used for different purposes in the first scenario. Second, the reports may have a different function and may thus have different contents which make the first scenario to be far reaching. Based on this, we assume that the second scenario might be the appropriate architecture to remove the duplicated information provided in the required reports by the faculties at ZUJ. We continue with the expected benefits and costs of DFS in the next section.

5 Expected Costs and Benefits of DFS

The cost of DFS system implementation can be classified into two categories: direct costs and indirect costs [14]. Direct costs, or tangible costs, are those costs explicitly related to DFS implementation that the hosting organization has to pay including the costs incurred due to software purchase or licensing, hardware and software costs and acquisition and instalment, procurement of external professional services, staff training, and maintenance costs and so forth [17]. On the other hand, indirect costs, or intangible costs, are elusive to define, identify and control [21]. Indirect costs of IT project (e.g. the overhead costs, business process reengineering and organizational restructuring, etc.). On the other hand, Dunne et al. [5] concluded that time and efforts to learn this technology, cost of filing software are inhibitors of XML adoption in the UK. Several studies found that XML technology is complex technology because it requires both special technical expertise to be adopted by large organizations in the UK [5], in the USA [16] and in Australia [4]. Pinsker and Li [16] argued that using complex XML taxonomy requires time and money which are barriers that negatively influence the decision to adopt it.

It is expected that XML technology will bring numerous benefits to filers and users in terms of time and costs saving and errors reduction during the filing process [6]. Dunne et al. [5] found that XML technologies offers several benefits: no re-keying of data, enhancing data comparability, accuracy, speed, re-usability of data without losing integrity, reliable source, easily integrated with other applications. Pinsker and Li [16] conducted interviews with business managers who are involved in XML adoption in Canada, Germany, South Africa and the U.S.A. They found that adopting this technology offers four main benefits, including competitive advantage; increasing the transparency of financial reporting; better compliance with statutory requirements; and saving time and costs. They concluded that "there is a reduction of 30% of postage and filing information costs and the time needed to generate financial statements is reduced from five to six days per statement to 15 min or less" after XBRL filing by organisations [16]

6 Conclusion

All faculties within ZUJ have to file multiple academic and administrative reports with different formats. Some of the reports contain the same information with different descriptions. Although the electronic mediums such CD-ROM, PDF have evolving and replacing paper based medium, they failed to drive out unnecessary or duplicated data descriptions.

This study examines the development of digital filing system (DFS) that is enabled by an eXtensible Mark-up Language (XML) technology as a new mechanism which aims at reducing the reporting burden on the filers and users of the information at ZUJ. This paper presents two scenarios for implementing DFS. DFS has revolutionised the electronic filing systems world like nothing before as it offers two major advantages over others systems: firstly, it removes the manual intervention of the information supplied by organisations, and secondly, it drives-out duplicated information that is filed to multiple users. By using this innovation, filers of information can report one set of information instead of having to report it repeatedly to the different (multiple) users and for the different purposes and using all the different forms. To conclude, this innovation can reduce the filing and reporting burden on faculties. DFR is enabled by a technology to develop an appropriate taxonomy. This helps by developing standard for all information descriptions and formats. The most widely used enabling technology is XML.

This study has several limitations which present avenues for future research. First, it based on literature review rather than analysis of the ZUJ case. Further case study needs to be used to specify the components of the DFS architecture at ZUJ. Future research can be useful to identify the filers and the users of required information within the ZUJ. Empirical research is recommended to explore the expected costs and benefits of digitising the filing system for the managerial and academic documents at ZUJ. The main focus might be on the views of the filers and those using the information by this innovation which will be enabled by XML.

Acknowledgements We are gratefully acknowledging the helpful comments and suggestions of the reviewers, which have improved the presentation.

References

1. Al-Yaseen H, Al-jaghoub S (2012) Success and failure of e-learning projects: alignment of vision and reality, change and culture. https://doi.org/10.1.1.646.736
2. Al-Zoube M (n.a) Enhancing Quality of Technology-Enhanced Learning at Jordanian Universities Baseline Jordanian TEL Quality Assurance Framework. Co-funded by the European Union. https://doi.org/10.1186/s41077-016-0025-y
3. Bharosa N, van Wijk R, Janssen M, de Winne N, Hulstijn J (2011) Managing the transformation to standard business reporting: principles and lessons learned from the Netherlands. In Proceedings of the 12th annual international digital government research conference: digital government innovation in challenging times, New York, USA, June 2011
4. Doolin B, Troshani I (2007) Organisational adoption of XBRL. Electron Mark 17:199–209. https://doi.org/10.1080/10196780701503195
5. Dunne T, Helliar C, Lymer A, Mousa R (2009) XBRL: the views of stakeholders, a research report. The Association of Chartered Certified Accountants, London. http://www.accaglobal.com/content/dam/acca/global/PDF-technical/technology-publications/rr-111-002.pdf. Accessed 25 Jul 2017
6. Eierle B, Ojala H, Penttinen E (2014) XBRL to enhance external financial reporting: should we implement or not? Case company X. J Account Educ 32:160–170. https://doi.org/10.1016/j.jaccedu.2014.04.003
7. El-Gharaybeh A (2004) Me and the university: fifty years of hope and despair. In Al-Rae Newspaper, Amman
8. Higher Education Accreditation Commission-HEAC (2015) Guides for quality assurance standards for academic programs and institutions. http://www.heac.org.jo/?page_id=2428. Accessed 25 Jul 2017
9. Holm-Nielsen LB (2001) Challenges for higher education systems. In International conference on higher education, Jakarta
10. Hulstijn J, Van Wijk R, De Winne N, Bharosa N, Janssen M, Tan Y (2011) Public process management: a method for introducing standard business reporting. In Proceeding of the 12th annual international digital government research conference: digital government innovation in challenging times. College Park, MD, USA, 12–15 June 2001. ACM, pp 141–150. https://doi.org/10.1145/2037556.2037577

11. ICAEW (2004) Digital reporting: a progress report (information for better markets). Institute of Chartered Accountants in England & Wales, London. https://www.icaew.com/technical/financial-reporting/information-for-better-markets/ifbm-reports/digital-reporting-a-progress-report. Accessed 25 July 2017
12. Jones A, Willis M (2003) The challenge of XBRL: business reporting for the investor. Balance Sheet 11:29–37. https://www.pwc.com/gx/en/xbrl/pdf/pwc_balsheet.pdf. Accessed 25 Jul 2017
13. KPMG (2006) Administrative burdens: HMRC measurement project. KPMG, London. http://www.hmrc.gov.uk/better-regulation/kpmg1.pdf. Accessed 25 Jul 2017
14. Lim N, Perrin B (2014) Standard business reporting in Australia: past, present, and future. Australas J Inf Syst 18. https://doi.org/10.3127/ajis.v18i3.895
15. Lymer A (1999) Business reporting on the internet. In: International Accounting Standards Committee, editor. International Accounting Standards Committee, London. https://www.icjce.es/images/pdfs/TECNICA/C02%20-%20IASB/C208%20-%20IASB%20-%20Estudios%20y%20varios/BusinessReportingInternet-IASC.pdf. Accessed 25 July 2017
16. Pinsker R, Li S (2008) Costs and benefits of XBRL adoption: early evidence. Communications of the ACM, presented at the annual conference of the mid-South instructional technology. Murfreesboro 51:47–50. https://doi.org/10.1145/1325555.1325565
17. Remenyi D, Money AH, Sherwood-Smith M (2000) The effective measurement and management of IT costs and benefits. Elsevier. https://www.elsevier.com/books/the-effective-measurement-and-management-of-ict-costs-and-benefits/remenyi/978-0-7506-8328-9. Accessed 25 Jul 2017
18. Roohani S, Zheng X (2011) Determinants of the deficiency of XBRL mandatory filings, applied finance, p 502. http://connection.ebscohost.com/c/articles/92528908/determinants-deficiency-xbrl-mandatory-filings. Accessed 25 Jul 2017
19. Sabri HA, El-Refae GA (2006) Accreditation in higher business education in the private sector: the case of Jordan. J Mark High Educ 16:47–76. https://doi.org/10.1300/J050v16n01_03
20. Wimmer MA (2002) Integrated service modelling for online one-stop government. Electron Mark 12:149–156. http://citeseerx.ist.psu.edu/viewdoc/download?doi=10.1.1.617.5236&rep=rep1&type=pdf. Accessed 25 Jul 2017
21. Zeng Y, Lu Y, Skibniewski M (2012) Enterprise resource planning systems for project-based firms: benefits, costs and implementation challenges. J Adv Perform Inf Value 1;4(1). http://file:///C:/Users/Esraaalkhatib/Downloads/73-75-1-PB%20(1).pdf. Accessed 25 Jul 2017

The Impact of the Adherence to Basel Rules on Banking Risk Management: Jordan Kuwait Bank Case Study

Abdul Razzak Al-Chahadah and Maha Ayoush

1 Introduction

The growth of the economic and financial globalization, the financial market opening, competition and the development in the electronic communication means lead to an increase in the banking institutions' services and in the complication degree of the banking operations. To meet this development and related risks, it has become a necessity to pay much attention to the banking risks management as the frequent financial crises in the past years have confirmed the necessity to manage the banking risk effectively for its role in rooting the concept of the banking institutions continuity. Accordingly, this study aims to shed a light on the banking risks that threaten the continuity of the banking institutions and the role of Basel standards of banking supervision in reducing or avoiding them.

2 Problem of the Study

The basic motive behind Basel convention was the fear of the central banks in the greatest tenth countries of decline of the capital to critical levels especially after the debt crisis that hit Latin America [6]. The Basel convention (I) sets minimum limits to the capital in light of the credit risk where all the banks need to commit to a minimum of 8% of its capital to the total of its risky assets after being weighted by credit risks.

When the capital increases, the bank's ability in facing the unexpected risks gets higher and accordingly its ability to manage these risks and maintain capital

A. R. Al-Chahadah (✉) · M. Ayoush
Al-Zaytoonah University of Jordan, Amman, Jordan
e-mail: abdulrazzaqsh@zuj.edu.jo

S. Şahin (ed.), *8th International Conference on Engineering, Project, and Product Management (EPPM 2017)*, Lecture Notes in Mechanical Engineering,
https://doi.org/10.1007/978-3-319-74123-9_31

adequacy will be stronger. Therefore, the following questions summarized the problem of the study:

1. To which extent does the Jordan Kuwait Bank abide by the implementation of Basel (II) of the international banking supervision?
2. To which extent does the Jordan Kuwait Bank's adherence to Basel (II) rules of the international banking supervision affect banking risk management and its future continuity?

3 Hypotheses of the Study

1. The Jordan Kuwait Bank abides by the implementation of Basel (II) standards of the international banking supervision.
2. The Jordan Kuwait Bank's adherence to Basel (II) rules of the international banking supervision affects banking risk management and its future continuity.

4 Significance of the Study and Its Objectives

The importance of this study arises from the definition of the banking industry, which is the art of dealing with the risks management. This has attracted the attention of bankers all around the world especially the Bank of International Settlements and the International Monetary Fund, in addition to monetary authorities' officials in the ten major industrialized countries.

Therefore, this study aimed at revealing the impact of Basel convention on banking risks management and identifying the process of addressing capital adequacy standard, which was confirmed by Basel (II) of the banking supervisory.

5 The Concept of Banking Risks and Its Types

The Basel Committee defines banking risks as a set of changes in the institution's market value. This concept reflects the view that claims that risks management is working on achieving the ideal return by making a balance between return level and risk degree [10]. The banking risks can also be defined as: the banking institution's possible exposure to unexpected and not planned losses, or fluctuation in the expected return on a specific investment which causes negative results that affect the bank's goals and the successful implementation of its strategies [9].

There are many types of risks facing the banking industry as shown below:

1. Financial risks: they include all the risks related to the assets and liabilities of the banking organization. According to Karasnah [8] the financial risks includes:
   (a) Credit risks that result from the customer's failure to fulfill his commitments towards the bank on time.
   (b) Liquidity risks that causes losses due to the banks' inability to meet short-term obligations on their due date.
   (c) Exchange Rate Risks are risks resulted from dealing with the foreign currency and the volatility in the prices of these currencies.
   (d) Pricing risks that result from the changes in the assets prices, particularly the financial investments' portfolio.
2. Operational risks are risks related to the direct and indirect losses arising from the internal operations, human factor and information technology problems, or due to external circumstances [2, 4].
3. Interest rate risks: they are the risks arising from the bank's exposure to losses attributed to changes in the interest rate fluctuations in the market [8].
4. Legal and organizational risks: Legal risks result from the banks' lack of commitment to the laws organizing the work while the organizational risks result from the bank's violation of the Control 's Authorities' standards and laws [10].
5. Reputation risks are risks resulting from the negative impression about the bank that motivates the customers to deal with other banks [1].

6 The Role of Basel Convention in Reducing Banking Risks

6.1 Basel (I) Convention

The banking capital has significant role in maintaining the safety and stability of the banks. It is the barrier that forbids any unexpected loss which the bank may face. There are two types of losses attributed to the banking risk [3]. The first type is the expected losses that any bank may incur repeatedly and these losses can be covered by the operational expenses. The second type is the unexpected losses that rarely happen and affect the bank greatly. Unexpected losses are covered from the bank's capital, but incase it was not sufficient then the customers' deposits are used to cover these losses. This is illustrated in Fig. 1.

It is clear from the previous chart that Basel (I) convention was the first international step in facing the banking risks and enhancing the capability in banking management and strengthening the confidence in its activities and role as an economic and financial median between the savers and the investors and customers.

1. Basel Committee has divided the capital in the banks into two tiers [11]. The first tier, which is called the primary capital, reflects the bank's self-capabilities

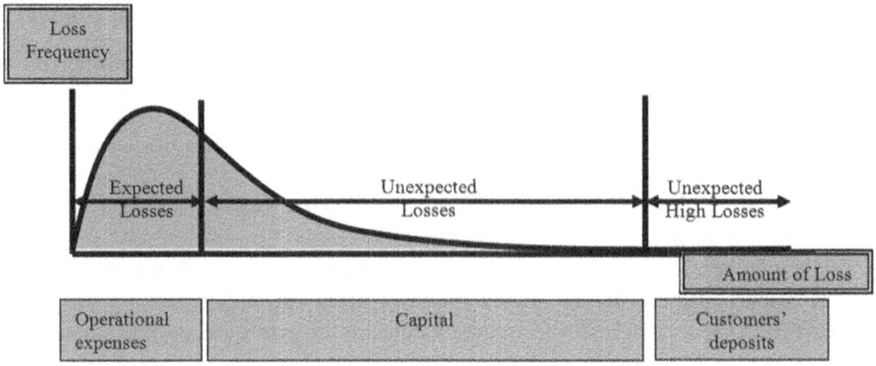

Fig. 1 Distribution of bank losses [3]

of profits and capital and its ability of continuity by relying on itself and protecting its depositors. The second tier, which is called supplementary capital reflects the bank's readiness for the risks through unpaid reserves and preservations.

2. Risk Weight System: The Basel Committee classified the assets into categories; each category is given a risk weight that suits the degree of risk which is exposed to Abo-Mohemid [1]. They set five weights (0, 10, 20, 50, and 100).
3. Target ratio: the committee sets minimum limit of ratio that reflects the relation between the capital base in one hand, and the ***Risk-weighted*** assets and the accidental commitments on the other hand. It was agreed on the minimum limit (8%) and so the ratio of the capital adequacy according to Basel (I) is calculated as follows [1]:

$$\text{Capital} \div \text{risks weighted assets} \geq 8\%.$$

6.2 *Basel (II) Convention*

Basel (II) convention is based on three main pillars: minimal capital requirements, regulatory supervision and market discipline.

First pillar: Minimal Capital Requirements:

$$\begin{aligned}\text{Capital Adequacy Ratio} = \text{Capital}/(&\text{Credit Risks} + \text{Market Risks}\\ &+ \text{Operational Risks}) \geq 8\%.\end{aligned}$$

It is noted that the ration remains the same 8% and the concept of capital does not change consisted of three tires: primary capital, supplementary capital and supplementary loans. However, the most important change was inserting operational risks to reduce the banking risks and support banks' continuity.

Second pillar: Regulatory Supervision:

The goal of regulatory supervision according to Basel (II) is not only to ensure capital adequacy in facing the risks but also to motivate the banks to use the best methods to manage its risks [11]. The Central Bank regulations require that paid in capital should not be less 100 million dollars and the shareholders equity ratio to assets should not be less than 6% while the regulatory Capital should not get less 12% compared to risk-weighted asset and market risk.

Third pillar: Market Discipline:

This pillar requires the availability of more information for the participants in the market to enhance and strengthen market discipline through increasing the banks' discourse and transparency, since the sufficient disclosure is considered a necessity to ensure the awareness and understanding of the dealers with the banks risks.

6.3 Basel (III) Convention

The amendment of the Basel (II) convention includes the following three pillars that should be implemented from the beginning of 2014 until 2019:

First pillar: minimum of capital adequacy, this amendment includes raising the minimal of capital reserve ratio from 2 to 4.5% in addition to margin reserve of ordinary share with 2.5% to be used in facing crises and so the total becomes 7%. Also, raising the capital adequacy rate to 10% instead of 8% and the new convention proposed two ratios to meet the liquidity requirement [5].

Second pillar: regulatory supervision:

The Basel committee recommended that the banks need to set technologies to determine and measure risks concentration using Stress test to examine the bank's capability in tolerating results under hard circumstances and therefore possible points of weakness could be identified through virtual exceptional events as sudden withdrawal operations. The committee also recommended applying the confirmation test to find out whether the actual daily returns equal the expected daily returns that are resulted from a specific trust level and this is the core of continuity.

Third pillar: Market discipline:

The amendments include more disclosures by the banks, in addition to disclosure requirements of securitization and re-securitization operations, which give a holistic image of the banks' risks, and thus there will be indirect pressure on the banks with insufficient capital compared to risks degree.

7 An Application Study on the Jordan-Kuwait Bank

The Jordan-Kuwait Bank is a Jordan limited public shareholding Company that was established in 1976 by the Jordan companies Law, number [13] for the year 1964. The shares of this bank are listed on Amman Stock Exchange, and the bank has 61 branches inside Jordan and 4 branches abroad, in addition to two companies. This bank is chosen for the study for many reasons such as the availability of data, the good reputation, the lack of problems and the availability of at home and abroad branches which means facing different kinds of risks.

The bank's annual statements and reports for the years (2011–2015) are examined in order to extract the needed information concerning the bank's commitment to Basel (II) implementation.

7.1 Methodology

The ratio of capital adequacy was calculated according to Basel (II) convention during (2011–2015), and risks are measured in a way that reflects the expected losses that aroused in the ordinary circumstances and the unexpected ones. This is based on the estimation of the total of the actual losses using advanced statistical methods that rely on the possibilities based on the previous experiments and they are adjustable to suit the economic circumstances. The results were illustrated in Table 1.

Table 1 showed the following:

1. The Jordan-Kuwait Bank has implemented Basel II concerning capital adequacy ratios, which should not be less than 8%, and this is an indicator of the bank's ability in facing possible banking risks.
2. The Bank has committed to the Jordan Central bank's instructions concerning the systematic capital that its ratio to the Risk Weighted Assets and risks market should not get less than 12% of the total of regulatory capital.
3. The bank's management did not distribute any profits to the Shareholders from the regularity capital factors as such distribution can lead to lack of commitment of the minimum limit of the required capital.

8 Results

1. The bank has a strategy adopted by the board council of the banking risks management including identifying the risks facing it in addition to the maximum limits of risks that are acceptable to the bank and the limits of the capital adequacy, which should not exceed 14% of the primary and regularity capital.
2. The bank is committed to the implementation of Basel II requirements through disclosure of the capital formation and through having provisions to face different risks and these provisions should be disclosed in the financial data that ensures its continuity.

Table 1 Ratios of capital adequacy in the Jordan-Kuwait Bank according to Basel (II) convention during (2011–2015)

Items	Year				
	2011	2012	2013	2014	2015
Items of primary capital (JD)	1000	1000	1000	1000	1000
Subscribed and paid up capital	100,000	100,000	100,000	100,000	100,000
Legal reserve	52,702	59,011	65,623	71,918	77,480
Optional reserve	93,858	106,476	119,699	132,235	143,357
Retained earnings (after deducting proposed profits and deferred tax assets)	63,818	67,997	74,929	79,370	80,149
Non-controlling rights	5375	–	–	–	–
Subtract					
Deferred allocations with the approval of the Central Bank of Jordan	14,000	25,234	23,663	17,625	10,687
Goodwill and intangible assets	1097	2227	3197	7362	7065
50% of investments in insurance, financial companies and investments of interest in other companies	12,817	7420	6351	11,024	–
Properties that were expropriated over than 4 years or more than 2 years but less than 4 without the approval of the Central Bank	561	551	613	6896	–
Deficit in additional capital	2036	–	–	–	
Total of primary capital	285,243	298,052	326,426	340,616	376,248
Supplementary capital items					
General banking risk reserve	10,554	12,116	12,141	12,982	13,353
Periodic fluctuation reserve	–	78	186	228	228
The reserve of financial assets' assessment with fair value through comprehensive income with 100% of the positive change and 45% of the negative change	226	270	2298	2720	1926
Additional capital deficit	2036	–	–	–	–
50% of investments in insurance، financial companies and investments of interest to other companies	12,817	7420	6351	11,024	6986
Group of systematic capital	285,243	303,546	334,701	345,521	384,769
Risk weighted assets	1,747,241	1,861,252	2,037,468	2,071,236	2,111,023
Regularity capital adequacy ratio (%)	16.33	16.31	16.43	16.68	18.23
Primary capital ratios (%)	16.33	16.01	16.02	16.45	17.82

3. The bank is committed to keep the capital adequacy rate stated in Basel II. It is noted that the bank has achieved higher rates during years of the study which revealed a strong and trusted basis for the capital that is capable to avoid possible banking risks.
4. Based on what has been mentioned previously, it is clear that the bank is committed effectively to the implementation of Basel II and therefore the study's hypotheses are accepted. Accordingly, the bank's commitment of the supervisory criteria affects positively risks management and the continuity of its future activities.

9 Recommendations

1. The Central Bank has to make a decision concerning the working banks in Jordan to implement Basel II requirements gradually with limited period of time that suit the recommendations of Basel committee so as to support the Jordan banks' continuity and efficiency in banking risks management and to increase the financial and economic society's trust of these banks.
2. The bank has to improve the tools used in managing banking risks including control and risk self-assessment and the international control procedures where the major indicators of the risks and its types were controlled monthly.

References

1. Abo-Mohemid M (2008) Islamic funding risks and its relation to Islamic Banks' capital adequacy through Basel II standard. Dissertation, Arab Academy for Banking and Financial Sciences, pp 26–27
2. Al-Biltaji M (2005) A proposed model to measure the risks in the Islamic Banks: field study. The Banking Institute, Saudi Arab Monetary Fund, p 12
3. Al-Sheikh Hasan M (2005) Measuring Islamic Banks solvency in the new frame of capital adequacy. A paper presented to the international conference of the third Islamic conference, Umm Al-Qura University, Makkah Al-Mukarramah, p 8
4. Al-Zo`bi T (2008) Developing a model of calculating Islamic Banks' capital adequacy based on Basel committee instructions: application study on Islamic Arab Bank and the Palestinian Islamic Bank at Gaza sector. Unpublished thesis, Islamic University, faculty of Commerce, Gaza, pp 83–90
5. Bo-Bozian M et al (2011) The Islamic Banks and new reservation standards and systems: status of Basel III application. A paper presented to the international eighth of Islamic economy and finance, Qatar, pp 10–26
6. Egyptian Ahli Bank (2001) Systems of banking supervisory in the developed countries. Econ Bull 54(4):24
7. Hashad N (2004) Your guide to Basel II. Union of Arab Banks, Part II, p 30
8. Karasnah I (2006) Basic and modern frame of the banks' control and risks management. Arab Monetary Fund, Institute of Economic Policies, Abo-Dhabi, p 39

9. Keegan M (2004) Management of risk, principals and concepts, H M treasury. The Orange Book, Working Papers, p 9
10. Nasr A, Abo-Salah M (2007) Operational risks according to Basel II requirements: a study of its nature and ways of managing them in the operating banks in Palestine. A paper presented to the fifth annual scientific conferences, Philadelphia University, pp 10–12
11. Sha`bo FB (2012) Effect of Basel II standards' application on Islamic Banks' liquidity and profitability. Thesis, University of Halab, pp 9–42)
12. Soliman N (2010) Do the Islamic banks need Basel III? http://www.cibafi.org
13. van Gestel et al (2009) Credit risk management, basic concepts: financial risk components, rating analysis, economic and regulatory capital, p 352

The Use of Capital Budgeting Techniques as a Tool for Management Decisions: Evidence from Jordan

Mohammad Ebrahim Nawaiseh, Hala Al-nawaiseh, Moh'd Attar and Azeez Al-nidawy

1 Introduction

No doubt that capital budget is one of the most important areas of the company's financial management. It is considered as an imperative tool in the process of planning, execution and control. This area encourages companies to focus on techniques that are beneficial in response of interest in these applications, and call for researchers to focus on this subject. Studies on this topic were conducted in developed countries, i.e. Sweden (Sven-Olov et al. 2014), Canada [14, 15], and Netherlands [19]. Other researches were accomplished in eastern and central Europe, which have a lesser degree of development compared with Western countries, and USA [30]. Other studies were conducted in Asian countries such as Malaysia and Philippine [30]. Many companies still rely on the traditional techniques of control budgeting for evaluating their projects [5]. Different capital budgeting methods are generally accepted in the business; Net Present Value (NPV); Payback Period (PBP); Profitability Index (PI); Internal Return Rate (IRR); Accounting Return Rate (ARR); Discounted payback period (DPP), and Economic Value Added (EVA). These techniques are used to make investment decisions; new projects; Expansion of extending operations; Leasing of assets; merger and acquisition. The choice, however, is not arbitrary, textbooks on financial management often encourage the use of net present value method (NPV); while discouraging the use of other techniques, such as PBP method [4]. These methods, such as (IRR), and (PBP) are often criticized; because they do not consider time value of

M. E. Nawaiseh (✉) · M. Attar · A. Al-nidawy
Department of Accounting, Al-Zaytoonah University of Jordan,
Amman, Jordan
e-mail: Nawaisehmohd@yahoo.com

H. Al-nawaiseh
Al-Quds College, Amman, Jordan

S. Şahin (ed.), *8th International Conference on Engineering, Project, and Product Management (EPPM 2017)*, Lecture Notes in Mechanical Engineering,
https://doi.org/10.1007/978-3-319-74123-9_32

money [14, 15, 3]. IRR can be misleading because of so-called multiple rates of return [25]. Discounted payback period ignores cash flows after the maximum payback point (Sven-Olov et al. 2014). Other studies investigated the use of capital budgeting methods in different countries. [23] indicates that NPV (69%); IRR (46%); and PBP (23%) were the most common techniques used in South Africa. All researches came to an agreement that the proportion of using each CBT is different. From this context; this paper aims at analyzing the practice of CBTs in Jordanian industrial companies, and to know what the most used CBTs are, the degree of application of different techniques, and the obstacles that might undermine the use of CBTs? The remainder of this paper proceeds as follows; in the next section, we address previous research on this topic; next, methodology; empirical results are presented; lastly, conclusions and recommendations are drawn. Many Studies on CBTs practice were conducted. Examples include [20] defines capital budgeting as "long-term planning for making and financing all investments that affect the financial results over the financial period longer than just the next period." Companies in emerging economies such as Rwanda [24]; Jordan [22]; Pakistan [21]; and India [29] are shifting towards more advanced techniques, but that will be in terms of large companies. Small sized companies in these countries are still using PBP as a key tool for evaluating the investment. There has been almost no research on CBTs in Jordan. Only one study [22] has been conducted. Hartwig [18] found a positive correlation between company size and use of NPV method. Other studies show that PBP, and IRR to be the preferred tools for managers in most companies in Canada [2], Japan [26], China [19], Table 1 provides a historical appraisal of the development of capital budgeting research.

Table 2 shows twelve CBT recommended or not recommended by textbooks.

Table 1 Most commonly used capital budgeting techniques as a result of prior research surveys

No.	Authors	NPV	PI	PBP	DPBP	ARR	IRR	Others
1	Arnold [1],	√		√		√	√	
2	Hermes et al. [19]	√						
3	Chen [8]	√					√	
4	Sridharan [27]	√		√				
5	Truong et al. [28]	√						
6	Bennouna et al. [3]	√						
7	Khamees et al. [22]		√	√				
8	Shinoda [26]			√				
9	Correia [9]	√					√	
10	Hall [16]	√					√	
11	Andres et al. 2014	√		√			√	√
12	El-Daour [13]	√	√	√		√	√	
13	Rachunkowosci [31]	√	√		√		√	
14	Shakila Yasmin [32]	√		√			√	

Table 2 Capital budgeting techniques recommended or not through textbooks

No.	Method	*Situation	No.	Method	*Situation
1	Net Present Value	√	7	Value-at-risk (VAR)	√
2	Annuity	√	8	Real options	√
3	Adjusted NPV	√	9	Payback period	×
4	Profitability index	√	10	Discounted PBP	×
5	Sensitivity analysis	√	11	Accounting Return Rate (ARR)	×
6	Earnings multiple	√	12	Internal Return Rate	×

*Recommended (√); Not-Recommended (×). *Source* Graham and Harvey [14, 15]

2 Methodology

This research is exploratory based on a review of the existing literature. In addition to the financial data compiled from the annual reports in 2016; Moreover, we developed a questionnaire circulated by hand to 50 people who are in charge of CBTs at 50 listed industrial companies; the response was just the opinion of one individual. Data do not represent the overall opinion of the company. We received 42 questionnaires. However, only 8 questionnaires were dropped from the analysis, leaving an adjusted response rate of 84% out of the sample, and 79.4% out of the population. Our Questionnaire focus on the following questions: To what extent the CBTs are used; How frequently companies use each of the different CBTs, response were on a five point scale of 0 to 4; with 0 meaning "never", 1 "rarely", 2 meaning "sometimes", 3 meaning "frequently" and 4 meaning "always; What are the obstacles that industrial companies might face in applying CBTs? The study posted a question requesting the respondents' characteristics such as experience, the professional Certificate level attained, and respondent's age.

3 Empirical Results and Discussion

3.1 Data and Summary Statistics

The collected responses were coded, and entered in E- views for analysis. Proper statistical analyses were computed, Level of significance chosen in this study is 0.05.

H1 Capital budgeting techniques used are similar to the methods recommended by textbooks. Respondents were asked to indicate what kind of CBTs they use. A list of different nine methods was given to provide their own. In some cases, respondents selected several methods. To verify this hypothesis, we observe that companies use these techniques as follows: Most popular recommended method used is NPV with (69.05%) which is indicated as a primary technique; the current use of NPV in the Jordanian companies seems far less common than in the USA, the NPV was used by 74.9% of USA Firms; and 75.7% use IRR as primary method [14, 15]. The second most popular method not recommended used is IRR (63.81%), followed by PI (62.86%). Despite its theoretical problems [9], PBP was the fourth most popular method (60.48%), together with ARR (60.48), whilst (37.62%) of the companies surveyed use SA. In fact, a limited companies use EVA, with (25.24%), and ANPV with (33.81%). Some statistical analyses for the used CBTs are shown in Table 3. The results are depicted graphically in Fig. 1 to indicate ranks in the use of these methods. It should be mentioned that combined proportion of companies use each of these two groups of techniques are different; not recommended compared to recommended methods; (47.60 < 50.83%). For testing H1, we use a One Sample t-test at significant level 0.05; results are shown in Table 3. Thus, we have an evidence to conclude that companies mostly use CBTs recommended when choosing among other appraisal techniques. This analysis enables us to accept hypothesis 1. These findings suggest that (60.48%) of respondents ignore PBP weaknesses which does not consider the time value of money; ARR also have been employed by (60.48%), because it is more flexible, agile, ease of calculation. EVA and SA were never used by participants with (74.76%) and (62.38%) respectively.

H2 large firms with younger; more educated and experienced staff use more DCF of CBTs. We split the companies' sizes into small versus large; these companies range from small (73.80%) of the sample have assets of less than total JD

Table 3 Frequency of companies that use each different method

No.	Method	*R/N	Mean	Use (%)	No use (%)	PV test	t test
1	NPV	R	3.45	69.05	30.95	.000	38.181
2	PBP	N	3.02	60.48	39.52	.000	5.441
3	ANPV	R	1.69	33.81	66.19	.000	4.287
4	PI	R	3.14	62.86	37.14	.000	6.632
5	IRR	N	3.19	63.81	36.19	.000	7.964
6	ARR	N	3.02	60.48	39.52	.000	6.987
7	DPBP	N	1.4	28.10	71.90	.000	7.765
8	EVA	N	1.26	25.24	74.76	.000	10.749
9	SA	R	1.88	37.62	62.38	.281	1.094
All recommended		R	2.54	50.83	49.17	.000	11.364
All not-recommended		N	2.38	47.60	52.40	.000	6.596

(R) Recommended, (N) not-recommended in textbooks, The % of use, means all respondents who frequently or always use a particular technique. The relative usage of CBT can be depicted by the following Fig. 1:

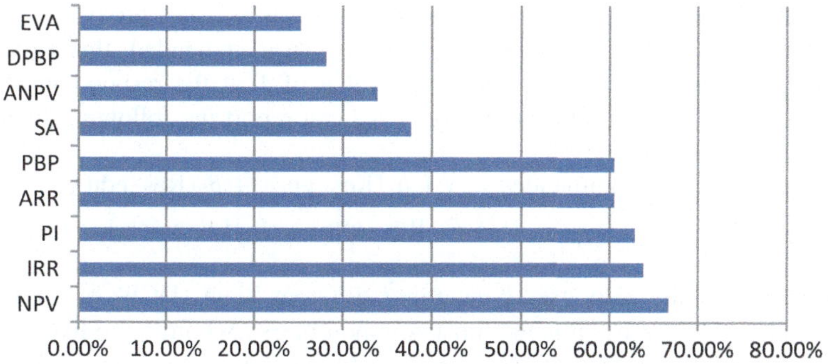

Fig. 1 The relative usage of CBT

10,0000,000 (log. assets <7) to large firms (26.20%) have total assets more than JD 10,000,000 (log. assets ≥ 7). We scored responses; nearly 61.9% of the respondents had been working with the company for more than 15 years in their jobs. While 81% of the respondents do not have any professional certificate; only 19% have JCPA; CMA; and CPA. It was noted that nearly (85.7%) are over age 35 years; Table 4 presents a summary about these characteristics.

We estimate the study model; using proxies for the expected influential factors which may affect the use of capital budgeting method; to proxy for these, we include a dummy Res_age in full years, which takes the value of 1, if the respondent's age is less than 40 years and 0 otherwise. As an alternative to professional education; we use the value of 1 as a dummy variable of Res_edu, if he has a professional certificate and 0 otherwise. Finally, we include Res_exp as a dummy in full years, which has a value of 1, if the respondent's experience is more

Table 4 Shows a demographic characteristic for respondents

Data	Variables	Frequency	%
Experience	5–10	5	11.9
	10–15	11	26.2
	15–20	18	42.9
	20 or more	8	19
Qualification	CMA	3	7.1
	CPA	2	4.8
	JCPA	3	7.1
	Others	34	81
Age	30–35	6	14.3
	35–40	4	9.5
	40–45	17	40.5
	45 or more	15	35.7
Size	Small less than 10,000,000	31	73.8
	Large 10,000,000 or more	11	26.2

than 20, and 0 otherwise. We include the log. of assets as a size proxy (Ln_assets), The dependent variable will focus on DCF_ as_ Recommended, this variable is measured by a binary variable, taking the value of 1, if the respondent chooses NPV, SA, ANPV, or PI; and 0 otherwise. We can report the following model:

$$\begin{aligned}\text{DCF\_as\_Recommended} = \alpha + \beta_1(\text{Res\_age})_\text{i} + \beta_2(\text{Res\_edu})_\text{i}\\ + \beta_3(\text{Res\_exp})_\text{i} + \beta_4(\text{Ln\_assets})_\text{i} + e,\end{aligned} \tag{1}$$

We regressed the responses concerning different main (DCF) of CBTs, and focused on use it; because it is the most common basis. According to hypothesis 2; we also use multivariate probit regression to estimate the model in line with Hermes et al. [19]. We expect positive signs for logarithm of assets as a proxy of Company size (LN_ASSETS); more age of the respondent (RES_AGE), professional education attainment (RES_PROF), and more specific company's staff experience (RES_EXP), will lead to use more recommended methods in the model as independent variables. Pearson correlation coefficients among the independent variables were examined (Table 5); multicollinearity found to be limited.

Table 6 shows that size of company is a significant determinant of the DCF based CBTs recommended; the expected sign is positive for LN_ASSETS. Our result for size supports earlier findings on the use of DCF based CBTs, means that use of these techniques is more common in larger companies. We test the effect of each respondents' characteristics on the use of DCF based methods recommended separately (Table 6); both RES_EXP and RES_EDU obtain the expected signs; these variables are significant. The use of DCF is insignificantly and positively related to RES_AGE. We conclude that age, experience and professional background of people who are in charge of CBT position influence the CBTs; this is consistent with Graham and Harvey [14, 15]. Results are similar to an indication in a study reported by Du Toit and Pienaar [12] that Companies undertake relatively large capital projects tend to select IRR and NPV in South Africa. Our results confirm that lager companies tend to use recommended methods more often when deciding on evaluation of capital investments; which is similar to study reported by Hartwig [17].

Table 5 Pearson correlation matrixes of independent variables

Variable	Data	Size	Experience	Profession	Age
Firm Size	Correlation	1			
Experience	Correlation	−0.095	1		
	Sig.	(0.551)			
Profession	Correlation	−0.031	−0.072	1	
	Sig.	(0.844)	(0.649)		
Age	Correlation	0.175	−0.456[a]	0.141	1
	Sig.	(0.269)	(0.000)	(0.372)	

[a]Correlation is significant at the 0.01 level (2-tailed)

Table 6 Results of regression analysis/ML—binary probit regression

Variable	Coefficient	Z-Statistic	Prob.
Constant	−0.482	−3.448	0.001
RES_EXP	0.38	2.653	0.008
RES_PROF.	0.405	2.984	0.003
RES_AGE	0.186	1.364	0.173
LN_ASSETS (Size)	0.424	3.006	0.003
Prob. (LR statistic)			0.000

Table 7 Obstacles undermine the use of capital budgeting Recommended

#	Obstacles	0	1	2	3	4	M	Sig.	t
1	Lack of knowledge	0	0	2	17	23	3.5	0.000	10.898
2	Unavailability of adequate data	0	2	8	17	15	3.07	0.000	4.274
3	Lack of specialized accountants	0	0	0	25	17	3.4	0.000	11.803
4	Lack of confidence in the use of DCF	3	1	13	25	0	2.43	0.593	−0.539
5	Lack of applicability to the business	0	2	4	10	26	3.43	0.000	7.002
6	Too sophisticated to apply in practice	0	3	11	10	18	3.02	0.002	3.396
7	No trust in DCF	8	4	6	0	24	2.81	0.229	1.222
8	Difficulty of estimating inputs	2	5	5	13	17	2.9	0.035	2.175
9	Requires many internal resources	1	1	4	15	21	3.29	0.000	5.545
10	Lack of top management support	0	0	5	17	20	3.36	0.000	8.025
11	Encourages too much risk taking	0	1	6	18	17	3.21	0.000	5.92
Average of relative importance %		62.60%					3.13	0.000	11.362

H3: Industrial Companies face obstacles to undermine the use of budgeting techniques. Results are summarized in Table 7. The most important obstacle is the Lack of expertise and knowledge, followed by Lack of specialized accountants. Combined mean of the result is 3.1 with P. value 0.000; we accept H3; and conclude that there are many obstacles companies face to impede the use of capital budgeting techniques.

3.2 *Findings and Recommendations*

Our study concluded that majority of companies use (50.83%) out of (4) methods recommended; NPV (69.05%), PI (62.86%, budgeting techniques were not all equally used. It is recommended that companies in Jordan adopt more capital budgeting methods like: NPV; ANPV; PI, and SA, in addition to give more attention to use of traditional techniques, particularly for the smaller investments. The study recommends that Companies should handle with strict care CBT, because of its impact on the future of the company. Two methods (EVA) and

(SA) should be supported and given more attention. The obstacles have scored (62. 6%). Future research may focus on the merit is to explore the effect of capital budgeting methods on performance.

References

1. Arnold GC, Hatzopoulos PD (2000) The theory–practice gap in capital budgeting: evidence from the UK. J Bus Finance Account 10(5):603–626
2. Baker HK, Dutta S, Saadi S (2011) Management views on real options in capital budgeting. J Appl Finance 21(1):18–29
3. Bennouna K, Meredith GG, Marchant T (2010) Improved capital budgeting decision making: evidence from Canada. Manage Decis 48(2):225–247
4. Brealey RA, Myers SC (2003) Principles of corporate finance (International Edition), 7th edn. McGraw-Hill, New York
5. Brealey RA, Myers S (2005) Principles of corporate finance, 8th edn. McGraw-Hill, New York
6. Brounen D, De Jong A, Koedijk K (2004) Corporate finance in Europe: confronting theory with practice. Financ Manage 33:71–101
7. Brunzella T, Liljeblomb E, Vaihekoski M (2013) Determinants of capital budgeting methods and hurdle rates in Nordic firms. Account Finance 53:85–110
8. Chen S (2008) DCF techniques and nonfinancial measures in capital budgeting: a contingency approach analysis. Behav Res Account 20(1):13–29
9. Correia C (2010) Capital budgeting practices in South Africa: a review. S Afr J Bus Manage 43(2):11–29
10. Daunfeldt S, Hartwig F (2011) What determines the use of capital budgeting methods? Evidence from Swedish listed companies. Soc Sci Res Netw, 1–37
11. Daunfeldt SO, Hartwig F (2014) What determines the use of capital budgeting methods? Evidence from Swedish listed companies. J Finance Econ 2(4):101–112
12. Du Toit MJ, Pienaar A (2005) A review of capital budgeting behaviour of large South African firms. Meditari Accountancy Res 13(1):19–27
13. El-Daour JI, Shaaban MA (2014) The use of capital budgeting techniques in evaluating investment projects: an applied study on the Palestinian corporations working in Gaza Strip. J Al-Quds Open Univ Res Stud Part 2(32)—October 2014
14. Graham JR, Harvey CR (2001) The theory and practice of corporate finance: evidence from the field. J Financ Econ 60(2):187–243
15. Graham JR, Harvey CR (2001) The theory and practice of corporate finance: evidence from the field. J Financ Econ 60:187–243
16. Hall J, Millard S (2010) Capital budgeting practices used by selected listed South African firms. SAJEMS NS 13(1):85–97
17. Hartwig F (2011) The use of investment appraisal methods and cost of capital estimation techniques in Swedish listed companies. Paper presented at the 34th EAA Congress, Rome 20–22 April 2011
18. Hartwig F (2012) The use of capital budgeting and cost of capital estimation methods in Swedish-listed companies. J Appl Bus Res 28(6):1451–1476
19. Hermes N, Smid P, Yao L (2007) Capital budgeting practices: a comparative study of the Netherlands and China. Int Bus Rev 16(5):630–654
20. Horngren C, Sundem G, Stratton W (2005) Introduction to management accounting, 13th edn. Person Education Inc, New Jersey
21. Hussain Asif, Shafique Imran (2013) Capital budgeting practices in Islamic banking: evidence from Pakistan. Euro-Asian J Econ Finance 1(1):9–23

22. Khamees BA, Al-Fayoumi N, Al-Thuneibat AA (2010) Capital budgeting practices in the Jordanian industrial corporations. Int J Commer Manage 20(1):49–63
23. Maroyi V, Poll H (2012) A survey of capital budgeting techniques used by listed mining companies in South Africa. Afr J Bus Manage 6(32):9279–9292
24. Mbabazizie PM, Daniel T (2014) Capital budgeting practices in developing countries: a case of Rwanda. Res J Finance 2(3)
25. Ross SA, Westerfield RW, Jaffe J (2005) Corporate finance (International Edition), 7th edn. McGraw-Hill, New York
26. Shinoda T (2010) Capital budgeting management practices in Japan: a focus on the use of capital budgeting methods. Econ J Hokkaido Univ 39:39–50
27. Sridharan UV, Schuele U (2008) Budget size and risk perception in capital budgeting decisions of German managers. Int Rev Bus Res Pap 4(3):213–221
28. Truong G, Partington G, Peat M (2008) Cost-of-capital estimation and capital-budgeting practice in Australia. Aust J Manage 33(1):95–122
29. Verma S, Gupta S, Batra R (2009) A survey of capital budgeting practices in corporate India. Vision: The J Bus Perspec 13(3):1–17
30. Wnuk-Pel T (2014) The practice and factors determining the selection of capital budgeting methods—evidence from the field. Procedia Soc Behav Sci 156:612–616
31. Wnuk-Pel T (2015) Factors determining the selection of capital budgeting methods in companies operating in Poland. Zeszyty Teoretyczne Rachunkowosci 84(140):217–240
32. Yasmin Shakila (2015) Capital budgeting in practice: an explorative study on Bangladeshi Companies. Int J Eng Bus Enterp Appl 11(2):158–163

Evaluating the Need to Use Integrated Project Delivery (IPD) Approach as a New Alternative Implementation System in Developing Countries

Farimah Noghli, Ehsan Saghatforoush and Zahra Forghani

1 Introduction

The progress achieved through social and economic development in today's world, results in expanding activities in the field of engineering projects [1]. One of the factors of economic development in each country is success of its civil infrastructure projects [2]. In recent years, the government of Iran has welcomed using technology in the civil and infrastructure section, in order to develop the country and create economic boom [3]. Considering the increase of complexities in the building industry, it is faced with several challenges. The building industry has learned many ways to face these challenges through offering diverse systems for project implementation. Moreover, due to the increasing need of societies and more importance of time, cost, and quality in evaluating the projects, systems of project implementation need presenting new approaches [1]. For this purpose, a number of practitioners in this industry have proposed a new delivery system under the title of Integrated Project Delivery (IPD). IPD is the result of changes of the traditional project delivery models which gathers all the project team members at an early planning and design time in order to reduce time, avoid wastes, improve project performance, and to create a win-win result for all those involved in the project (Glick and Acree Guggemos, n.d.). This approach improves project objectives through a participatory system [26]. It optimizes project implementation system by enhancing cooperation among different project stakeholders [4]. In fact, increasing coordination and participation of members of project team are considered as the major IPD target [5]. Studies have shown that IPD can be used in all kinds of projects, but as it is not used widely, it is supposed that it is not still approved in Iran as a developing country; whereas this approach has been used in many projects

F. Noghli · E. Saghatforoush · Z. Forghani (✉)
MehrAlborz Institute of Higher Education, Tehran, Iran
e-mail: forqanizahra@gmail.com

S. Şahin (ed.), *8th International Conference on Engineering, Project, and Product Management (EPPM 2017)*, Lecture Notes in Mechanical Engineering,
https://doi.org/10.1007/978-3-319-74123-9_33

abroad [6]. In this study, due to incompatibility and inability of the common civil contracts in Iran in using the experiences of other stakeholders and inappropriate distribution of profit and loss among them, duplications and other similar problems in them, IPD has been introduced as a new system for conducting the civil projects and a response to most of these problems. According to lack of understanding of IPD in Iran and other similar developing countries and consequently using previous traditional approached, it is required to introduce this approach as a new approach in implementing the civil projects.

2 Literature Review

The major civil projects, because of their complexities, are always faced with several challenges all over the world. These problems either occur while signing a contract, during implementation or even after project completion. Some of the reasons of occurrence of these challenges are the existence of multiple contracts and consequently lack of integration among project stakeholders, lack of prediction of proper solutions for possible claims and generally incomprehensive views when starting the project planning. It results in extra project time and cost and reduction of its quality.

In this section, at first we investigate the evolution of project implementation systems, then different kinds of currently used project implementation systems are studied, and following that IPD systems are examined particularly and after that, it will be introduced as the most modern system in this field.

2.1 Project Implementation Systems

Project implementation system is a process, during which the project is designed and constructed. Moreover, it is a method for project preparation, through which the employer transfers the risk of delivering project and performing design and construction to other parties [7]. At the beginning of 20th century, most of the projects were performed through contracts with fixed prices that now they are known as systems for design- tender- construction or traditional ones. The concept of performance management was introduced in 1960s, the concept of design and construction consulting in 1970s, and plan management in 1980s. They appeared in response to the search of employers to find better solutions to conduct complex projects [1]. Figure 1 shows revolution of project implementation systems.

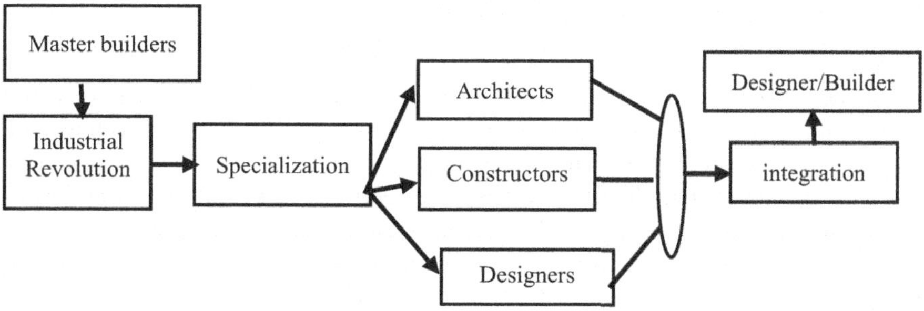

Fig. 1 Revolution of project implementation systems [7]

2.2 Types of Project Implementation Systems

Types of project implementation systems are divided into six categories of in house system traditional (Design-bid-built) systems, design-build systems construction management systems, program management system, and integrated project delivery system. Below, each of project implementation systems are described briefly. In house system or full implementation is a mechanism, in which almost no contract is signed, and in fact the employer has all of the responsibilities of the project [8]. Conventional systems, in which at first the employer provides project design from external resources, and then assigns it to another external resource [7]. This method is common in Iran [8].

Design-build systems are in which the employer provides design and construction services of the project from an external resource. In this type of systems, the employer should have sufficient knowledge about design and construction process, which usually an independent consultant is employed to help the employer [9]. In fact, the employer reduces his/her risks and responsibilities by transferring a large portion of his/her responsibilities to the contractor [10]. Types of design-build systems can be considered as turnkey system, types of construction- operation- transfer systems, bridging system, and buyback method [7]. According to the studies of American Design and Construction Institute, almost 87% of the construction projects are done through this method [3]. construction management system, this kind of system can be considered as a type of conventional systems, in which the owner employs an external source to coordinate design and construction phases and project management to reduce his/her responsibilities and risks [1]. Construction management system is used in two forms as agency CM, and at risk CM [7]. Program management system principally is not considered as a project implementation system, but it is an overall service system that covers multiple needs of the employer. This system is called "Professional Project Management System" that can use a combination of implementation systems for design and construction. In fact, a wide range of services provide feasibility studies for project implementation systems [1]. There is another type of project implementation system known as Integrated Project Delivery. The next section will describe this method in detail.

2.3 *Integrated Project Delivery (IPD)*

Integrated project delivery is one of the project implementation systems, in which project stakeholders, tools, methods, and structures are integrated to achieve the objectives of the project. The aim of this newly born system is creating a process that focuses talents and capacities of all of the project stakeholders towards optimizing the outcomes, increasing the earned value by the owner, reducing wastes, and maximizing efficiency in the design and implementation phases based on multilateral cooperation. This system has eight conceptual phases, such as basic design, detailed design, providing project documents, reviewing documents, buy and check out, implementation, and completion [11].

The aim of IPD implementation system is integrating processes and phases of the project that is shown in the above picture. Standard forms of contract for integrated delivery of implementation were provided and published by some institutes such as Architectural Institute of America. 3xPT Strategy Group has summarized the basics of integrated delivery system in 5 clauses of processes and structure of project organization, scope, implementation issues, tools and methods, and treaty agreements [12]. Participatory decision making and design integration are among necessities of IPD contracts [13]. Figure 2 shows communications in IPD system.

The most important success of IPD can be considered as clear definition of contractual relations, project objectives and also team building at the beginning of the project [14]. Research methodology and data collection methods applied are presented in the following section.

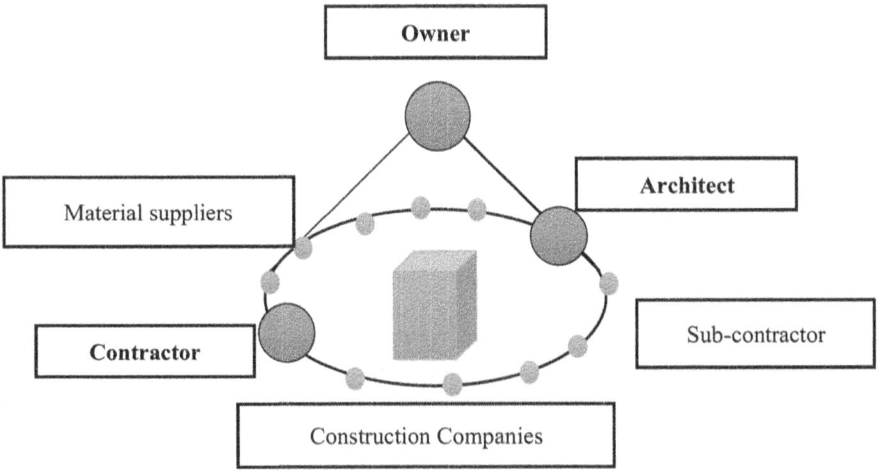

Fig. 2 Communication in the IPD system [7]

3 Research Methodology

Method means initiative or determining steps that are taken to realize a special objective [15]. Selecting a research methodology is usually implemented according to the research objectives, nature and available equipment. Consequently, one can decide about research methodology when its subject, scope, nature and the equipment of its implementation are clear enough [16]. Given the novelty of the IPD approach and insufficient information in this field in Iran, literature review method has been used in this research. One of the most familiar software for managing data in the qualitative analysis is Nvivo [17]. In this study, for qualitative analysis of the collected data, the Nvivo Software has been used. In the next section, data obtained from the literature review method are analyzed.

4 Data Analysis

As stated in the previous section, in this study, the Nvivo Software has been used for qualitative analysis and classification of data collected from the literature reviews. Analyzing this study addresses advantages and disadvantages of the current systems of implementing the civil project and the necessity of applying IPD as a new delivery system in implementing the civil projects, specifically in developing countries.

4.1 Conventional Systems of Implementing the Civil Projects in Iran: Advantages and Disadvantages

In the civil sector of Iran, like most of other developing countries, conventional implementation systems are used in quite all civil projects. Among existing systems introduced in the Sect. 2.2, in house, traditional (Design-bid-built), design-build, and construction management systems are used more than others in Iran. The differentiating factor among project implementation systems is the scope of responsibilities and the way of distributing risks among project stakeholders [7]. According to this fact that the amount of integration is one of the main characteristics of each implementation system, integrated decision making can be a differentiating factor. The concept of integrated decision making that is stated against linear decision making, is such that at first project stakeholders integrate their experiences with each other and accelerate the overall process of design and construction [1]. In the in house system, the employer has the advantage of eliminating tracking costs due to lack of signing contracts in this system. In this regard, time is saved significantly as well. But this system is applicable in small projects [7]. Owner in this system either should use his/her trained forces, or should outsource different sections of the project. For this reason, he/she doesn't have to focus decision making and integrated coordination of the project necessarily [18]. In the in house system, the owner has all of the responsibilities, and all of the jobs are done inside the organization [7]. So it can be concluded that the in house system has poor performance in the way of distributing risks and integration. Traditional (Design-bid-built) system is common for normal projects in many countries including Iran. Accordingly, this method is completely known and its procedures have been identified. The employer has a clear idea about the final costs of the project. The contractor is selected impartially through holding tender. One of the disadvantages of this method is delay in the completion time because of the separation of design and implementation stages, and also hostile relations of designers, contractors, and employer due to being in different positions [7]. This issue shows lack of integration in this system [1]. In the traditional (Design-bid-built) system, the responsibility of design and construction is assigned to the outside of the organization, and the employer has financial responsibility [7]. In this system, we are faced with inappropriate distribution of risks. One of the advantages of construction management system is better cooperation, reducing time of design and construction processes, saving costs through reducing time, and less disputes. Besides these advantages, like other systems, there are some disadvantages too, such as more bureaucracy between the employer and construction manager, management and administrative duplications, and reducing the control on contractors [1]. From two kinds of construction management systems, agency CM is more common in Iran. In the performance construction system, performance manager has the responsibility of management and coordination of design and construction stages [7]. In this system, like traditional (Design-bid-built) system, we are faced with lack of integration and inappropriate distribution of risks. In design-build systems, the employer is responsible for design and construction and his/her risks are minimized and the projects proceeds faster, and most importantly, the builder is involved in the design process [9]. Value engineering and investigating constructability are improved by integrating design and construction processes [1]. In this system, the contractor accepts risks more than any other construction system [7]. Two types of contracts, EPC and BOT, are more common in the construction industry in Iran. Moreover, in the design and construction system, the responsibility of project design and construction is completely assigned to the designer-builder. But a part of the responsibility of feasibility studies and initial designing are assigned in the EPC contract [7]. In the design-build system, in comparison with the previous systems, integration has been enhanced, but still the contractor has the responsibility of project risks, which indicates inappropriate distribution of risks. Project construction system, which has two factors of integration and appropriate distribution of risks together, is IPD that has been examined in the next section.

4.2 The Necessity of Utilizing IPD as a New Alternative System in Implementing the Civil Projects

The overall performance of IPD is such that the key selected stakeholders for implementing the project form a central team composed of owner, architect, consultant engineer, contractor, and other members. After that, a general and comprehensive contract is signed among the owner and other members. After signing the contract, several management groups will be formed to determine the objectives of the project. Project profits, awards and risks will be shared by the same central team. In this method, it is emphasized on using Lean Construction, BIM, and PMIS techniques [19]. BIM is caused better understanding of design and costs. As it is found from the name of IPD, it is based on integration concept and is known as a solution for integrating implementation of the building projects [20]. Writing a contract as win-win, causes distributing project risks appropriately among contract parties. In most of the civil project implementation systems, one-way view benefits of the owner prevails win-win attitude [21]. In the controversial systems of project implementation, inappropriate distribution of risks is seen, while in IPD, contract clauses are set somehow to be a stimulus for maximizing cooperation in the projects, and consequently risks and opportunities are distributed appropriately [12]. In other words, in IPD system contractual clauses, the existence of a common profit and loss and win-win attitude is considered. Early participation of key stakeholders and project integration lead to saving time and cost [22]. In in house systems, conventional systems, and construction management, we are faced with lack of integration. Whereas one of the principles of IPD is that the employer should lead the project with perfect obligation to project integration and total responsibility [11]. In order to compare design-build system with IPD system in terms of integration, it can be said that despite integration has been improved in the design-build system in comparison with the traditional systems, it is not implemented seamlessly. In contrast, implementing projects in IPD system is defined in an integrative way. Early participation of project stakeholders in IPD is one of the advantages of this system in comparison with design-build systems [23]. Due to the universal acceptance of utilizing this method in the civil projects, still there are few studies about this issue in Iran and other similar developing countries. According to the report of Plan and Budget Organization [3], 90% of the civil projects in Iran are faced with an increase of time and cost, and 60% of unfinished projects need 15 more years to be completed. Accordingly, lack of using modern contracts, is a factor for increasing cost and time of the civil projects in Iran [24]. So it is expected that IPD can solve the problems of the civil projects in Iran.

5 Discussion and Conclusion

Studies show that selecting the best project implementation system reduces project's costs up to 5% and its implementation time, up to 30% [25]. During recent years, the IPD approach has been grown in some developed countries through integrating processes, stakeholders, and components of project implementation. Generally, it can be concluded that participation of project stakeholders in the early planning, design and construction phases, and also the presence of common profit and loss in the IPD approach, enhance the possibility of achieving success in the civil projects [23]. It is hoped that the results of this study can make project owners familiar with this newly born approach, and encourage them to try using this implementation system idea in order to make it more and more popular in civil projects. For this aim, it is suggested that contractual principles of IPD be localized according to the culture and laws of Iran, the required infrastructure to create a culture of collaboration be provided among stakeholders of the project and an integrated view by the related entities be organized to quicker replacement other available systems with the IPD approach. Moreover, some studies can be conducted to examine the potential problems in implementing IPD in Iran.

References

1. Dorsey W, R (2009) Project delivery systems in construction industry. University of Tehran Press, Tehran
2. Talkhabi H, Jalal MP, Golabchi M (2014) Analyzing the causes of contracting claims in design and build underground projects in the country. J Eng Tunnels Undergr Spaces
3. Shakeri E, Ghorbani A (2006) Project management and construction the major couse of contracting claims in civil project. Project Management Institute (PMI), Tehran
4. Mollaglu-kormaz S, miller VD, Sun W (2014) Assessing key dimensions to effective innovation implementation in interorganizational project teams: an integrated project delivery case. Eng Proj Organ J
5. AIA, AIA Minnesota, School of Architecture-University of Minnesota (2012) IPD case studies. AIA, Minnesota
6. Collins W, Parrish K (2014) The need for integrated project delivery in the public sector. AIA California Council, California
7. Golabchi M, Faraji A (2010) Project strategic management. University of Tehran Press, Tenhran
8. Latifi Rostami M, Shirazi Rostami G, Hajizadeh Rostami F (2005) Examine and evaluate the advantages and disadvantages of different methods of implementation of the project and compare them. Development Finance Institution, london
9. Levy M, S (2009) Design-build. University of Tehran Press, Tehran
10. Shakeri E, Sajadi SV (2012) Comparative three general condition of the three factors, design and construction of civil industrial construction projects (5490) within the scope of the project changes. METKA, Greek
11. AIA National/AIA California Council (2007) Integrated project delivery: a guide. AIA, California

12. 3xPT Strategy Group (2007) Integrated project delivery: first principles for owners and teams. 3xPT Strategy Group, America
13. Lee WH, Anderson SM, Kim Y-W, Ballard G (2014) Advancing impact of education, training, and professional experience on integrated project delivery. Integrated Project Delivery, Canada
14. Kent DC, Becerik-Gerber B (2010) Understanding construction industry experience and attitudes toward integrated project delivery. J Constr Eng Manag
15. Khaki G (2015) Research methods in management. Foojan, Tehran
16. Hafeznia M (2005) Introduction to research methods in human sciences. Organization of research development, Tehran
17. Blekster L, Hiuz K, Tiat M (2012) How do research?. PNU, Tehran
18. Hendrickson K, Aow T (2009) Construction project management. Unversity of Tehran Press, Tehran
19. Shahhossieni V, Hajaralasvadi H, Nozhan Naderi A, Joshaghani A (2013) The use of building information modeling technology in integrated project items delivery. Building Information Modeling (BIM), UK
20. Nida A, Youngcheol K, Irtishad UA (2014) Factors influencing integrated project delivery in publicly owned construction projects: an information modelling perspective. Integrated Project Delivery (IPD), California
21. Bidi A (2008) Risk management projects by applying conventional considerations. Project Management Institute, Tehran
22. Ghassemi R, Becerik-Gerber B (2011) Transitioning to integrated project delivery: potential barriers and lessons learned. Lean Constr J
23. Pishdad-Bozorgi P, Moghaddam EH, Karasulu Y (2013) Advancing target price and target value design process in IPD using BIM and risk-sharing approaches. Building Information Modeling (BIM), Uk
24. Safavi SA, Shayanfar MA, Nasr Azadani SM, Eshteharian EA (2011) Investigate the causes of delay in the implementation of urban development projects according to the project. Project Management Institute, Tehran
25. Golabchi M, Hosseini SZ (2012) Principles of project management. University of Tehran Press, SL
26. Becerik-Gerber B, DDes, Kent DC (2009) Implementation of integrated project delivery and building information modeling on a small commercial project
27. Glick S, Guggemos AA IPD and BIM: benefits and opportunities for regulatory agencies. Building Information Modeling (BIM), Uk

Evaluating Risk Management in Jordanian Construction Projects: An ISO 31000-2009 Implementation Perspective

Naser Abuyassin, A. S. H. Yousif and Najm A. Najm

1 Introduction

Risks are the syndrome of businesses as the warriors have to tolerate hazardous situations to achieve triumph. To realize the maximum possible level of profit, business organizations must afford a very high level of risk. Many management thinkers are deeming competition as a war, but without arms. This type of comparison which, always raises by businesses, may be the main reason that has motivated military strategist to consider war as a case of risk management and therefore, calls on his warriors to act as a risk manager [4]. Primarily it might be necessary to say that risk management is not a pessimistic perspective of company's projects rather than it is actually a conscious managerial action. As planning is one of the major functions of management, so one of its essential requirements is the exploration of the future to differentiate between adequate circumstances (i.e. Favorable events and opportunities) and inadequate ones (i.e. unfavorable events and threats) that might occur in the future.

Risk actually, is a part of any type of business, because there is no achievement of good return without risk. Therefore, management is highly concerned about risks that might face in the future as much as it is concerned about the achievement of total return and projected profit margin. The main reason behind the increasing importance of risk management is the state of information shortage which produces inappropriate procedures that required eliminating the negative impact of risk.

Risk is essentially represented a type of effect on the projected objectives of a business organization. This effect can be a positive or negative deviation from the expected performance to meet the company's objectives that may take different aspects (such as financial, quality specification and delivery date). These objectives

N. Abuyassin · A. S. H. Yousif · N. A. Najm (✉)
Al-Zaytoonah University, Accounting Department, P.O. Box 130,
Amman 11733, Jordan
e-mail: najimnajim@zuj.edu.jo

S. Şahin (ed.), *8th International Conference on Engineering, Project, and Product Management (EPPM 2017)*, Lecture Notes in Mechanical Engineering,
https://doi.org/10.1007/978-3-319-74123-9_34

could be adopted at different levels (such as strategic, organization-wide, project, product and process). As far as, risk in this context is concerned, risk is often interrelated with the potential future events, its occurrence likelihood and its expected consequences. Finally, risk is closely connected with uncertainty which is the state, of information deficiency (totally or partially) concerning the understanding or knowing of, a future event, its consequence and its occurrence likelihood (ISO 2009, pp 1–2, ISO Guide 73:2009).

According to (ISO 31000-2009) risk is usually represented a critical influence of meeting business organization objectives. This influence could be of positive or negative deviations from what is projected. This deviation comprises different levels the strategic, project, process or product, levels of the financial, environmental and health & safety aspects. Therefore, it can be concluded that risk is closely related to a probable incident and its consequences or both. Finally, risk is the outcome of uncertainty due to information shortages (totally or partially), which leads to poor understanding of the future events and its consequences and likelihood of occurrence. (ISO 2009, pp 1–2, ISO Guide 73:2009).

Despite the availability of useful guides and standards for the project management filed, it (PM) remains a highly problematical endeavor and risk management is still the most critical activity due to the fact that any project may face many risks during its executing process [11]. while more organizations are becoming project-based, risk management becomes one of the most important management to effectively address project risks. According to Caroll and Webb (2001), there are four types of risk: Imminent peril (the Sword of Damocles), Invisible risk (Pandora's box), Cost—benefit report (the Scale of Athena) and Voluntary Risk (The Myth of Hercules or Risk for the risk's sake). Biebold et al. (2004) presented another classification of risks: Known (there is a model that explains the risk), Unknown (more than one model explain the risk) or Unknowable risks (no specific model explains the risk). These diverse and wide-ranging risks in project management require specialized management that is risk management.

2 Risk Management

ISO (21500) [5] documents define project management as the application of methods, tools, techniques and competencies to a project. It includes the integration of all the phases of the project life cycle, which covers the period from the start up to project completion. IT also was defined by PMBOK guide (2008, p 6) as the application of knowledge, skills, tools, and techniques to a project in order to meet its general requirements. The UK- Institute of Risk Management (IRM), pointed out that project management considers risk management as the essential part of organization's strategic management (IRM 2002, p 3). It is clear that the latest definition focuses on the strategic dimension as, a range of risks that can be classified as strategic risks such as: competition risk, change in the loyalty of customers, and changing in customers' needs and wants.

Oxford dictionary defined risk as a chance or possibility of danger, loss, injury or other adverse consequences. In this meaning, risk can cover the losses of resources, the angible and intangible damages or any other negative consequences that could happen in the future. Therefore, risks are some types of probabilistic uncertain events. Hence, the certainty of risk occurrence is unacceptable aspect; because it simply means that, business organizations are targeting danger, which is an illogical practice business world. Risk is a future event, while the historical records are sources for lessons to be learned and a database for future forecasting. Paxson and Wood (1998) defined risks as "facing unpleasant coincidence". Harris (2006), defined risks as the likelihood of threat that creates damages to business.

The British institute of risk management (BIRM) has defined risk management as "the central element or function of company's strategic management, which usually concerns with all types of risks that have a negative effect on the company's general activities. It is playing a distinctive role to eliminate harmful consequences of risks and facilitating the way of achieving all the expected benefits of the project (BIRM 2002). It is quite clear that this definition is focusing on the strategic dimension as some types of risks which are characterized as strategic risks relying upon the nature of the projects. As projects usually, characterize by their level of newness and uniqueness, this makes them always facing a high probability of failure, which simply means, losing resources, opportunities, and creating new types of threats for the business and its projects. Risk is the most appearing business attribute nowadays.

The ISO organization has defined risk in tis directory of risk management reformist language as "the effect of uncertainty on objectives". It might be relevant to mention here that the ISO definition of risk indicates, that risk is not a product of coincidence or incident that cannot be overcome, but it is essentially the effect of uncertainty. According to loader [8], Risk management seeks to achieve many objectives such as: identify what the risks are, know the frequency of occurrence of the risk, understand how and where the risk will potentially impact, measure the impact of the risk, and reduce the controls that will manage the risk.

Risk management has been considered as sub-administration within project management. The American institute of projects management has considered risk management as one of the mine basic aspects of project management knowledge according to its project management directory (APMI 2009).

3 ISO 31000-2009

In risk management, there are many approaches to manage and control the project risks such as: risk-based dependability framework [12], Failure Mode and Effects Analysis: FMEA [1]; the two-dimensionality of project risk [14].

The ISO (International Standardizations Organization) has issued the (ISO 31000) standard for risks in 2009. These developments were the main elements behind the current consideration of risk management as an essential aspect of scientific knowledge and very important professional specialization of value creation.

Relying upon ISO literature it can be concluded that ISO (31000-2009) is consisted of five main components (), these are:

3.1 The Strategic Vision of Risk Management: Some risks are strategic and others are operational. The context of risk management includes a board scope of all those types of risks [3, 13, p 6]. The context of risk management covers a wide range of aspects extends from operational risks to market risks including competitive risks. This broad context requires a strategic vision of risk management which can contribute efficiently to the achievement of the company objectives.

3.2 The Scientific and Statistical Definition of Risk: Risk is, no longer, be considered as, just a, coincidence, an unexplained or unpredictable event that happens without any warning signals. Risk is simply the effect of the state of uncertainty due to deficiencies in information, attention and focus. That is why risk can be treated like any other business phenomenon, which can be investigated and analyzed to determine its patterns of occurrence and thus determining its occurrence probability.

3.3 Five Attributes of Risk Management in Construction Companies: These attributes are: company has an organizational unit, adapting a continuous improvement approach related to risk management, team of Individuals with the required qualifications and skills in risk management to handle all kinds of potential risks, applying of a risk management process system for all kinds of risks regardless of the type or level of importance, and finally, the adoption of a transparent governance process that covers all types of risks through a periodical reporting system to all stakeholders and other concerned parties.

3.4 The Risk Management Principles: According to ISO 31000-2009 standard, effective risk management is guided by a set of principles.

3.5 The Implementation of Risk Management Process: This process represents systematic structured and integrated steps to manage and control the context of the company.

4 Methodology

The Variables of the study: the independent variables of the study are the main five components of ISO 31000-2009 risk management process. These components are: Establishing the context, Risk assessment, Communication & Consultation, and Monitoring & Review. The dependent variables are the performance criteria: Project financial budget obligation, quality specifications, and delivery date.

These three criteria are the essential determents of any project management effectiveness (i.e. extent to which the company goals are attained).

4.1 Study Hypotheses

This study intends to determine the effect of the main components of risk management on project management performance. This objective has been addressed by the following three hypotheses:

- H1: ISO 31000-2009 risk management components are positively affected the project financial budget obligation.
- H2: there is a positive effect of ISO 31000-2009 risk management components on the quality specifications commitment.
- H3: there is a positive effect of ISO 31000-2009 risk management components on the delivery date commitment.

4.2 Questionnaire

Questionnaire as one of the most commonly used tool for data collection was adopted to gain the required data. A three sections questionnaire was designed and evaluated. These three sections include demographic and functional characteristics, the ISO 31000-2009 components of risk management, and project management performance criteria (i.e. budget, quality and delivery date).

Participants: The study based on the convenience sample that comprised four construction companies, Greater Amman Municipality engineering department responsible for executing Amman Municipality engineering projects, DAMAC Real Estate Development Limited in Jordan, Arab Towers Contracting Company (Jordan), and Al-Yacoub Contracting Est. 75 questionnaires were distributed 51 were returned 9 of which were not totally completed.

The sample participants are managers, engineers and supervisors who are directly involved in project management and risk management activities.

4.3 Sample Characteristics

Table 1 displays the distribution of the sample by gender, age, educational level, working experience, etc.

Table 1 Sample characteristics

Character type	Data	Freq.	%	Character type	Data	Freq.	%
Gender	Male	36	85.7	Marital status	Single	11	26.2
	Female	6	14.3		Married	31	73.8
Total		42	100	Total		42	100
Age	<30	10	23.8	Specialty	Manager		
	30–39	20	47.6		Engineer (non-manager)		
	40–49	6	14.3		Technician		
	≥ 50	6	14.3		Other		
Total		42	100	Total		42	100
Education	Sec.	1	2.4	Working experience	<5 years	8	19.0
	Bach.	32	76.2		5–9	11	26.2
	High Dip.	3	7.1		10–15	9	21.5
	Mas.	6	14.3		>15	14	33.3
Total		42	100	Total		42	100

4.4 Tests

- Construct validity: In order to examine the construct validity of the data collection tool to ensure that the questionnaire items are relevantly measuring what they are designed for, and factor analysis loadings was used for this purpose. Table 2 illustrates the results of this test which are plainly indicate that all the percentages of the study variables and their items are larger than 50 percent, as well as KMO percentage. (Kaiser 1981) or (60%) (Mulaik 2010). KMO value was also greater than (50%) which means that the sample of the study is adequate. These results also ensure that all the items of risk management

Table 2 Factor analysis loadings

Items	No. of items	Factor 1	Extraction	KMO
Risk management components				
– Establishing the context	6	0.899	0.808	0.879
– Risk assessment	5	0.894	0.799	0.787
– Risk treatment	5	0.908	0.825	0.717
– Communication and consultation	5	0.803	0.645	0.753
– Monitoring and review	5	0.644	0.414	0.793
Performance criteria				
– Financial budget	3	0.585	0.343	0.601
– Quality specifications	3	0.860	0.763	0.680
– Delivery date	3	0.874	0.874	0.618

Table 3 Inter-correlation matrix

	Risk management components				
	EC	RA	RT	CC	MR
EC	1.000				
RA	0.785	1.000			
RT	0.739	0.605	1.000		
CC	0.667	0.583	0.725	1.000	
MR	0.515	0.500	0.459	0.334	1.000

Independent variables: EC: establishing the context, RA; risk assessment, RT: risk treatment, CC: communication and consultation, and MR: monitoring and review

components are homogenous and factor loadings are appropriate to maintain the construct validity.

- Reliability: Reliability test was used to determine the extent to which the items of risk management components are internally consistent. Cronbach's alpha is one of the most recommended measures for internal consistency. This test was implemented and the generated outcomes are: 0.916 for establishing the context, 0.824 (risk assessment), 0.784 (risk treatment), 0.798 (communication and consultation), 0.844 (monitoring and review).

Variables relationships: To specify the intensity of relationship between independent variables, the inter-correlation matrix was used for this purpose. Table 3 shows that all correlation coefficients values are larger than 0.30.

5 Hypotheses Testing

To test the hypotheses of the study and determine the relationship between independent and dependent variables, the statistical correlation test was used where the correlation and determination coefficients were calculated to. Multiple regression analysis was also conducted to determine the effect of independent variable on dependent variables in a causal model.

H1: ISO 31000-2009 risk management components are positively affect the project financial budget obligation. The Value of the determination coefficient was (0.218) which indicates that there is a positive relationship between risk management components and project financial budget obligation. Table 4 shows that three components of risk management (the context, risk assessment and risk treatment) have a positive effect on c project financial budget obligation at a significant level (p-value <0.05). The other two components of risk management (communication and consultation and Monitoring and review) have no effect on project financial budget obligation and they were ignored.

H2: there is a positive effect of ISO 31000-2009 risk management components in on the quality specifications obligation.

Table 4 Multi-regression: effect of risk management components on performance criteria

Variables	R	R^2	T	Beta	Sig.	Result
H1						
Establishing context	0.446	0.218	2.835	0.522	0.007	Accepted
Risk assessment			2.004	0.092	0.042	Accepted
Risk treatment			2.220	0.200	0.032	Accepted
Comm. and consult			1.007	−0.095	0.096	Rejected
Monitoring and review			0.048	−0.243	0.804	Rejected
H2						
Establishing context	0.332	0.110	0.311	1.098	0.030	Accepted
Risk assessment			−0.755	−2.068	0.046	Accepted
Risk treatment			0.464	1.168	0.012	Accepted
Comm. and consult			−0.174	−0.578	0.567	Rejected
Monitoring and review			0.098	0.459	0.049	Accepted
H3						
Establishing context	0.327	0.107	1.723	0.175	0.023	Accepted
Risk assessment			1.899	0.192	0.009	Accepted
Risk treatment			2.010	0.237	0.003	Accepted
Comm. and consult			0.467	−0.092	0.982	Rejected
Monitoring and review			0.897	−0.233	0.653	Rejected

Results of hypothesis testing reveal that there is a positive relationship between risk management and quality specifications. The regression analysis also showed that four components of risk management (communication and consultation Monitoring and review) have an effect on quality specifications at (p-value < 0.05) significant level. While the communication and consultation components do not have any effect on quality specifications (see Table 4).

H3: there is a positive effect of ISO 31000-2009 risk management components on the delivery date obligation.

For this hypothesis testing, the value of the coefficient of determination was 0.107, which indicates that there is a type of a positive relationship between risk management components and the delivery date. In general, the regression analysis results exposed that there is an effect of the three components of risk management on the delivery date at a significant level of (p-value <0.05). The main results are presented by Table 4.

6 Discussion

In This study has identified three types of risk: the strategic, financial, and technical risks. The related literature combines many types of risk classification [3, 10]. The broad identification of risk in the "Design and Build projects" indicts that there are thirty-five types of risks. These risk can be classified in several groups according to related project management dimensions (time, cost and quality) [9].

This study has adopted four broad types of risk that were ranked according to their importance. First in the list is the strategic risks, second type represented the technical risks, third type was the legal risks, and the fourth was the financial risks. These types of risk are in consistent with the classification of Build-Operate-Transfer model. (BOT) [15].

This study sought to determine the effect of ISO 31000-2009 five components (establishment of context, risk assessment, risk treatment, communication and consultation and monitoring and review) on the three performance criteria (financial budget, quality specifications, and delivery date). The statistical analysis results revealed that the three components (establishment of context, risk assessment, risk treatment) have a positive effect on the three performance criteria. It might be relevant to mention the benefits of adapting the ISO 31000-2009 framework for risk management such as:

- Developing applicable approach to risk management.
- Improving the risk management process in its five dimensions.
- Achieving an effective response of risk management to the corporate strategy in general and to project management in particular.

Improving the performance of the company, especially the parts relate to the financial budget, quality specification and delivery date.

7 Conclusions

Results discussion suggests that there is a urgent need for developing a new approach of risk management for Jordanian construction sector. This approach can be formulated relying upon the five components of ISO 31000-2009 risk management framework (i.e. establishment of context, risk assessment, risk treatment, communication and consultation and monitoring and review). The suggested approach also requires a revised classification of risks. This study has developed the required classification that comprises four categories of risk. These categories are the strategic, financial, legal and technical risks. The statistical analysis results of this study indicated that the main three components of risk management (establishment of context, risk assessment, risk treatment) have a positive effect on the three criteria of performance (financial budget, quality specifications, and delivery date). The results also confirm that there is no significant effect of the other two components on performance criteria. These results might urge Jordanian construction companies to take the necessary actions to integrate the two components into risk management.

References

1. Antoniolli PD, Argoud ART, Benevides G, Pires SRI (2013) FMEA as a tool for managing Risks in ICT projects, based on the PMBOK. Asian J Bus Manage Sci 3(12):1–24
2. Diebold FX, Gunther TA, Tay A (2004) Evaluating density forcasts with applications to financial risk management. Int Econ Rev 39:863–883
3. Evans J, Ganegoda A, (2012) Classification of risks and management implications. Risk management today, October, 66–68
4. Heng Y (2006) War as risk management. Routledge, London
5. ISO 21500 (2012) Guidance on project management. International Organization for Standardization, Geneva
6. ISO 31000 (2009) Principles and Guidelines on Implementation. International Organization for Standardization, Geneva
7. Joao V, Joao C, Silvac H (2012) ISO 21500:2012 and PMBoK 5 Processes in Information Systems Project Management, Computer Standards and Interfaces (Accepted manuscript), available at http://www.rcolomo.com
8. Loader D (2006) Advanced operational management. Sons, Chichester, John Wiley
9. Ogunsanm OE, Salako OA, Ajayi OM (2011) Risk classification model for design and build projects. J Eng Proj Prod Manage 1(1):46–60
10. Porrini D (2015) Risk classification efficiency and the insurance market regulation. Risks 3:445–454
11. Rehacek P (2014) Standards ISO 21500 and PMBoK guide for project management. Int J Eng Sci Innovative Technol (IJESIT) 3(1):288–295
12. Norrbin PSP (2013) Risk-based dependability approach to maintenance performance measurement. J Qual Maintenance Eng 19(3):316–329
13. Economist The (2015) Holistic risk management. The Economist, London
14. Williams TM (1996) The two-dimensionality of project risk. Int J Project Manage 14(3):185–186
15. Xenidis Y, Angelides D (2005) Identification and classification of risks in a new modeling process for build—operate—transfer projects. In: Khosrowshahi F (ed) 21st Annual ARCOM Conference, 7–9 September 2005, SOAS, University of London. Association of Researchers in Construction Management, vol 2, pp 803–812
16. Zandhuis A, Stellingwerf R (2012) ISO 21500. Van Haran Publishing, Zaltbommel

Outsourcing Projects and Achieving the Organizational Goals: Applied Study in Greater Amman Municipality (GAM)

Mohammad Salameh Alhmeidiyeen

1 Introduction

In recent years, the government has used outsourcing widely to deliver high quality, cost-effective, services and projects. Turner and Muller [38] defined projects in terms of temporary organizations: A project is a temporary organization, to which resources are assigned to undertake a unique, novel and transient endeavor managing the inherent uncertainty and need for integration in order to deliver beneficial objectives of change. Hällgren and Wilson [19] also argued that project is special groups are set up to accomplish the task at hand.

In another words, projects deal with producing an events that have not been accomplished before. The accomplishment of these projects by another party called outsourcing. Troaca and Bodisolav [37] argued that concept of outsourcing came from the terminology "outside resourcing", which means to get resources from the outside. In other words, outsourcing when handed over to providers in the same country is called domestic outsourcing, and when handled by providers in offshore locations is called offshore outsourcing [18]. The general objectives of outsourcing projects do not simply include the purchasing of products and services, as all organizations need to purchase things from external sources. Outsourcing can be an important strategic decision that entails a series of performance effects in different parts of an organization, furthermore; Organizations are seeking to outsource any process that is not deemed as core to their business [40]. Outsourcing can require top management decisions involving organization policy and may change the boundaries of an organization.

Meanwhile, Mutiangpili [28] argued that governments look to outsourcing for the progressive transformation of the overall operation of their agencies. The public

M. S. Alhmeidiyeen (✉)
Alzaytoonah University of Jordan, Amman, Jordan
e-mail: mhmeidiyeen2012@gmail.com

S. Şahin (ed.), *8th International Conference on Engineering, Project, and Product Management (EPPM 2017)*, Lecture Notes in Mechanical Engineering,
https://doi.org/10.1007/978-3-319-74123-9_35

agencies' bottom-line of providing quality service and optimizing taxpayers' money are the primary drivers for government sector outsourcing.

Outsourcing in Greater Amman Municipality (GAM) are commonly used in many projects that usually accomplished by other party. In 2015 more than 52 projects were executed by outsourcing with value about 28 million JD, and in 2016 about 74 projects with a value about 35 million JD [15].

Major outsourced projects in these two years, executed for GAM are:

1. Accomplishment of hot mix asphalt in different places
2. Rehabilitation of King Abdullah II's stadium phases 1, 2, 3
3. Maintenance of King Husain Gardens
4. Establishing retaining walls in different places
5. Supply of base coarse gravels
6. Establishing of water drainage lines in different places
7. Rehabilitation of Amman Slaughterhouse
8. Establishing a new garden in different places

Although outsourcing is not new to project management, it is done through two lines of business, in construction; projects are mainly run by a general contractor, whose job is to select and manage a battery of contractors and subcontractors. The second line that is depends on outsourcing is government business. The great share of government work is conducted through contracts. Governments rarely build bridges, highways, or computer systems using their own employees. Such work is usually done by private outside contractors [14]. GAM usually executes the bridges and tunnels by private contractors through outsourcing projects.

2 Literature Review

Outsourcing as a concept has been around for decades; its roots go to the 1970s when manufacturers began contracting out the production of components to smaller specialized suppliers, outside their organizations to manufacture some or the entire product [40]. In this context, the lack of researches in the area of outsourcing in public sector can be a problematic for making comparisons. However, it does provide an opportunity to explore a gap in the literature and to discover if anything can be learnt from how outsourcing is managed in government's organizations [8]. On the other hand, outsourcing decisions including make or buy can be based on a single factor, such as, costs and financial evaluation, lead-time and delivery reliability, cost capability, product quality and technical capability [36]. Outsourcing is one of the management strategic tools and should be used by an organization to deliver its services to community and achieving business goals. The outsourcing contract, whether for the first time or as a renewal, the first step is to determine the reasons for the outsourcing.

There are many reasons for outsourcing, the main are:

- Unavailability of service in-house.
- Focusing on core services such as equipment and roads.
- Risk diversification.
- To reduce time.
- Access to information, skills and technology.

In this context, outsourcing can make research and development (R&D) feasible, when government does not have the financial or human capacity to invest in innovation. In addition to the above benefits, outsourcing promises to reduce costs, improved flexibility, provide new capabilities those economies of technological scale offer, and achieve efficiencies in response to higher levels of global competition [7, 8, 16, 28, 37, 40, 43].

Meanwhile, in a study by Wilding and Juriado [42] aims to identify the customer perceptions on outsourcing, the study overviews the main reasons as established by five previous studies [4, 12, 22, 31, 33]. The studies found that cost reduction, improvement of service levels, increasing in operational flexibility, focusing on core competencies, improvement of asset utilization and change management were the most common reasons for outsourcing.

On the other hand, Outsourcing has a potential risk to the organization, following some of them

- Dependency on supplier.
- Loss of control over technology.
- Loss of technical skills.
- Risks due to incompatible organization culture.
- Vulnerability of information and intellectual property [5, 8, 9, 20, 37].

Although the use of outsourcing has increased in recent years, therefore the definitions of outsourcing vary according to its use. Accordingly, many definitions and descriptions of outsourcing have been argued by many authors. Child [6] describes outsourcing as the contracting out of activities that are needed to be undertaken on a regular basis, which otherwise would be conducted within an organization. Below some selected definitions of outsourcing available in the literatures are shown in Table 1.

For the purpose of this study, outsourcing can be defined as "the practice of having certain projects done outside GAM instead of doing it by its departments or employees, and these projects can be outsourced to either an external organization or supplier or an individual".

On the other hand, outsourcing takes different forms or arrangements, Frame [14] argued these forms as follow:

(a) Consultants, the simplest form of outsourcing is employment of a consultant to do needed work, may be hiring one person to execute a single job. This is done sometime in the areas of accounting, training, design, and software

Table 1 Some selected definitions of outsourcing

Author	Definition
Parkhe [32] In: Moreira et al. (2015)	Moving activities that were previously developed in-house to external organizations
Kavcic [20]	A transfer of some activities, which were previously carried out by the company, to an outsourcer
Deliotte [10] outsourcing handbook	The contracting out of a business function to an external supplier, involving the transfer of people, processes and assets
Mitra [26]	The procurement of material inputs or services by a firm outside the original firm
Ross [34] In: Wang (2012)	The contracting of services or products to another independent supplier organization as a way of achieving the desired supply or as a way of cutting costs
Kern and Willcocks [21] In: Cox et al. (2012)	A decision taken by an organization to contract out or sell the organizations IT assets, people and/or activities to a 3rd party supplier, who in exchange provides and manages assets and services for monetary returns over an agreed time period
Belcourt [3] In: Hansen et al. (2011)	A contractual relationship for the provision of business services by an external provider
Overby [30] In: Marvin (2011)	Outsourcing is viewed as involving the contracting out of a business function—commonly one previously performed in-house- to an external provider
Vitasek et al. [40]	Contracting with an outside supplier, this may or may not involve moving the work offshore
Milecová et al. [25]	Outsourcing means providing the company activity by an external provider
Olausson [29]	The extent a legal entity has contracted out manufacturing to another legal entity
Fitzgerald and Willcocks [13] In: Dalcher (2005)	The commissioning of a third party (or a number of third parties) to manage a client organization's IT assets, people and/or activities to required results
Schaaf [35] In: Kavcic (2014)	A concept, which represents a contractual transfer (long-term or constant) of activity, which had been carried out by the outsourcing company, but has been outsourced to an external supplier
Frame [14]	An important trend in recent times has been the growing use of outsiders to carry out an organization's business

development, as well as in specialized technical and business areas, this offers the outsourcing company more flexibility.

(b) Organized suppliers of specialized services, employers of contract employee are not required to pay benefits, but these are often provided by organizations that have been set up to provide specialized services to other organizations, such as security, foods services, supply hot mixed asphalt, building an executing bridges and tunnels as in the case GAM.

(c) Suppliers of parts and materials, outsourcing involves having outsiders supply organization with spare parts and materials it needs to produce its goods and services, such as supplying spare parts for vehicles in GAM.
(d) Personal services contracts; contracting employee is a growing phenomenon in organizations. It is common to see a contracted secretary to execute some work in an organization.

3 Method

3.1 Study Variables

The outsourcing is a complex phenomenon that affects the private as well as public sector. Outsourcing is no longer just about cost saving; it is a strategic tool that may power the twenty-first century global economy. GAM as a public sector adopted this strategic tool in its projects, in order to serve the citizens of Amman. Study variables are two kinds of variables, Independent variables representing three dimensions of outsourcing [outsourcing strategy: (the decision of GAM to use an outside organization to perform or execute the outsourcing project), speed of implementing project: (the time required to execute the outsourcing project), and project cost: (the cost that required to execute the project successfully)], and dependent variables that represent the total citizens' satisfaction [citizens' satisfaction: (how extent the citizens of Amman are satisfied with works done by GAM) and serving larger number of citizens: (the larger number of citizens that GAM can be served due to executing the project)].

3.2 Questionnaire

A questionnaire survey was used to reach a broad range of managers and non-managers in the field to collect data. Questionnaire data analysis was used to compare the results from the completed questionnaires and to identify significant issues. The five-point Likert scale answers on the questionnaires could be converted into numerical values for quantitative testing. The questionnaire consists of three main sections;

- Personal and informational data of the Sample which is comprised of six statements.
- Questionnaire phrases, this section insisted of (14) which covered three dimensions of outsourcing.
- Organizational objectives (total citizens' satisfaction). This section insisted of (7) phrases, which covered citizens' satisfaction and serving large number of citizens.

3.3 Sample

This study was conducted with a sample of (70) respondents which were distributed at two levels: (25) questionnaires were distributed for managers of GAM and (21) were recovered, and (45) questionnaires were distributed randomly for the second level non managers; and (39) were recovered. Total questionnaires recovered are 60, which represents 85% of the questionnaires distributed. The study sample represented (21%) of total sum managers and non-managers in the targeted departments. Table 2 demonstrates sample characteristics.

3.4 Hypotheses of the Study

H1: There is a positive effect of outsourcing dimensions (outsourcing strategy, speed of implementing project, and project cost) on citizens' satisfaction in GAM.
H2: There is a positive effect of speed of implementing project on citizens' satisfaction in GAM.
H3: There is a positive effect of project cost on citizens' satisfaction in GAM.
H4: There is a positive effect of outsourcing strategy on serving large number of citizens in GAM.

Table 2 Characteristics of respondents (n = 60)

Characteristics		Frequency	%
Sex	Male	30	50
	Female	30	50
Age	<30	2	4
	30–39	24	40
	40–49	20	33
	50–59	14	23
	>60	0	0
Social status	Single	12	20
	Married	48	80
Education	Bachelor	47	78
	Master	13	22
	Doctorate	0	0
Occupation	Managers	21	35
	Non manager	39	65
Experience (years)	<5	0	0
	5–10	9	15
	10–15	27	45
	>15	24	40

H5: There is a positive effect of speed of implementing project on serving large number of citizens in GAM.
H6: There is a positive effect of project cost on large number of citizens in GAM.

3.5 Study Validity and Reliability

To testify the validity of the study, a number of questionnaires were distributed to a number of referees, of a specialized academic staff, where their notes have been taken into account to develop the questionnaire. To examine the harmony of questionnaire statements, Cronbach's alpha was used for this purpose. Validity test: the statements of the questionnaire were tested by professors from Alzaytoonah University of Jordan to ensure that the content represents what needs to be tested and meets the research variables. The draft questionnaire was returned and adjusted based on the recommendations from the reviewers to build the final version that was used in the research. Reliability analysis was used to determine the extent to which the questionnaire items are realizing the internal consistency. Cronbach's alpha was calculated for the outsourcing dimensions (outsourcing strategy, speed of implementing project and cost of project) in the questionnaire. The value of Cronbach's alpha was (0.699, 0.644 and 0.623) respectively which are larger than (0.60). The result indicated that there is a good effect on these dimensions and the questionnaire was good fit to be used in the study.

3.6 Sample Characteristics

Table 2, shows demographic and informational characteristics of respondents.

3.7 Hypotheses Testing

To test study's hypotheses, the determination and the regression coefficients were used, to determine the effect of independent on dependent variables. In Table 3

Table 3 Effect of outsourcing dimensions on citizen's satisfaction (n = 60)

Dimensions	Total citizens' satisfaction	R	R^2	ß	T	sig
Strav	Citizens' satisfaction (Czsav)	0.586	0.344	0.474	3.280	0.002
Spdav				0.127	0.981	0.331
Cstav				0.071	0.576	0.567

Czsav citizens' satisfaction av., *Strav* outsourcing strategy av., *Spdav* speed of implementing project av., *Cstav* project cost av., *Slnav* serving large number of citizens

Table 4 Effect of outsourcing dimensions on serving large number of citizens (n = 60)

Dimensions	Total citizens' satisfaction	R	R^2	ß	T	sig
Strav	Serving large number of citizens (Slnav)	0.397	0.157	0.054	0.332	0.741
Spdav				0.143	0.971	0.336
Cstav				0.314	2.257	0.028

coefficients of determination (R2) indicated that there is a positive relationship between dimensions of outsourcing (outsourcing strategy, speed of implementing project and cost of project) and citizens' satisfaction. In order to determine the effect of outsourcing dimensions, the regression coefficient used. The results of analysis showed that the values of beta (ß) which represent the regression coefficient are; outsourcing strategy (0.474), speed of implementing project (0.127) and cost of project (0.071). Also in this table, values of the calculated-t were 3.280 for strategy dimension of outsourcing which is higher than the tabulated-t; therefore, the hypothesis is accepted. The value of the calculated- t was 0.981 for speed of implementing project and for cost of project were 0.576 which are both lower than the tabulated-t (at $p < 0.05$ and n = 60 is 1.67). Therefore; the hypotheses are rejected and the alternative null hypotheses are accepted. The results of regression coefficient (value of ß parameter at the table) indicated that there is a significant effect of outsourcing strategy on citizens' satisfaction. (The result for outsourcing strategy is in line with the results from other studies by Alexandrova [1], McIvor et al. [24], Wang [41].)

In Table 4 coefficients of determination (R2) indicated that there is a positive relationship between dimensions of outsourcing (outsourcing strategy, speed of implementing project and cost of project) and Serving large number of citizens. In order to determine the effect of outsourcing dimensions, the regression coefficient used. The results of analysis showed that the values of beta (ß) which represent the regression coefficient are; outsourcing strategy (0.054), speed of implementing project (0.143) and cost of project (0.314). Also in this table, values of the calculated-t were 0.332 for outsourcing strategy, 0.971 for speed of implementing project which are both lower than the tabulated-t (at $p < 0.05$ and n = 60 is 1.67), therefore; the hypotheses are rejected and the alternative null hypotheses are accepted. While the value of calculated-t for cost of project was 2.257 which is higher than the tabulated-t. Therefore; the hypothesis is accepted. The results of regression coefficient (value of ß parameter at the table) indicated that there is a significant effect of cost of project on serving large number of citizens. (The result for cost of project is in line with the results from other studies by Alexandrova [1], MacCormack et al. [23], Mutiangpili [28], Wang [41] and Wilding and Juriado [42].)

4 Discussion

Outsourcing is an increasingly popular method of achieving performance improvement. The evolution of outsourcing into value-creating global collaboration provides a significant competitive advantage for those who employ it effectively [23]. Public sector outsourcing's first driver was the need for expertise to manage the evolving needs of the citizens it serves. It has been made more relevant by the realized cost savings. In public sector, managers adopt a way to outsourcing that based on goals and values other than cost efficiency, whilst being further constrained by other accountability [39]. Outsourcing is a powerful approach to help organizations keeping up with rapid environmental changes and the fast-paced technological advancements in various fields. Certainly, the adoption of outsourcing by organizations and government institutions such as GAM was associated with many difficulties and challenges, such as the pressure of Public Auditing Bureau. The main aim of GAM is to concentrate on its core objective; serving the citizens of Amman. The results of this study confirm that outsourcing dimensions (strategy of outsourcing, cost of project) have a positive effect on the citizens' satisfaction and serving large number of citizens respectively (Tables 3 and 4). The same conclusion reached by other studies [1, 11, 23, 24, 28, 41]. The other dimension (speed of implementing project) has no significance that is due to concentrate GAM on costs and financial issues, that it works according to the tight governmental regulations and faces pressures of internal and external auditors. On the other hand, the changes in orders and design of the projects and weather conditions, these could lead to a significant disruption and delay in the projects, therefore, a reschedule of plans are required which consequently result in time delay. This conclusion reached by other studies [2, 17, 18].

References

1. Alexandrova M (2009) International outsourcing: incentives, benefits and risks for the companies in SEE countries. In: 4th international conference of ASECU development cooperation and competitiveness, The Bucharest Academy of Economic Studies, pp 11–16
2. Alsendi MAY (2015) Studying the effect of decision making on delayed construction projects. Dissertation, The George Washington University
3. Belcourt (2006) In: Hansen ZNL, Ahmed-Kristensen S, Rasmussen LB, Conrad F (2011) On outsourcing and offshoring: challenges facing management and engineering. Dissertation, Kgs. Lyngby: DTU Management
4. Boyson et al (1999) In: Wilding R, Juriado R (2004) Customer perceptions on logistics outsourcing in the European consumer goods industry. Int J Phys Distrib Logistics Manage 34(8), 628–624
5. Chiasson M, Dexter Al, Wotherspoon D (2005) Systems development outsourcing: lessons from litigation, book chapter. In: Brudenall P (ed) Technology and offshore outsourcing strategies. Palgrave Macmillan, New York

6. Child J (2005) In: Cox M, Roberts M, Walton J (2012) IT outsourcing in the public sector local government: experiences of the management and selection of IT service providers. Electron J Inf Syst Eval 15(3)
7. Choi K (2008) Serving the community by using the private sector, a general guide to outsourcing, 3rd edn. Efficiency Unit
8. Cox M, Roberts M, Walton J (2012) IT outsourcing in the public sector local government: Experiences of the management and selection of IT service providers. Electron J Inf Syst Eval 15(3). Available online at www.ejise.com
9. Dalcher D (2005) From fixed contracts to dynamic partnerships: successful outsourcing in a changing world, book chapter. In: Brudenall P (ed) Technology and offshore outsourcing strategies. Palgrave Macmillan, New York
10. Deliotte (2013) The outsourcing handbook: a guide to outsourcing. United Kingdom: Deliotte MCS limited
11. Eger III RJ, Knudson DA, Marlowe J, Ogard L (2002) Evaluation of transportation organization outsourcing: decision making criteria for outsourcing opportunities. TRB 2003 Annual Meeting CD-ROM
12. Fernie (1999) In: Wilding R, Juriado R (2004) Customer perceptions on logistics outsourcing in the European consumer goods industry. Int J Phys Distrib Logistics Manage 34(8), 628–624
13. Fitzgerald G, Willcocks (1994) In: Dalcher D (2005) From fixed contracts to dynamic partnerships: successful outsourcing in a changing world, book chapter. In: Brudenall P (ed) Technology and offshore outsourcing strategies, New York
14. Frame JD (2002) The new project management: tools for an age of rapid change, complexity, and other business realities. Wiley, New York
15. GAM tender's department (2017) GAM's outsourced projects in 2015 and 2016, January 2017
16. Goldsmith S, Rendell E, Fernandez S (2014) Government outsourcing: a practical guide for state and local governments, Indiana University
17. Grüter K (2007) Outsourcing tasks and new methods of financing-the challenges facing external auditors. In: Speech at the 6th EURORAI CONGRESS 21–23 October in Crans-Montana, Switzerland, Director of the Federal Audit Office, Bern
18. Hansen ZNL, Ahmed-Kristensen S, Rasmussen LB, Conrad F (2011) On outsourcing and offshoring: challenges facing management and engineering. Dissertation, Kgs. Lyngby: DTU Management
19. Hällgren M, Wilson TL (2007) Mini-muddling: learning from project plan deviations. J Workplace Learn 19(2):92–107
20. Kavcic K (2014) Strategic management of outsourcing. Published by University of Primorska, Faculty of Management
21. Kern, Willcocks R (2000) In: Cox M, Roberts M, Walton J (2012) IT outsourcing in the public sector local government: experiences of the management and selection of IT service providers. Elect J Info Sys Eval 15(3). Available online at www.ejise.com
22. Laarhoven et al. (2000) In: Wilding R, Juriado R (2004) Customer perceptions on logistics outsourcing in the European consumer goods industry. Int J Phys Distrib Logistics Manage 34 (8), 628–624
23. MacCormack A, Forbath T, Brooks P, Kalaher P (2007). From outsourcing to global collaboration: new ways to build competitiveness, HBS Working Paper 07-080
24. McIvor R, Humphreys PK, Wall AP, Mckittrick A (2009) A study of performance measurement in the outsourcing decision. Elsevier/CIMA Publishing, Oxford, vol 4, issue 3
25. Milecová M, Grznár M, Szabo L (2010) Outsourcing: decision making in the case of outsourcing the company activity and the management of outsourcing project. Agric Econ Czech 56(8):387–396
26. Mitra RM (2013) The information technology and business process outsourcing industry: diversity and challenges in Asia. Working Paper, Asian Development Bank, Philippines

27. Moreira MR, Andrade SR, Sousa PS (2015) International outsourcing: a process approach for apparel industry. Rev Bus Manage 17(58):1444–1463
28. Mutiangpili J (2010) Government sector outsourcing transforming public service with outsourced IT services. THOLONS Global offices, New York, www.tholons.com
29. Olausson D (2009) Facing interface challenges in complex product development. Dissertation, Linköping University, Department of Management and Engineering, SE-581 83 Linköping, Sweden
30. Overby (2007) In: Marvin KT (2011) Global trends in outsourcing and their impact, an interactive qualifying project report submitted to the faculty of worcester polytechnic
31. P-E International (1994) In: Wilding R, Juriado R (2004) Customer perceptions on logistics outsourcing in the European consumer goods industry. Int J Phys Distrib Logistics Manage 34(8):628–624
32. Parkhe (2007) In: Moreira MR, Andrade SR, Sousa PS (2015) International outsourcing: a process approach for apparel industry. Rev Bus Manage 17(58):1444–1463
33. Penske Logistics (1999) In: Wilding R, Juriado R (2004) Customer perceptions on logistics outsourcing in the European consumer goods industry. Int J Phys Distrib Logistics Manage 34(8):628–624
34. Ross (2006) In: Wang JJ (2012) Examination of outsourcing of components and finished products from Australia to companies in China—inter-firm business problems, solutions and business success factors. Dissertation, faculty of business and law, Victoria University
35. Schaaf (2004) In: Kavcic K (2014) Strategic management of outsourcing. Published by University of Primorska, Faculty of Management
36. Tayles M, Drury C (2001) In: Wang JJ (2012) Examination of outsourcing of components and finished products from Australia to companies in China—inter-firm business problems, solutions and business success factors. Dissertation, faculty of business and law, Victoria University
37. Troaca VA, Bodisolav DA (2012) Outsourcing, the concept. Theor Appl Econ XIX 6 (571):51–58
38. Turner, Muller (2003) In: Hällgren M, Wilson TL (2007) Mini-muddling: learning from project plan deviations. J Workplace Learn 19(2):92–107
39. Vilvovsky S (2008) In: Cox M, Roberts M, Walton J (2012) IT outsourcing in the public sector local government: experiences of the management and selection of IT service providers. Electron J Inf Syst Eval 15(3)
40. Vitasek K, Ledyard M, Manrodt K (2010) Vested outsourcing: five rules that will transform outsourcing. Palgrave Macmillan, New York
41. Wang JJ (2012) Examination of outsourcing of components and finished products from Australia to companies in China—inter-firm business problems, solutions and business success factors. Dissertation, faculty of business and law, Victoria University
42. Wilding R, Juriado R (2004) Customer perceptions on logistics outsourcing in the European consumer goods industry. Int J Phys Distrib Logistics Manage 34(8):628–624
43. Willcocks L, Oshri I, Kotlarsky J, Rottman J W (2011) Outsourcing and offshoring engineering projects: understanding the value, sourcing models, and coordination practices. IEEE Trans Eng Manag 58(4):706–716. http://doi.org/10.1109/TEM.2011.2128873

Author Index

S. Şahin (ed.), *8th International Conference on Engineering, Project, and Product Management (EPPM 2017)*, Lecture Notes in Mechanical Engineering,
https://doi.org/10.1007/978-3-319-74123-9

The manufacturer's authorised representative in the EU is Springer Nature Customer Service Centre GmbH, Europaplatz 3, 69115 Heidelberg, Germany. If you have any concerns regarding our products, please contact ProductSafety@springernature.com

Printed and bound by CPI Group (UK) Ltd, Croydon, CR0 4YY
30/09/2026
02225238-0007